CAN'T DECIDE WHICH CHAPTER TO READ FIRST?

THIS DICE WILL HELP YOU

Natalie Denk, Alesha Serada,
Alexander Pfeiffer, Thomas Wernbacher (Editors)

A LUDIC SOCIETY

Editors:
Natalie Denk, Alesha Serada, Alexander Pfeiffer, Thomas Wernbacher
Cover, Illustration: Constantin Kraus

Publisher: Edition Donau-Universität Krems
Print: tredition GmbH, Halenreie 40-44, 22359 Hamburg

ISBN Paperback: 978-3-903150-72-0
ISBN e-Book: 978-3-903150-73-7

Contact:
Center for Applied Game Studies
Department for Arts and Cultural Studies
Donau-Universität Krems
www.donau-uni.ac.at/ags
ags@donau-uni.ac.at

Contents

PREFACE

Natalie Denk, Alesha Serada, Alexander Pfeiffer, Thomas Wernbacher

Everybody plays. We find play in every corner of our world in all its multi-layered manifestations. Whether in digital or analogue form, games have long since become an important part of our society. As economic products, they are at the core of an industry that generates a higher turnover worldwide than the film and music industries combined. However, the importance of video games for society goes far beyond their mere entertainment value. As communication and learning tools, games can make even the most difficult topics tangible, experienceable and easy to master. Game-based learning, the methodical use of games in a pedagogical framework, has been established in the educational sector throughout its entire history.

So far, teachers, psychologists and game scholars have collected enough evidence to demonstrate that playful approaches can make positive changes to people's behaviour. As this book will demonstrate, such changes may happen in such diverse fields as health, environment or mo   Some companies use game-based methods to increase the moti   ı and satisfaction of their employees, to initiate innovation processes or to make training measures more effective. At the same time, however, many negative developments in our society cannot be understood without thinking about their often hidden but all the more determining game character: the distribution of fake news is just as much an example of this as cyber-bullying or the clever hunt for user data by major marketing companies and 'information brokers'. The good news is that we also have serious games that teach players about privacy, trust and informational hygiene. Besides, many beloved popular games in the 'cyberpunk' setting openly discuss these problems in their narratives, warning about the dangers of cyberspace in the most fascinating manner.

Last but not least, games are a cultural phenomenon. In the current mode of production, they become essential goods as well as legitimate means of artistic expression. Video games shape our identities, offer new ways of communication and contribute to our symbolic vocabularies to perceive and explain the world. Gaming practices create social ties and personal bonds that underlie vibrant online communities with their own unique 'game cultures', opening up new spaces for negotiating ideas about society and ourselves. Of course, there is always the dark side to such abundance of expressions, not necessarily prosocial, which further stresses the importance of game research and education in the times of global 'virtualization'.

Since Covid-19 turned our world upside down, more than ever before, many things in our lives take place in virtual space. Thus, the participants of the 14th Vienna Games Conference "FROG - Future and Reality of Gaming" in 2020 came together for the first time in virtual space. The international conference brings together scholars, players, students, game designers, developers, educators and experts from various disciplines to discuss the Future and Reality of Gaming. In 2020 the conference was hosted by the Center for Applied Game Studies (Danube University Krems) in cooperation with the Austrian Federal Chancellery and was dedicated to key challenges of a "Ludic Society". With this anthology, we invite our readers to engage in discussions between game scholars who present a multitude of viewpoints on games, learning, society, identity, and change.

ACKNOWLEDGMENTS

We would like to dedicate this book to the ever-growing FROG community, to all those who attend the "FROG - Future and Reality of Gaming" conference year after year. We thank you for your loyalty, commitment, interest, discourse and support. Especially in the Covid-year 2020, the cohesion of the community was a crucial factor for the success of the conference and is always present even in the virtual space. Special recognition goes here to Herbert Rosenstingl, who has been committed to the conference since its very beginning.

INTRODUCTION

Contemporary game scholarship offers a broad palette of theories and methods inherited from such fields as sociology and communication studies, experimental sciences, literary analysis, educational sciences and cultural critique. At large, this inherently interdisciplinary research aims for a holistic perspective on the 'ludic society', which is also our goal here. With that in mind, this book is organized into four sections that present related and often intertwined ideas and observations about the ways we manifest ourselves in games and play, how games represent us in the present and in the past, how games and play change us, and what it all may mean for contemporary society.

GAME, PLAY & IDENTITY

Video games provoke a full spectrum of emotions in their players, from passion and jubilation to hostility and fear. Whether these emotions and attitudes can bleed into real life situations remains the subject of active research, even though the 'media panic' around violent video games has been relatively stagnant recently. Today's research places a special importance on emotional development of gamers, as well as the potential of serious games for nurturing compassion and promoting prosocial behavior. For the reader to catch up with the current state of related research and game design, **Ricarda Goetz** provides a thorough review of identity, empathy and the Proteus Effect in video games. Teaching empathy for often most marginalized groups, such as immigrants (the intended mission of *Papers, Please!* (2013)) or transgender people, which is the case of *The Missing* (2018), autoethnographically analysed for this book by **Josephine Baird**. She questions the prescriptive 'queer pedagogy' of LGBTQ+ games and comes to the conclusion that the success of this particular game partially relies on its intentionally 'misdirecting' strategy to present the most traumatic experience of exclusion.

But how authentic is the identity that can be acquired in a video game? This is not a simple question: therefore, in their chapter, **Mona Khattab, Tanja Sihvonen and Sabine Harrer** reveal the Orientalist gaze on ancient Egypt in the game and connect it to the idea of 'identity tourism'.

Video games can help us understand who we are, in new and fascinating ways, as they require active involvement of the player. In his work in progress, **Steve Hilbert** explores the potential of the game Gris in self-reflection and emotional growth of a person, particularly in relation to the self-shattering experience of depression. **Doris Rusch** and **Andrew Phelps** propose the project for existential games - "games of the soul" that appeal to myths and symbols at the deeply personal level. Finally, **Frank Pourvoyeur** provides a Jungian perspective on user experience in games by introducing synchronicity and verisimilitude as analytical categories for the experiential journey of the player.

GAME, PLAY & HISTORY

History in games has everything to do with the history of games. **Eugen Pfister**'s work bridges the gap created by the Anglo-American focus of most published game histories by turning to the most turbulent years of the Austrian games industry instead. Furthermore, the author connects notable events in local game history to the political and economic atmosphere of these times: one important discovery that he makes is the proliferation of business simulators in Austria in the 1990s and 2000s.

How do we connect with (or, in some cases, disconnect from) our historical past? To many gamers, the first lessons in world history came from the settings of their favourite games, and this is also true for the first generation of digital gamers. In their chapter, **Wilfried Elmenreich and Martin Gabriel** study representations of colonial history and international trade in three critically and commercially successful video games released in the 1980s, played on the Commodore 64. This unique gaming experiment provides the backdrop for critical reflection about how historical storytelling developed since then.

Representations of history become powerful weapons in political arguments. Historical accuracy is a particularly troublesome concept, as it often relies on 'selective authenticity'. There is no shortage in representation of World War II in videogames, but they rarely provide realistic depictions of that time's society: an authenthic interaction with a Nazi would not be so much fun for consumers of digital entertainment. To further explore this paradox, **Benjamin Kirchengast** has studied negative reviews of the controversial game *Through the Darkest of Times* (2020). Kirchengast applies qualitative content analysis to negative reviews to discover several directions of criticism, which can be seen as symptoms of the current political atmosphere. Meanwhile, **Pascal Wagner** deals with another similar controversy in his own study, and active involvement into the antifascist network "Keinen Pixel den Faschisten!", uncovering and confronting extremist tendencies in some of the German-speaking video gaming communities.

Since the first installment of *The Oregon Trail* (1971-2021), serious games have been created with the direct aim to teach history to their players. In this book, **Michael Black, Jared Derry, Kathryn Friesen, Josey Meyer and Montserrat Patino**, a group of students from Texas A&M University, share the development process for the game that they have developed about the prominent event in the contemporary history of the United States - the 2020 presidential election.

GAME, PLAY AND A BETTER FUTURE

The educational potential of games and play has been discovered early at the beginning of pedagogy. Nevertheless, the implementation of digital games into the study process has been comparatively slow. In her chapter, **Daniela Hau** presents an extensive empirical study of digital game-based learning that summarises the results of 13 teaching projects in Luxembourg. 17 different games were used to teach social studies, media literacy, mathematics, sports, foreign languages and other subjects in formal school context.

What is video game culture? Is it a culture of inclusion or exclusion? Video game culture inevitably enters educational spaces such as schools, being brought there by the most curious and progressive generations of younger gamers. In their summarizing chapter, **Natalie Denk, Alexander Pfeiffer and Thomas Wernbacher** place video game culture in the perspective of Cultural Studies and focus on its societal situatedness. They suggest the directions for research in the pedagogical potential of video games in schools, such as gender equality, career orientation, social inclusion, and more. Two practical projects are introduced, already running in Austrian schools.

Siding up with Jane McGonigall's powerful statements, many believe that games can help solving large-scale real-life challenges. In their presentation of the gamified Ride2Park project, **Constantin Kraus, Simon Wimmer and Thomas Wernbacher** present the study of a smart incentive system for car pooling that can potentially contribute to slowing down climate change. **Thomas Wernbacher, Alexander Pfeiffer, Alexander Seewald, Mario Platzer, Constantin Kraus, Simon Wimmer and Dietmar Hofer**, the research group around the Cycle4Value project, is developing a gamified reward system based on blockchain technology to increase the attractiveness of cycling.

However, gamification is not a universal solution, even though it can be a powerful tool in the right context. **Mario S. Staller and Swen Koerne**r call for the critical assessment of its commonplace understanding, and present their concept of 'non-defining gamification' for pedagogical practice. This concept has been tested in a gamified learning environment already in use for educating police recruits.

GAME, PLAY AND SOCIETY

Is the virtual world for real? It is most certainly so to cybercrime researchers and lawyers. Unregulated internet activity may result in 'mixed reality crimes', as **Alexiei Dingli** names them. Dingli takes a closer look at the mischievous adventures of children on the dark web in his chapter "Children in an online world, victims or perpetrators? - A collection of case studies". Although the younger generations become better accustomed to the virtual realms provided by digital technologies, they still easily fall victim to their dangers and even turn to the dark side themselves. Based on his summary of unlawful deeds, Dingli calls for the new approach to ensure safety of underage dwellers of the internet.

It remains the question what exactly constitutes violence and crime in a digital world. However, there is no question that verbal abuse, 'doxing' and online bullying can lead to psychological suffering and long-lasting traumas, withdrawal from social activities and, eventually, lower quality of life. As an example, **Sonja Gabriel** extensively covers the topic of hate speech in digital games and outlines the possible measures to deal with this problem.

The state of lockdown has drastically influenced ordinary life and the wellbeing of people all around the world. The feeling of helplessness can scale up to the state of 'anomie', when social ties are undone, and society may slip into chaos. For this reason, it becomes particularly important to claim one's political and social agency, and video games can offer the space and the tools for practicing it. In his chapter, **Tobias Unterhuber** carefully examines the situations of the loss of agency in three games and comes to the conclusion that game mechanics are particularly important and meaningful for mapping the limits of one's political agency, not only in games, but in society and culture in general.

The portrait of a typical male gamer as an asocial competitive achiever is often found to be a stereotype. Yet another evidence may be found in the study of **Bastian Krupp** on the effect of games on the development of emotional intelligence (EQ) in the context of media effects research. Based on the results of his online survey, there was hardly any difference between genders in terms of emotional intelligence of gamers, apart from one specific aspect, which was empathy. As a valuable addition, Krupp presents an interesting breakdown of preferred game genres preferred by male and female participants.

Virtual spaces gained special importance during COVID-19, providing comfort, the sense of community and the experience of adventure that became unavailable to many in real life. In their study, **Wilfried Elmenreich and Mathias Lux** compare user activity during the lockdown in three games: *CS:GO*, *Drawful 2* and the old-school MMORPG *Eternal Lands*. All these saw the influx of old and new players during the pandemic times, but they demonstrated different patterns of player activity.

COVID-19 has pushed us to create completely new formats to reach and connect people - despite (or often because of) local distance. **Swen Koerner and Mario S. Staller** are using gamification to facilitate online lessons in Krav Maga during the COVID-19 pandemic at the German Sport University Cologne. The story of "Sneak Gaming" is yet another example of overcoming the challenges of the global pandemics with a new format of gaming events, carried out by **Simon Wimmer, Natalie Denk and Jogi Neufeld**.

GAMES, PLAY
& IDENTITY

GAMES, PLAY AND IDENTITY

EMPATHY AND INCLUSIVITY IN GAMES AND THE PROTEUS EFFECT

Ricarda Goetz-Preisner

Games have become more varied and inclusive. The hero and protagonist is not only a 30-something white heterosexual man anymore. Protagonists in games, often represented by playable avatars, look and behave differently nowadays. Women-avatars can be warriors, protagonists in games can represent different ethnicities or abilities, and avatars can have same-sex relationships and follow new narratives. "The Sims" is one game that has offered these possibilities for a long time. Players can create virtual characters with or without any physical attributes. This inclusive attitude toward the appearance of gender, visual identity traits and sexuality, once a rarity in video games, is becoming more common as games take on more diverse and weightier subject matters. There are different reasons for creating these sometimes called 'serious games' or 'empathy games'. Many of the reasons can be linked to the Proteus effect. The Proteus effect proposed by Yee and Bailenson (2007) suggests that the human embodiment in digital avatars may influence the self-perception of the player both online and offline, based on their gaming avatar's aesthetics or behaviors. Different studies (e.g. Fox et. al 2013) since then focused on how players can be influenced by their avatars. She found that women may be at risk for experiencing self-objectification when their avatars wore revealing clothing (ibd.) or that participants responded better to avatars modelled closely on their real appearances (Fox 2009). Future studies need to clarify the extent of these and show how different avatars can be created to elicit positive changes in attitudes, game play and self-image. This paper provides a literature review and different close readings of games within the concepts of inclusivity, diversity and the Proteus effect in and of games.

Keywords: Games, Inclusivity, Diverse Characters, Proteus Effect, Influence

---

EVOLUTION OF GAMES AND CHARACTERS

Video games have been in a constant state of evolution. From the early beginnings where pixilated graphics and simplistic story lines were the norm, the aesthetics of games have developed into crystal-clear high definition pictures with multi-varied narratives. One example where we can clearly see the development of games is a comparison of the graphics of *The Legend of Zelda* in 1986 and Zelda: Breath of the Wild in 2017. The later game provides 3D graphics with a detailed appearance of the main character *Link*, and even though the game is a fantasy adventure game, it creates the illusion of really exploring imaginative forests and landscapes. The world and perspective that players find themselves now is utterly emerging with the 3rd person perspective and abilities one has, nothing close to the rudimentary graphics and bird perspective from the game in 1986.

Games have also changed in the way gamers are able to interact with them and with their game characters. Characters that players control now have gone from simple forms like squares and circles to hyperactive realistic human-looking avatars. „In early games like Asteroid or Pac-Man, player representation was quite simple (…) as technology advanced, player representation became more detailed" (Graner Ray 2004: 94). Game characters now display unique traits in their appearance and embody different genders, groups of ethnicity or body and ability types. Playable characters also offer unique personalities that easily create an empathic experience for the gamer. According to Graner Ray, the better a game character, she uses the word avatar, is created, the more players feel comfortable and the longer they play (cf. ibd.). A long and intense engagement in a game is in that regard the goal of every game producer. That might lead to the assumption that inclusive player representation in avatar design is benefitting all parties involved in a game.

In this paper, the concepts of game characters and avatars are used synonymously, however in different research the latter is often referred to as a digitized image of the player itself. As different studies use these concepts within their respective definitions in more or the same way, this author will refer to avatars and game characters following a definition by Ahn et al:

"Broadly defined, any form of representation that marks a user's entity can be considered an avatar. (…) Over time, avatars have become more complex creations, rendered in three-dimensional forms with an extensive range of animated movements that aid in the expression of the avatar's personality and supplement various social interactions. Options for individual customization of avatars have increased significantly as well, allowing users to modify a number of physical features including eye color, hair style, height, body shape, clothing, and even facial expressions. Using these diverse features, users have great freedom to build not just a graphical marker of themselves, but virtual humans with distinctive personalities, unique appearances, and individualized behavioral patterns" (Ahn et al 2012).

Another change has become noticeable in long-time beloved characters like *Lara Croft* from *Tomb Raider* or *B.J. Blaskovics* from *Wolfenstein*. These game characters have changed drastically over time with both new technical possibilities as well as feedback from the side of gamers. *Lara*, still appealing to a broad audience as a strong attractive fighter, now wears slightly more clothing when she fights evil and has more realistic body features. *B.J.* is still muscular and hyper-masculine in his appearance but also resembles more or less a real person with realistic facial features and body shape. *Lara Croft* has been subject to different academic papers, focusing broadly on her role, her importance for players, as one of the first women avatars or her sexualized aesthetics. Kennedy (2002) gives an extensive insight in different research about the game character *Lara Croft* and sums up that:

"it is impossible to securely locate Lara within existing feminist frameworks, nor is it entirely possible to just dismiss her significance entirely. These readings demonstrate the range of potential subversive readings, but there exists no real 'extra-textual' evidence to back this up – hence the focus on the text itself, which is on its own inadequate to explore the range of pleasures available from playing as Lara – we can only conjecture." (Kennedy 2002)

Figure 1. © Crystal Dynamics [https://whatculture.com/gaming/10-legendary-video-game-heroes-you-wont-recognise-now?page=9]

METHODOLOGY

This paper provides a literature review of both theoretical game studies works as well as reviews of game experiments that are concerned with identity and empathy in digital games focusing on the so-called *Proteus effect*. Additionally different close readings of games will give empirical examples. The focus lies on so-called triple AAA games, playable on PC and consoles who feature human looking gaming characters. Triple AAA games refer to games with high production value, extensive budgets for advertising and marketing their games and a high number of players. Compared to other media texts they can be referenced to Blockbuster-movies (cf. Demaria & Wilson, 2002). Rather than independent movies or games that are done both as the name suggests, by either independent producers or smaller development teams, big game production companies have more means to create more elaborate games which differ in the way a game is then playable. Games that will be discussed in this paper range in the genre of Role Playing Games (RPGs) which on the one hand offer rather realistic game opportunities, and on the other feature mostly human looking characters.

These games represent different opportunities to explore aspects of our own or different identities and stories. They are also story-driven games that put more emphasis on the narrative within the game, contrary to other genres where the emphasis lies strongly on game play itself (how the game is played). Story-driven games like RPGs also spend more time and resources developing the protagonists of their games with regard to aesthetics and personality traits and pay greater attention to details such as dialogues. Sometimes these games also become other media texts in form of movies or TV shows, like for example *The Witcher*, *Tomb Raider* or *Assassins Creed* have been franchised into high grossing movies.

That said, a (video) game is still a game, so even though these games try to feature realistic narratives or characters, they still provide a some-what out of this world experience where real-world physics, human abilities or stories are not the main objective per se. Nonetheless these games present means to broaden our own empathy towards the lives of others and let the players experience different worlds.

INCLUSIVE GAMES

Inclusivity in games with regard to this paper means to show and include representations of different identities of humanoid characters, as well as the game story shows varied narratives and realities. The term identity describes the way individuals and groups define themselves and/or are defined by others based on their gender, orientation, ethnicity or ability. Identification, as a concept has been studied in the context of interactive and traditional media, and functions to understand how attitudes and stereotypes about groups can manifest themselves in virtual environments. Identification is especially important in the context of video games, because players act "as", as opposed to "with" a game character (Cohen, 2001 qt. in Ash 2015: 3). As Bayliss explains, the terms avatar and character are often used interchangeably to describe the player's means of engagement with a game (Bayliss, 2007). A game character however, "exists in a world where meaning is always-already present, perhaps even more so since both the character and the game-world that they inhabit are designed as part of a greater system. Simply put, both the character and the game-world in which they operate are consciously designed things, created to fulfill the specific purpose of providing a gameplay experience to the player." (ibd.). Players may or may not act with several game characters in a game. An avatar can be most of the times customized to the player´s design wishes (within the game logic), if there is an avatar, players mostly play like that embodiment throughout the game. Avatars are in that logic modeled after real people (sometimes also after celebrities). A thorough analysis of the differences of the distinctions or means of creation of playable characters will not be provided, in this paper avatars and game character are used more or less without distinction except when the references literature distinguishes between these concepts in the way it influences the research.

When we look at inclusivity in games, more often than not the question will be raised of: "Who is represented as lead protagonist?" which leads to another question "In a game, can you really play anyone?" Or do you mostly play muscular, brown haired, slightly-bearded, able bodied 30 year old white heterosexual men who display little emotions?

Figure 2. © https://kotaku.com/brooding-white-male-video-game-protagonists-ranked-1762597481

The answer to that lies in the inherent history in and of games. Historically, young white men have been perceived as the main demographic for the gaming industry (Soukup, 2007). This reflects what type of people are traditionally thought of as gamers and what kind of games and game characters are developed to satisfy these player´s desires and in that regard identities. Game developers tend to cater to that perceived audience, which in turn leads to viewer representation for other people who play games who do not fit this identity (cf. ibd.). Other gamers are then perceived as nontraditional gamers who simply do not fit into this narrow-minded target audience. Nontraditional gamers is a term used to describe anyone who falls outside the established gamer stereotype, such as women, older gamers, people of color, LGBTQ communities (ibd.). Richard (2017) explains that video games and in that regard computing and technology "have significant and historically documented diversity issues, which privilege whites and males as content producers, computing and gaming experts (…)" (Richard, 2017: 36). With the words of Leonard, "just as video games are an area about and for males, they are equally a white-centered space" (Leonard, 2003, p. 3). More often than not, when for example women or POC were included in games, they would reproduce sexist or racist stereotypes. When for example black women appear in games, Dall`Asen (2020) says, more often than not they are depicted in a hyper-sexualized way and held to white beauty standards. She notices that in black female game characters like *Jade* from *Mortal Kombat*, *Sheva Alomar* from *Resident Evil 5*, and *Purna* from *Dead Island*, the avatars are portrayed as having straight hair, light skin and sometimes light-colored eyes (cf. ibd.).

Furthermore queer (or LGBTIQA+) characters in games are still not regularly included. The main protagonists represent heterosexuality. In his master thesis, Wilberg (2011) focused on how race, gender, and the lesbian, gay, and bisexual communities are represented in video games. He analyzed 1.083 video game characters and found only 29 of them to depict queer characters. Of those, 12 were female,17 were male and all were white. The author mentions that certain instances of queerness are more accepted, such as female bisexuals and lesbians, due to some appeal to male fantasies. He also found that white women were consistently shown as fearful, while women of color mostly appeared angry at first, then fearful, with no emotion from lesbian or bisexual women (Wilberg, 2011).

Video game characters are also mostly able bodied and rarely show any disabilities. Carr (2014) points out that ability is so natural in games that "it hides in plain sight when it comes to critique or reflection." While discussions of the need for more diverse characters in terms of gender, sexual orientation, and ethnicity are increasingly common, little attention has been paid to the presence of disabled characters in games (Disturbing Shadow, 2013) (qt. in Gibbons, 2015: 32).

But, do we actually need more inclusive games?

If we just consider a recent study from Austria (ÖVUS, 2019), that shows that 5.3 million people play games and that the average gamer is around 35 years old and of all those gamers 46 % are women, we do need more inclusive games. It makes sense that game characters have to get more inclusive. Because in the end game developers want to sell their games to a wide audience. There is also a 2019 survey conducted by Electronic Arts, one of the biggest game producers, of popular games such as *FIFA*, *Star Wars* or *Battlefield*, which states that 56% of the 2,252 survey participants consider that it important for companies to make their games more inclusive. Only 13% of the people surveyed felt that inclusivity in gaming was explicitly not an important topic to them (Gravelle, 2019).

If you analyse online comments about new (inclusive) game characters you will however find the typical misogynist or racist slurs next to praises for these changes. In a 2019 *YouTube* video with more than 110.000 clicks and the title "*The Desire for Representation in Games - An Honest, Open Conversation*" for example, you can find comments such as:

"'The whole point of diversity is that it should feel natural' (…) As a mixed-race woman, I always enjoy seeing this kind of character on screen, and though I think representation is important - when it's well-done and natural - as long as the character is well-written and the movie/game/series is good, I don't mind if I don't "see" myself. I have nothing in common with John Marston or Joel and yet they're ones of my favorite video games characters. If diversity is used to make more money, one can tell, it doesn't ring true." (Simbi)

"People are too concerned about 'relatability', when actually that doesn't matter if you enjoy the content. That only matter if you are an extremist or a blind activist, if you are normal person the only thing that actually matter is 'is it good?'. (…)." (Isori)

It is interesting that the commenter with the nickname Isori references "normal people" because the discussion in video games as well as other forms of representation in media surrounds the question of what is the norm and the need to represent it and/or others, which leads to another discussion of "othering" when talking about any form of representation that differs from the inherent logic of media codes that have been reproduced for a long time as mentioned above in the inherent history of game production.

Shaw (2017) describes the responses to her study about the need for representation that can be summed up in three categories: Some participants did express the possibility that representation had a great effect on others, particularly on children and young adults. "However, they rejected market logic-driven forms of representation that only represent groups well when they are being targeted as consumers." (2017: 55) The second group of participants insisted that they did not necessarily need to see one specific aspect of their identity (like sex, gender, sexuality, ethnicity, etc.) represented in their media texts, "because their identities and relationship to media characters were complex enough to let them feel like they could connect with a wide array of media characters" (ibd.). But what the third category needed, was to see people like them, in an expansive intersectional sense, to be seen. Representation matters because it makes their identity legible (cf. ibd.).

Inclusive Game Characters

Games from the perspective of this paper can be seen as rather influential on gamers. Gamers often infer gender, age, race, and personality traits from human-like avatars, just as they would when meeting another person (Guadagno et al., 2011 qt. in Fox et al., 2013: 930). Game scholars like Kaye and Bryce (2012) say, that games exist not just as entertainment but powerful tools of social integration. When different people are able to share their experiences with others through games, some sort of integration and in wider sense cultural capital is built. That cultural capital permits the feeling of inclusion in society. Gibbons (2015) quotes Cross (2014) that there is a vital need to address exclusion and harassment in gaming communities, arguing that we need to embrace new cultural scripts that will allow for greater inclusion of women, people of color and LGBTQ people in gaming culture and design. Representation of different identities and realities can strongly contribute to these new scripts. Shaw (2017) argues against the constant justification and perspective of defense for inclusivity, or as she calls it representation in games. She claims that she has "yet to hear a good articulation of why diversity in representation is bad for anyone. All of the arguments I have seen dismiss the discourses in favour of "representation does matter" out of hand, but they never actually make the case that diversity in media is a bad idea. At most, they point to specific failed commercial examples, but even then fail to take into account marketing explanations for those failures. New arguments for representation can offer new possibilities for what representation and critiques of representation can look like" (2017: 56).

Games in that sense and in recent years have definitely become more varied and more inclusive. The hero and protagonist can be strong female warriors who wears non-revealing clothing and does not have to be saved by their male counterparts. Male characters can show feelings and play in narratives that are more diverse and avatars can have same-sex relationships, be of color or differently abled. The difficulty for these depictions as already mentioned is not to recreate stereotypes. Inclusive games can be achieved more easily, when people from diverse backgrounds are included or heard in the respective game development. Some recent examples of inclusive game characters shall be mentioned at this point.

Women game characters

Rather than assets, sexy booth babes or damsels in distress (Sarkeesian, 2013), female protagonists do not have to be tropes anymore. There are different examples nowadays of strong female characters who wear body-covering clothing and do not represent sexist stereotypes about women. Such as game characters like *Aloy* from *Horizon Zero Down* who is a fierce fighter that wears armor that covers her whole body. The same game also features more than one women as vital characters; women are represented in different ways and have different roles in the game. None of the depicted women serves to decorate the game or fulfil pseudo-erotic desires.

Hillier sums up the discussion about abandoning over-sexualized avatars in games:

„The industry's reliance on over-sexualised, impossible female design is somewhat insulting to those who've grown past the point of getting erections from passing bra stores. (...) Surely most gamers don't play video games because they're looking for female companionship. They play games because they want to blow something up, or drive a fast car, or be enthralled by a beautifully realised fantasy world" (Hillier, 2012).

With *Battlefield V*, a well beloved combat game, you finally can play female warriors. Even though this release has provoked also criticism of some of the gamers that it is ´unrealistic` to play female fighters, they seem to be okay with jumping out of helicopters and surviving, reanimating lost team members or fighting tanks with guns. Also *FIFA 16* finally features women´s leagues, which was long overdue. The popular battle royal game with younger players, *Fortnite*, included different female and male heroes from the start, they also seem to represent more diverse body types – there are muscular women, petite men, android characters or curvy ones. In *Assassins Creed Odyssey* you can choose to play either *Kassandra* or *Alexandros* which makes none or little difference to the story line. This fact is important because the skills of a game character, or in other words, how powerful and vital the character is to the story, makes it enjoyable to play a character.

Queer game characters

The Sims was one of the first video games where characters were able to be in same sex relationships. Next to binary options of character choice, games can also offer gender fluid characters. *The Sims* has offered that possibility since 2016. According to Shaw (2014) many teens who come out as a member of the LGBTQ community often feel alone and they may not have the same support systems as those who identify as heterosexual. Often media texts like games can function as key channels for these youths to observe and learn social roles, rules and norms and feel represented. There are different stories of players that show that these possibilities helped people to find their identities as well as with their sexual orientation. Rachel Franklin, the vice president of Electronic Arts and general manager of Maxis, the studio behind *The Sims*, said: "It has always been important to us to provide our players with powerful ways to express themselves and tell a wide range of stories — whether they're customizing their Sims' age, skin color or gender." (qt. in Parker, 2016). Jansz et al. (2010) explain that *The Sims (I and II)* also became a very popular game with girls and women due to the developer's desire to build a game for everyone. So to speak an inclusive game. They game did not market to a particular group of players and with these goals in mind, the game avoided gendered stereotypes in content and marketing.

Other LGBTQ characters that were featured in older games such as *Fallout 2* in 1998 and *Fable* in 2004 allowed same-sex marriage between characters. The role-playing game *Mass Effect* in 2007 and *Dragon Age* in 2009 introduced LGBTQ characters as more vital characters. Since then, many games like the very successful action adventure survival horror game *The Last of Us I* and II introduced LGTBIQ characters as protagonists. The latter game has also received numerous awards and praises for its depiction of very 'human-like' diverse characters, story lines and featuring more than one LGBTIQ character.

Game characters of color

Game characters of color have been included in a variety of games, especially game characters depicting Asian ethnicities have been featured in fight games and games produced by and for the Asian gaming market. Other avatars of color such as black characters were more or less displayed as sidekicks to the protagonists and displayed more often than not in racist stereotypes. Ash (2015) finds that there is different evidence of the stereotype that links African Americans to aggression which can be found in different American media such as games. Although gamers of color constitute a large part of the digital gaming player base, especially in the US, the stereotypical representation in games is rather persistent.

However more and more games feature lead characters in big games such as *Uncharted: The Lost Legacy* which was released in 2017 (cf. Dornieden, 2020). The game gained a lot of attention as it was set in India, had two lead women, and one of them, *Nadine Ross*, is a black South African. Other big releases include *Assassins Creed Origins*, which is set in Egypt with an African protagonist, while *Star Wars Battlefront II* used the likeness and voice of Janina Gavankar, an actress with part-Indian heritage (ibd.). Chella Ramanan, a journalist and game developer tells in an in interview that diversity in games is:

"about including a broad range of voices and experiences in your games so that people from different backgrounds see themselves represented. (...) That is important because it might make them think that they can also come up with a cool story for a game with people in it who look like them. (…) There are more black leads coming through but we just want to encourage even more diversity and more voices of young people who may otherwise think the games industry is not for them." (qt. in Sheikh, 2017).

Also other big game series such as the action-horror game *The Walking Dead: The Final Season* feature two protagonists of color in the latest addition: *Clementine and AJ.* Even though the game is rather violent, the characters in their actions and conversations show different feelings and make the story that much more relatable.

Figure 3. © https://www.theguardian.com/games/2019/mar/26/telltale-the-walking-dead-the-final-season-review-playstation-xbox-pc-nintendo

Another good example is the already mentioned game *The Last of Us Part II* which features inclusive characters besides the protagonists. Dornieden (2020) says that this game features characters that represent a variety of people in a way that seems natural and well placed. There are characters of Asian descent like *Jesse, Lev* and *Yara, Dina,* a Jewish woman who is the protagonist's (*Ellie*) girlfriend, *Isaac,* an African American leader and *Manny,* a Latinx member of the fictional Washington Liberation Front. Dornieden (2020) explains what makes this diversity special: "None of these characters' race or ethnicity serves as an important plot point, they're all just normal people trying to survive in a mushroom zombie world. What makes representation effective is not only visibility but ensuring that characters are included without forcing them to be tied to whatever stereotypical trauma their demographic endures."

Differently abled game characters

Disability studies with regard to games can explore whether the content of a game allows for multiple understandings of disability (Gibbons, 2015: 27). Another way in which disability can be represented in a game is through simulations that allow players to explore alternative sensory or personal experiences that let you embody life from a different perspective. However these games are more or less independent games and rarely done by big game production companies.

One famous example by a big gaming company however is *Lester Crest* from *Grand Theft Auto*, who has an unspecified disease, which gradually wears down his motor skills as he ages. He also has asthma. He is nonetheless a vital and beloved character in the *GTA* series. Other examples can be found in the new *Marvel´s Avengers*, where there is character named *Cerise*, who is a superhuman in a wheel chair, however which is unfortunate, she is not a playable character. *BJ Blazkowicz* is also wheelchair-bound for the early parts of *Wolfenstein II: New Colossus*, and the game also lifts the veil on his mental health. Ability studies and in that regard ability game studies are also concerned with mental ability. The game *Dys4ia* by Anna Anthropy and Zoë Quinn's *Depression Quest* encourage players to empathize with experiences similar to those that the developers have personally experienced. *Dys4ia* allows the player to experience many of the frustrations that accompanied Anthropy' s experience of hormone replacement therapy. Quinn's *Depression Quest* is an interactive game that allows players to make decisions from the perspective of a person who experiences depression. People that have experienced the game narratives first hand created these games and are therefore able to share these experiences with players to create an empathic game experience.

PROTEUS EFFECT

Inclusivity with regard to the appearance and behavior of characters is becoming more common, as games also take on weightier subject matters. There are different reasons for creating these sometimes called 'serious games' or 'empathy games'. Many of the reasons are linked to what is known as the Proteus effect. The Proteus effect proposed by Yee and Bailenson (2007) suggests that the human embodiment in digital avatars may influence the self-perception of the player both online and offline, based on their gaming avatar's aesthetics or behaviors.

"The Proteus effect occurs when a user's self-representation is modified in a meaningful way that is often dissimilar to the physical self. The user then embodies the self-representation, observes him or herself behaving in this virtual form, and draws inferences regarding his or her internal beliefs or attitudes based on these observations." (Yee et al., 2013: 932).

The Proteus effect has been studied in several studies conducted in different virtual settings (Yee and Bailenson, 2007, Konjin et al., 2007, Fox and Bailinson 2009, Eastin et al. 2009, Yao et al. 2010, Fox 2013, Ash 2015). The study by Yee and Bailenson (2007) determined that participants using an attractive avatar disclosed more personal information when asked questions later (Yee and Bailenson 2007). Participants also showed more intimacy in the real world than those assigned to a less attractive avatar (cf. ibd.). Similarly, participants who received cues in a virtual setting that they were embodying a tall avatar, behaved more confidently in a similar real-world setting (cf. ibd.). Konjin et al. (2007) found that players may experience similar feelings to the character like feeling distraught or happy and potentially even imitate their character's behavior.

Ash (2015) also interprets the Proteus effect in a way that it is likely that a player using a female avatar would behave less aggressively due to certain views about femininity. The effect she says can also be applied to assumed generalizations about other social groups, also known as stereotypes. Different studies have analyzed, how players can be influenced by their avatars. Jesse Fox has been studying how interactions with digital avatars influence people's offline attitudes since 2009. Fox et al. (2013) for example disagree strongly - due to their exploration of the Proteus effect - that games are merely there for entertainment. Especially the sexist portrayal of female characters in a game has real life effects on the players. In a 2013 study, they found that women may be at risk for experiencing self-objectification when their avatars wore revealing clothing:

> "This study supported the Proteus effect and demonstrated that there are psychological consequences associated with embodying sexualized avatars. The findings here added new insights on the effects of exposure to sexualized representations in new media and what happens when images of the self are incorporated. Women who were embodied in sexualized avatars that resembled the self demonstrated greater rape myth acceptance than women who were embodied in other avatars. Women in sexualized avatars reported more body-related thoughts than women in nonsexualized avatars, indicating that sexualized avatars may promote self-objectification" (Fox et al., 2013: 835).

The latter rape myth refers to the assumption that women´s clothing is somehow to blame for, if a woman gets assaulted. Fox et al. quote different studies who analyzed this effect while studying other media texts such as TV shows or pornographic materials, which all functioned to make violence against women more acceptable within the sampled participants. Yao et al. (2010) also found that stereotypical and especially sexist gender representations affect players well after playing a game. For instance, consistent representation of video game female characters as sexualized objects affected how participants perceived women in the real world.

Fox and Bailenson (2009) created avatars based on photographs of participants. They called them "doppelgängers" and compared their effectiveness to representations of unknown people. They found that participants responded better to avatars modeled closely on their real appearances, as opposed to those of unknown people or generic-looking 'perfect' avatars. Doppelgängers have also caused participants to replicate eating patterns from games in the real world (Fox et al. 2009). Ahn and Bailenson (2012) found out that players prefer brands or products endorsed by their gaming doppelgängers. Considering these findings, an avatar's resemblance to the physical self may influence the effects of the Proteus Effect stronger. Ash (2015) tests the Proteus effect in determining whether players ascribe stereotypes associated with aggression to black avatars when playing a video game and how that affects aggressive play. While focusing on a boxing game she found that the Proteus effect was higher, demonstrated in those players who experienced greater embodiment within their avatars. Pena, Hancock, and Merola (2009) found that participants whose video game avatar wore a black cloak even displayed more aggressive intentions and attitudes compared to those using white cloaked avatars in the game. Research by Eastin et al. (2009) how a Black gaming avatar would influence video game experiences. Their research demonstrated effects of the avatar race on postgame hostile thoughts, or aggressive cognition, and that White participants who played a violent game as a Black avatar showed higher levels of hostile thoughts after game play compared to White participants playing as a White avatar. The authors discussed these findings as support for the Proteus effect, concluding that the increase in aggressive thoughts for White players, playing as a Black avatar is explained that those participants manifested negative stereotypes about African Americans that associate Black avatars with aggression-related concepts, so they played more aggressively in the game.

However, Konjin et al. (2007) only looked at adolescent boys and the influences avatars have on them. Fox focuses a lot on first person shooters in her different studies or creates avatars that you do not play in an emerging fantasy, but in a scientific setting, which might help to find comparable results but rather misses the special traits of gaming. Ash (2015) only looks at one boxing game for his findings. Most of the studies presented have a rather limited scope – Fox et al. (2013) looked at 92 women in their studies about sexual objectification, which is a big sample in comparison to other studies. Other studies only look at one genre of games but make conclusions about all kinds and genres of games. The literature the studies draw on also focuses on other media texts than games, such as TV shows, film or other entertainment formats.

The extent of the Proteus effect also strongly depends on the player´s involvement in games, how much do they play, which games do they play, how long have they been playing and what other indicators define their personal and social life. Games are interactive, the way we play a game or how it is even playable – referring to game play – is of utmost importance because a game is nothing static, it only works with the player and then there are also big differences in how different players approach and interpret the same game. Fernandez-Vara (2015) says the analysis of games can´t simply be compared to that of other media texts such as film or literature, because games are an expressive medium, which do not simply tell a message to the player. The player becomes a necessary part pf the text. This statement follows closely what the *Material engagement theory* suggests, that in contrast to other media types, video games promote and ask for the direct involvement of the player (Jansz 2005).

Future studies need to clarify the extent of these effects while considering game play. They might also focus on how different avatars and narratives might elicit positive changes in attitudes and self-image as well as provide new (and at times empathic) experiences to players.

Serious Games

So why do we create inclusive games and game characters that try to elicit empathy when games should be mainly fun and depend on the player´s engagement?

Bourazeri and Pitt (2013) say that serious or empathy games create environments, where features such as thought provoking, informative or stimulating are as important as fun and entertainment. They can empower different groups and communities, function for teaching or raising awareness, enable users to develop new expertise and capabilities or even might show users the consequences of their behaviors. Eastin et al. (2009) define empathy in a game in the way that the act of the game playing activity with an avatar creates a mental connection between the self and the avatar, triggering particular behaviors related to the avatar with which the user identified.

In recent years, different kinds of these serious or empathic games were published. Two games that received a variety of both praise and critique were *Papers, Please*, which puts players in the role of an immigration officer at the border of a fictional country. The other game *That Dragon, Cancer* lets players feel the grief of family loss, based on the experiences of two of its creators, Ryan and Amy Green, whose son died of cancer in 2014 (Parker, 2016). Such games elicit a different empathy and involvement with the player. Järvinen (2009) uses Kubovy's categories of *pleasure of the mind* that he adapted to video games to explain that empathic category as one of "suffering". He says that the suffering category finds its realizations in the paradoxical nature of player motivations, that is, the player's willingness to play even in the face of potentially suffering loss or experiencing negative emotions. This paradox has been explained and applied to psychological theory with the concept of "metamood". The term accounts for a mental process where individuals experience unpleasant emotions on the object level, but also positive emotions and enjoyment on a meta-emotional level. This achieves other goals and purposes than simply being entertained (cf. Järvinen, 2009: 106).

One example for an empathic serious game is *Hellblade Senua´s Sacrifice*.The game despite being a formidable RPG with great fighting possibilities, is designed to provide an empathic experience about mental health. The female protagonist, *Senua*, is a Pict warrior with an anything but sexist look, who struggles with her mental health. *Senua* embarks on a journey in the 8th century after Vikings raided her village and killed her love *Dillion*. She hopes to redeem the soul of *Dillion* by going to the Norse version of hell, Hellheim, to confront the underworld goddess, *Hela*. She is plagued by her own inner darkness, and in the game this manifests as somewhat distracting visions.The voices in her (your) head never stop while you are playing, and she is constantly followed by the sinister entity she only knows as *the Darkness*. The game was designed with neuroscientists, mental health specialists, and people who experience different psychotic symptoms, which makes it even more relatable (Takahashi 2019). The game not only won many awards, but helped a lot of people. Paul Fletcher, a professor of health neuroscience from Cambridge University was a consultant for the game. He said the game changed many lives, as players with mental illness who played the game or knew people with mental illness contacted the production company by the hundreds to share what it meant to experience such a realistic portrayal of the illness (cf. ebd.).

Järvinen (2009) says there are specific game genres that elicit nurturing and caring actions among games, such as caring for virtual pets, maintaining social relationships and striving for the well-being of characters like in *The Sims*. Yet also player roles such as football managers and urban planners can be seen to afford the pleasures of nurturing. The pleasure of nurture can be elicited in a number of ways in games, but it is useful to point out the consequences of different game themes, that is, subject matter and metaphors for rules, in the elicitation of nurturing (Järvinen, 2009: 105). *Spiritfarer* is a somewhat serious game which falls into this genre of gaming. The design of the game is as colorful and friendly as possible and the game play works without any (fantasy) violence. It is nonetheless a very interesting game, combining caring, exploring, puzzle-guests and strategic playing. If one remembers the hand-held cult game *Tamagochi*, the whole game is one. You are a girl, owning and maintaining a boat, rescuing (fantasy) animals and helping them with their (last) desires and at the same time you are responsible for taking your guests on 'their last boat ride'. A player not only constantly worries about the wellbeing of all guests, through stories and conversations you get to know the dreams and sorrows of them, as well as your (the avatar's) own experiences with loss and sorrow. What makes a game a good game, is that it is still very enjoyable to play and that the protagonist character does not display any stereotypical gender-traits at all.

CONCLUSION

Games can show and let emerge you in fantastic but at the same time empathic experiences. Different studies were presented in this paper focusing on the Proteus Effect and how game avatars affect players. However most of these studies show a rather limited perspective on games and how players interact with them. One conclusion that can be made, is that games are played by a variety of players and therefore should also embody a variety of narratives and game characters to make all players feel included and to move away from stereotypical portrayals of in-game marginalised groups such as women, LGBTIQ persons, people of color or with different abilities. Recent popular games have followed this assumption more and more.

The overall conclusion of this paper follows a quote by Fernandez-Vara who says: "rather than limiting ourselves to thinking about games as a medium to convey messages, we can think of them as artefacts that encode certain values, which players decode and engage with as they play." Good games draw the player in, provide them with an experience that might be either completely different or very close to their own lives and identities to either help with the feelings the player experiences or provide new empathic ones. A game however must still create an enjoyable playable experience as the player engages with the game and is not a passive viewer.

While studies often focus on how games might affect experiences of identity and empathy, efforts to make inclusivity an inherent part of gaming can be noticed in different games. Both game players and developers have begun to acknowledge the desire for greater inclusion and representation of diversity. Efforts to include and represent different identities of people should consider a diverse display of these identities while including people from those groups in the production of said game. Without reinforcing stereotypes, games can provide different forms of empathic experiences to players and satisfy the diverse market of players. Characters that represent women or queer people do not have to be sexualized, people of color or with different abilities are created in an inclusive non-stereotypical way, they do not have to be side-kicks but take up active roles in the game. We have come far with games that embody empathic experiences and depict a variety of identities. The future will only bring more, so do not stop playing.

REFERENCES

Ahn, S., J. Fox and J. Bailenson (2012). *Avatars.* Leadership in Science and Technology: A Reference Handbook

Ash, E. (2015). *Priming or Proteus Effect?: Examining the Effects of Avatar Race on In-Game Behavior and Post-Play. Aggressive Cognition and Affect in Video Games.* In Games and Culture 11(4)

Bailenson, J. and N. Yee (2007). *The Proteus Effect: The Effect of Transformed Self-Representation on Behavior.* Human Communication Research, 271–290. Retrieved from URL https://vhil.stanford.edu/mm/2007/yee-proteus-effect.pdf

Bayliss, P. (2007). *Beings in the Game-world: Characters, Avatars, and Players.* Retrieved from URL https://peterbayliss.edublogs.org/files/2008/02/bayliss-ie2007.pdf

Bourazeri, A. and Jeremy Pitt (2013) *Serious Game Design for Inclusivity and Empowerment in Smart Grids. Imperial College London.* Retrieved from URL http://www.fdg2013.org/program/workshops/papers/IDGEI2013/idgei2013_6.pdf

Dall`Asen, N. (2020). *The Future of Diversity and Inclusion in Video Games.* 2020/10/18 Allure Magazine. Retrieved from URL https://www.allure.com/story/video-game-inclusive-beauty

Demaria, R., & Wilson, J. (2002). *High score!: The illustrated history of electronic games.* McGraw-Hill Osborne Media

Dornieden, N. (2020). *Leveling Up Representation: Depictions of People of Color in Video Games.* PBS 12/22/2020. Retrieved from URL https://www.pbs.org/independentlens/blog/leveling-up-representation-depictions-of-people-of-color-in-video-games/

Eastin, M., O. Appiah & V. Cicchirllo (2009). *Identification and the Influence of Cultural Stereotyping on Postvideogame Play Hostility*. Human Communication Research 35: 337–56.

Fox, J., and J. Bailenson (2009). *Virtual self-modeling: The effects of vicariousreinforcement and identification on exercise behaviors*. Media Psychology 12,1–25

Fox, J., J. Bailenson and J. Binney (2009). Virtual Experiences, Physical Behaviors: The Effect of Presence on Imitation of an Eating Avatar. Department of Communication Stanford University. Retrieved from URL https://vhil.stanford.edu/mm/2009/fox-pres-virtual-experiences.pdf

Fox, J. et al. (2013). *The embodiment of sexualized virtual selves: The Proteus effect and experiences of self-objectification via avatars*. Computers in Human Behavior 29 Retrieved from URL https://vhil.stanford.edu/mm/2013/fox-chb-sexualized-virtual-selves.pdf

Fernandez-Vara, C. (2015). *Introduction to Game Analysis*. Routledge

Gaming in Austria 2019 Österreichischer Verband für Unterhaltungssoftware (ÖVUS) Retrieved from URL https://www.ovus.at/news/ueber-fuenf-millionen-oesterreicher-spielen-videospiele/

Gibbons, S. (2015). *Disability, Neurological Diversity, and Inclusive Play: An Examination of the Social and Political Aspects of the Relationship between Disability and Games*. Loading…The Journal of the Canadian Game Studies Association Vol 9(14): 25-39

Gravelle, C. (2019). *Players Want More Inclusive Games, According To EA*. Retrieved from URL https://screenrant.com/ea-survey-inclusive-games/

Graner Ray, S. (2004). *Gender Inclusive Game Design: Expanding the Market*. Hingham, Massachusetts, Charles River Media

Hillier, Brenna (2012). Tameem Antoniades and the trouble with tits. VG 24/7. Retrieved from URL https://www.vg247.com/2012/03/22/tameem-antoniades-and-the-trouble-with-tits/

Jansz, J., C. Avis, & M. Vosmeer (2010). *Playing The Sims 2: An exploration of gender differences in players' motivations and patterns of play*. New Media & Society, 12, 235-251

Jansz, J. (2005) *The emotional appeal of violent video games for adolescent males*. Communication Theory 15 (3), 219-241

Järvinen, A (2009). *Understanding Video Games as Emotional Experiences*. The Video Game Theory Reader 2 (eds. E. Perron & M. Wolf). 85-109

Kaye, L. K., and Bryce, J. (2012). *Putting the fun factor into gaming: The influence of social contexts on the experiences of playing videogames*. International Journal of Internet Science, 7(1), 24–38

Kennedy, H. W. (2002). *Lara Croft: Feminist Icon or Cyberbimbo? On the Limits of Textual Analysis*. Game Studies, 2(2). Retrieved from URL http://www.gamestudies.org/0202/kennedy/

Konijn, E. et al. (2007) *I wish I were a warrior: The role of wishful identification in effects of violent games on aggression in adolescent boys*. Developmental Psychology, 43/4, 1038-1044

Leonard, D. (2003) *Live in your world, play in ours: Race, video games, and consuming the other.* Simile, 3(4), 1-10

Parker, L. (2016). *Video Games Allow Characters More Varied Sexual Identities.* New York Times Aug. 31., 2016. Retrieved from URL https://www.nytimes.com/2016/09/01/technology/personaltech/video-games-allow-characters-more-varied-sexual-orientations.html

Richard, G. (2017). *Video Games, Gender, Diversity and Learning as Cultural Practice: Implications for Equitable Learning and Computing Participation Through Games.* Educational Technology March-April 2017. 36-43

Sarkeesian, Anita (2013). Tropes vs Women in Video Games. Feminist Frequency. Retrieved from URL https://feministfrequency.com/video/damsel-in-distress-part-1/

Shaw, A. (2014). *Gaming at the edge: Sexuality and gender at the margins of gamer culture.* University of Minnesota Press

Shaw, A. (2017). *Diversity without Defense: Reframing Arguments for Diversity in Games.* Kinephanos, 54–76. Retrieved from URL https://www.kinephanos.ca/Revue_files/2017_Shaw.pdf

Sheikh, R. (2017). Video games: How big is industry's racial diversity problem? BBC Asian Network 09/20/2017 https://www.bbc.com/news/technology-42357678

Soukup, C. (2007). *Mastering the game: Gender and the ethical motivation system*

of video games. Women's Studies in Communication, *30:* 157-178

Pena, J., J. Hancock & N. Merola, N. (2009). *The priming effects of avatars in virtual settings.* Communication Research 36, 838–856

RobinGaming (2019). The Desire for Representation in Games - An Honest, Open Conversation. Retrieved from URL https://www.youtube.com/watch?v=405_YpDzL0U&t=224s

Takahashi, D. (2019). *How Hellblade: Senua's Sacrifice changed lives with its thoughtful portrayal of mental illness.* Games Beat Oct. 26., 2019. Retrieved from URL https://venturebeat.com/2019/10/26/how-hellblade-senuas-sacrifice-changed-lives-with-its-thoughtful-portrayal-of-mental-illness/

Wilberg, H. M. (2011). *What's in a Game? Race, gender, and the LGB representation in video games.* Master's thesis Roosevelt University

Yao, M. Z., C. Mahood & D. Linz (2010). Sexual priming, gender stereotypes, and likelihood to sexual harass: Examining the cognitive effects of playing a sexually explicit video game. *Sex Roles, 62,* 77-88

Yee, N., and Bailenson, J. N. (2007). *The Proteus effect The effect of transformed self-representation on behavior.* Human Communication Research, 33, 271–290.

THE MISSING'S MISDIRECT: OR HOW I CAME TO STUDY TRANSGENDER EXPLORATION, EXPRESSION AND EMBODIMENT IN VIDEOGAME-BASED-LEARNING

Josephine Baird

Hidetaka Suehiro's (SWERY) game, *The Missing: J.J. Macfield and the Island of Memories* (*The Missing*) (2018) presents an LGBTQ narrative by initially misdirecting the player with an implied, and quite different, LGBTQ narrative. The game is designed to suggest that the player-character's narrative goal is to seek out another (non-player) character who, it is implied by careful use of metaphor and mechanics, is the player-character's lesbian romantic partner. This careful construction of game design obscures an eventual reveal that the narrative was rather the dreamlike experience of a transgender (trans) woman's near-death exploration of her own identity, experience, and trauma. I present an autoethnographic close-reading of the game's intersectional design features in order to demonstrate how it does this, to consider why, and to examine its potential to achieve the designer's stated goal; which is to teach the player empathy for trans people (SWERY, 2019). I show how the game's design demonstrates an awareness of its wider socio-cultural context – a context that the game designer is in turn trying to impact with the game. I will show how this context is leveraged, which tropes and conventions the game deploys and subverts, and what presumptions of the player the game relies on to make this misdirect function. I present my analysis of *The Missing*'s (2018) design misdirect as an "assemblage" (Taylor, 2009) of game elements combined with recognisable LGBTQ narrative conventions (Shaw & Friesem, 2016); all with the goal to provide a specific pedagogical moment, performed partially through this misdirect. I compare my findings to commentaries and reviews by others, presentations and interviews with the designer himself and to some of his other game designs that feature LGBTQ references and representations. I position this reading as an introduction to my wider research into how games might provide an opportunity for the exploration, expression and embodiment of trans subjectivities through videogame-based-learning.

Keywords: Transgender, Videogames, Game Design, Game-based-learning, Close-reading

---

INTRODUCTION

I feel it is only reasonable to begin this article with a content note about some of the issues I will be discussing, not least because the game I analyse starts with a similar note but also because the subject of the game and this article reference trauma and experience that is all too common in the world and all too often under-represented. The game and this article feature discussions of self-harm, suicide, transphobic and homophobic abuse, depression and anxiety – and the representations of these are particularly harrowing in this game despite the fact that some of them are, at least initially, obscured by metaphor, implication and misdirection.

The Missing; J.J. Macfield and the Island of Memories (*The Missing*) (2018) is a videogame that represents the dreamlike mental experience of a transgender (trans) woman who is dying from a suicide attempt brought on by discrimination that she faced for being trans. It does so with an initial misdirection which is achieved through a combination of game mechanics, narrative conventions (and subversions thereof), musical and other artistic cues; as well as an acute awareness of the context within which the game is released and the experience of the player the designer wishes to influence. I explore this misdirect in terms of how it functions towards the designer's goal to create a pedagogical moment that could create greater empathy for trans people. I present this reading as an introduction to analysing games that are designed with the intention of educating on trans experience and to the consideration that trans subjectivities might be explored, expressed and embodied in videogames and through videogame-based-learning.

THEORY AND METHODOLOGY

I locate my reading of *The Missing* (2018) in theory and analysis of the last decade that sees games as both products and producers of socio-cultural discourse and as sites for social modelling, representation, and opportunity for change in relation to LGBTQ issues (Shaw, 2011, 2014; Anthropy, 2012; Murray, 2017; Marcotte, 2018; Ruberg, 2019). Specifically, I base my analysis in the argument that games might provide pedagogical opportunities and mental health tools to improve understanding, and experience, of LGTBTQ people in often hostile socio-cultural contexts (Allen, 2016; Kostopolus, 2017; Straus etc al, 2017; Egan, 2019) and provide a space to learn to challenge normative structures that lead to such a context (Flanagan, 2009; Klein, 2020). However, I do not do so uncritically, nor with the presumption that just because videogames might provide such an opportunity, that they necessarily will. Games-makers and analysts alike have challenged an unproblematised view that games can teach simple "empathy" for marginalised people and that this will necessarily lead to social change (Pozo, 2018; Ruberg, 2019).

To account for the multiple and intersecting considerations that go into game design and a determination of whether, and how, they might succeed at their queer potential, I contend a close-reading methodology (Bizzocchi & Tanenbaum, 2011; Keogh, 2014; Ruberg 2019) is appropriate. This works as part of a versatile "conceptual toolkit" (Keogh 2014) that can account for games as "assemblages" (Taylor, 2009); a perspective that sees videogames as products of multiple complex intersecting elements and socio-cultural influences that must be considered to get a fuller picture of their cohesive function and potential.

I believe this methodology is especially appropriate to a rigorous reading of *The Missing* (2018) as the creator himself, Hidetaka Suehiro's (SWERY), has said he approached making the game with a similar "cohesive" design philosophy in mind (2019); a game which he argues can teach people to develop an empathetic response to marginalised peoples (2019). I critically examine the game therefore as an attempt at a videogame-based-learning approach to the subject of trans experience and in relation to its potential to provide a site for the exploration, expression and embodiment of trans subjectivities.

I perform this close-reading with a reflective and autoethnographic approach (Marcotti, 2018; Rapp, 2018; Brulé & Spiel, 2019) to my analysis. Doing so, I argue, locates my own positionality and experience of the game in relation to the other sources that I present, including those provided by game industry commentators and journalists.

AUTOETHNOGRAPHY

I chose to play *The Missing* (2018) because I had heard it featured a "positive" LGBTQ narrative. Though queerness certainly can be found even in games that do not otherwise seem to make any reference to LGBTQ people (Shaw & Friesem, 2016; Ruberg, 2019), there still are few explicit LGBTQ narratives in mainstream games (Shaw, 2009, 2014).

The Missing (2018) starts with its queer narrative only being alluded to and far from explicit. Despite this, and from the outset, I thought I knew exactly how the story would eventually play out and what "the big reveal" at the end would be. Given the metaphors and clues the game was deploying from the very first messages displayed upon starting the game and the opening scenes, I was certain that the titular (player-)character, J.J., was hiding her romantic involvement with another (non-playable) character (Emily) and her sexuality from her family and friends. I was sure that the game as a whole functioned as a surreal metaphor of her sexuality's non-consensual exposure ("being outed") and the trauma of that experience. I recognised the game's framing from many examples that use similar metaphor and narrative "hints" instead of overt LGBTQ narrativization to tell such stories – a process which Shaw and Friesem (2018) describe in their extensive cataloguing of how LGBTQ people are represented in games, including metaphorically, which in turn draws inspiration from the likes of Russo (1987) and Epstein & Friedman (1995) who did the same with their examination of LGBTQ (metaphorical) representation in narrative film. Every aspect of *The Missing*'s (2018) cut-scenes, gameplay, text, and even "achievements" that pop up during certain sections, suggested to me that what was being alluded to throughout the game was a recognisable, if obscured, lesbian narrative. Only by the time the game was nearly finished, was I shocked and intrigued to finally realise the misdirect – that J.J. was instead dealing with the exposure of being outed as trans to her family and friends. On reflection, the misdirect seemed intentional because of the way genre and gameplay conventions had been deployed and then subverted.

After completing *The Missing* (2018) for the first time, I read reviews and commentaries and watched playthroughs of the game by others and found many experiences that were similar to my own (M., 2018; Blondeau, 2019; Cox, 2019; Dale, 2019). I read presentations by, and interviews with, the designer, which seemed to confirm some of what I had suspected about his design goals and that the misdirect was indeed intentional. I present here the resulting close-reading of this misdirect, how it functions as a result of the combination of the multiple and intersecting game design elements, the socio-cultural context of its production and reception, and the reported intentions of the designer.

SETTING THE (CUT-)SCENE

Upon loading the game, and following an initial graphics and sound options menu, the screen goes black and is filled with the message "This game is made with the belief that nobody is wrong for being what they are" in white bold letters. This message is then replaced with "This game contains explicit content, including extreme violence, sexual topics and depictions of suicide."

The first text is reminiscent of the messages that appear at the beginning of Ubisoft's *Assassin's Creed* videogames. The first *Assassin's Creed* game from 2007 features the introductory message, "Inspired by historical events and characters. This work of fiction was designed, developed and produced by a multicultural team of various religious faiths and beliefs." This message appeared at the beginning of every *Assassin's Creed* game in the ongoing franchise until 2015, with the release of *Assassin's Creed: Syndicate*; which included an amended version that reads, "Inspired by historical events and characters, this work of fiction was designed, developed, and produced by a multicultural team of various beliefs, sexual orientations and gender identities." Coincidentally, this game also featured the first woman player-character in the mainline series of games and the franchise's first trans (non-player) character (Phillips, 2015). This new message has remained the same for every *Assassin's Creed* game since.

Adrienne Shaw notes that the function of these messages is clearly as disclaimer, characterising them as "…the obvious attempt to curb representational critiques by anchoring their right to portray groups in their group's diversity" (2015: 11). *The Missing's* (2018) first message does not appear to have the same disclaimer quality. Rather, in my first playthrough it felt like it was positioned as a statement of belief or focus for the game. This is confirmed by SWERY in a GDC presentation in which he says that "[t]his is the most important message in this game" and functions as a core statement of what he is hoping the game communicates to the player (2019). He goes on to contrast the message with the *Assassin's Creed* style disclaimer, by suggesting that:

> "We had several meetings where we had discussed whether or not to include a message along the lines of 'this work was created by people of various races, faiths and genders.' But looking back at it now, all I can think about is how distasteful it would have been and almost sounds like an excuse." (2019)

Despite the importance that he places on the message, he notes that it almost did not make it into the game. SWERY recalls that, "[a]fter speaking with my advisor for this game, I decided that it was necessary to put this at the beginning." (2019). In an interview with Kazuma Hashimoto (2018) SWERY indicates that he worked with trans consultants during the making of *The Missing* (2018). Dale (2020) credits this consultation with LGBTQ advisors with what she sees as the quality of the trans representation within the game, and contrasts this with games that SWERY has produced which feature what she argues is problematic LGBTQ representation that were created without such consultation.

This statement of belief (rather than disclaimer) serves a second function, I argue, in that it gives the first clue as to the narrative (misdirect) of the game. No game that represents a normative narrative would require such a statement. This is obvious in the fact that such a message does not appear in front of any of SWERY's other games, nor in front of games and media in general. It implies to the player that the game they are about to play is making the overt declaration for a purpose – that the game is likely to challenge a presumably normative and problematic presumption. This demonstrates a socio-cultural awareness of discrimination towards marginalised people and suggests that this is likely to be addressed in the game. This supposition is made clearer in the message that follows, the content note. Such a note does not appear in front of most mainstream videogames (short of a PEGI or ESRB rating of all games sold commercially that functions as an "age rating" for the media), despite the prevalence of violence in many mainstream games. Here, the message functions as both a content note and as a clue as to what the narrative (misdirect) is to come. The note in this case specifically references "sexual topics" in relation to "extreme violence" and "depictions of suicide." Gender is not explicitly mentioned however, so the suggestion seems to be that there is a relation between the first statement and "violence" and "sexual" topics.

Once the messages fade out, an idyllic scene appears of J.J. Macfield, the player-character, standing on a pier with a plush toy (named F.K.) overlooking a sun-dappled lake. In the distance there are clear blue skies and a lush-looking island covered in green trees. Hitched to the pier is a row-boat and the gentle sounds of waves and seabirds can be heard. The game menu and title are superimposed over the scene. Upon starting a new game, the title and options fade and a cheerful call for, "J.J.!" can be heard from off-screen. Another figure – Emily – runs onto the pier and past J.J. before playfully hopping onto the rowboat. Both characters giggle. J.J. takes her plush-toy and walks towards the boat as the scene fades to black.

The scene fades into an ominous view of an island covered with fog. A new voice is heard, saying, "Memoria island. A remote island in North America, just off the coast of Maine…" The image shifts to silhouettes of trees and a foggy sky in the background, as ominous atonal music plays. The voice-over continues, "Filled with artefacts left over from the good old days, this island has the power to awaken the formative memories of all its visitors." The scene changes again to dark clouds over the island, rain in the distance and a lightning bolt striking the ground. The voice-over continues further with, "The island's ancient inhabitants called it 'A'Lapo Gymo." The scene shifts to a series of images flashing in quick succession – a set of letters and numbers, lightning, a clock spinning backwards, a tunnel being sped through, a set of old gravestones, flames, a writhing human form, and finally a video of blood dripping down and then rewinding to fall back up in reverse. The images fade to black as the voice-over concludes with, "which means 'place to find the lost' in their long-forgotten language…"

The juxtaposition of these two scenes is deliberate and references a clear series of horror genre tropes – that of carefree young people setting off an adventure to a secluded location in apparently idyllic circumstances which are portended to be supernatural in nature (Grant, 2010). The second scene with the voice-over confirms this with a series of other familiar horror-tropes being deployed, including a sudden shift to dark and stormy weather (fog, rain, thunder and lightning) as atonal music begins to play, eliciting an uneasy and trepidatious affect (Hayward, 2009). The voiceover from a mysterious narrator repeats a version of the problematic "Indian Burial Ground" (IBG) trope (Smith, 2014) by referring to an "ancient culture" that inhabited the island and knew of the island's supernatural qualities – going so far as to make mention of the name of the island in a "long-forgotten" language. This narration is the first and only time we hear of this ancient culture, the island's name or any of these supposed supernatural qualities it has; this is also the only time we hear the narrator at all. Also, the characters in the game are never aware of any of this information. I argue the voice-over is a part of establishing a recognisable narrative position for the player, rather than forwarding any specific plot-relevant details. By referencing these recognisable tropes, the game experience to come is positioned in a recognisable gothic horror narrative (Sage, 1990). This becomes more obvious in the following scene where the two are presented as intimate and negotiating what might be a romantic relationship.

The scene fades into another idyllic setting, now on the island, featuring a clear night sky full of stars framed by trees. Romantic country-style music plays whilst crickets and the sound of a camp-fire crackling are also heard. There is a tent pitched on the beach and a dock and dock-house are visible on the shore. The scene pans down to reveal the backs of J.J. and Emily sitting on a log bench in front of a camp-fire, warm light splashing over them and the rest of the scene. They are laughing gently and seem to be teasing each other good-naturedly. They then muse on the beauty of their environment and how they feel like they are the only ones in the world.

This scene features the second of only a few musical cues throughout the whole game. Whereas many videogames feature music throughout and even continuously (Zehnder & Lipscomb, 2006), *The Missing*'s (2018) musical cues happen rarely and only for a short amount of time. However, on each occasion, the music that plays seems to have the clear intent of guiding the player to a certain affect and narrative reading. In the preceding scene, the music was ominous and atonal, very much reminiscent of horror soundtracks (Hayward, 2009). The message was clear, this island is frightening and it has terrible supernatural characteristics – the music portents a mysterious, unsettling, experience ahead. In this scene however, the musical choice is light and melodic, intimate and simple.

The dialogue in the scene matches the music, in that it too implies a romantic moment is being witnessed by the player. Emily asks J.J. shyly, "We'll be together forever, right?" J.J.'s response seems to understand the implication, as she replies, "Whoa, Emily… What exactly are you asking?" Emily's reply continues to make allusions to the romantic quality of the moment, saying, "Well, it's just… The stars look so pretty…" as a shooting star trails across the sky. Finally, J.J. makes the point explicit by saying, "You're such a romantic." Emily responds in kind that "You're one to talk. You just hide it better. You love this kind of stuff, don't you?" to which J.J. admits, "I certainly don't hate it."

The conversation implies that there is a barely unspoken romantic connection here, either from Emily or both of them. In my initial playthrough, I came to believe this was the case, especially as the following moments seemed to confirm a narrative of two young adults awkwardly trying to see if feelings might be reciprocated in the other. I found a similar presumption was made from this moment by other commentators I read and playthroughs I watched (Nadia M., 2018; Cox, 2019).

After a pause, Emily scootches closer to J.J. and reaches out to take her hand. The moment is broken momentarily by a distraction off-screen and Emily retracts from J.J., seemingly disappointed or saddened. J.J. tries to peer around Emily to catch her gaze. The first point of player-interaction now occurs with a button-prompt appearing on the screen reading "Oh, fine…" On pressing the button, J.J. rests her head on Emily's shoulder silently. After a moment, Emily raises her head and so does J.J. and their gazes seem to meet. Another shooting star is seen crossing the sky behind them. A moment passes before J.J. reaches to hold Emily's hand and their heads seem to come together, perhaps in a silent kiss or some other form of intimate connection. It is impossible to see exactly what occurs from the reverse angle as the scene fades to black and the soft music continues to play.

Every aspect of the scene seems to be guiding the player to understand the two characters as (potential) romantic partners. As they are both feminine-presenting, the presumption would be that the relationship is lesbian. And given the messages at the beginning, the relationship can already be read in terms of a potentially discriminatory socio-cultural context. The ominous and portentous introduction to the island and its supernatural qualities sets up the pair for an impending horror narrative that somehow pertains to this. From a design perspective, this specific scene is clearly pivotal to SWERY as he uses it in his GDC 2019 presentation to introduce "the story" of the game, and it features as one of the few images chosen to be part of the press kit for the game (2018).

When the screen finally turns black, the music and the sounds of crickets and the fire stop. The only sound that can be heard initially is of a strong wind and waves. After a moment, Emily's voice-over says "… J.J." but is spoken now in a distorted way. If the subtitles are on, the text appears in the same font as before, but now the letters and punctuation shiver and jump in such a way as to suggest the unnaturalness of the speech. Emily's voice-over continuous distorted, "Hey! J.J.!" but louder and seemingly more insistent. Finally, and more urgently, Emily's distorted voice-over shouts, "J.J.!!!" The scene snaps back quickly to the same island view of the tent, log bench, fire (which is now out and smouldering), waterfront, and dock (though the dock house has vanished). The scene is now much darker, the horizon obscured by clouds and fog. It is raining, and the wind is moving the clouds in the background at an unnaturally high speed. The trees shake and lightning flashes whilst thunder is heard. J.J. stands from lying on the ground and immediately wraps her arms around herself clutching the plush close and calls out, "Emily?" At this point, full control is given to the player to move J.J. as an achievement pops up (depending on the system the game is played upon) reading "Where's Emily?"

YOUR PRINCESS IS IN ANOTHER CASTLE...

In order to progress, the player must traverse the environment from left to right, overcoming environmental obstacles, moving others, and avoiding more still. J.J. can jump from platform to platform to manoeuvre over certain environments and she can crawl under others. She is able to interact with puzzles by manipulating the environment or objects like machinery for example, or by redirecting electricity, water or fire to bypass doors or otherwise impossible to navigate rooms. There are optional collectibles, in the form of donuts which are starkly obvious in some places and hidden in others. The player is informed as soon as they collect the first one that there are a finite number throughout the game, which encourages those keen to complete such tasks, to find them all. All of these are features common to a "2-D puzzle-platformer" style of game (Smith & Whitehead, 2008).

Much like many such games, there is also a definite goal, Emily, who has fled to the right of the 2-D plane and must be sought after. Every time J.J. seems to get close to catching up to her, Emily always seems out of reach. So even if the player is able to guide J.J. to Emily's location, by the time she gets there, Emily has moved on to the next. This is reminiscent of the iconic line from Super Mario Bros. (1983) that is told to the titular player-character every time a castle has been defeated and it seems he might finally be able to rescue the princess - "Thank you Mario. But our princess is in another castle."

In this sense, the analogy is clear. The nature of the 2-D platformer, the mechanics of traversal, the puzzles to solve, the collectibles and the "princess" in another castle, are all short-hand that many who have ever played digital games before will recognise (Smith & Whitehead, 2008). SWERY (2019) makes a point of how he used these game design conventions in order to create a sense of recognition and comfort for the player. It also reproduces the "damsel-in-distress" trope (Dickerman, Christensen & Kerl-McCain, 2008) that even those who have not played games would likely recognise. And this matches with the metaphors presented before interactive gameplay had started – that J.J. is seeking her "princess," who seems, forever "in another castle" somewhere off to the right.

The Missing (2018) also features a definite antagonist, a tall, grey and feminine-presenting humanoid-monster with long hair that moves unnaturally, screeches and regularly runs on her hands and lunges at her would-be victims with claw-like feet. The creature is called the "hairshrieker," a name the player only learns in-game because an achievement pops up on the first encounter with her. The hairshrieker chases both Emily and J.J. intermittently, also always from left to right, appearing at set intervals. What is incongruent with her otherwise supernatural countenance is the antagonist's weapon of choice is an oversized box-cutter. In context, this weapon seems mundane at best and yet is colourful and brightly lit in a way that the hairshrieker is not, being monochrome and often cast in shadows. This is one of the examples of where the game breaks with convention, and begins to give some hint that things are not necessarily as they would otherwise have been suggested.

Despite the surreal quality of the horror and LGBTQ themes, as I have noted, the game follows many of the usual conventions such narratives take. However, there are also notable departures throughout, which eventually culminate in the reveal that these conventions were at least partially obscuring another narrative by inference, metaphor and implication.

Perhaps the most obvious departure with convention is the gameplay mechanic in which the player can only progress by controlling J.J. to brutalise herself. By interacting with certain hazards such as wire-fences, spikes, fire, electricity, wrecking-balls, amongst others, J.J. is severely physically harmed but not killed. She is however, able to use her altered state to her advantage, and reform herself by the player pressing a dedicated button for this. Her brutalisation goes through a series of stages depending on the nature and quantity of interaction with the hazard; and can include loss of limb or indeed entire body, massive head trauma, and being burned. Each phase of brutalisation, however, can be used (and in some cases are the only way) to solve certain puzzles, move obstacles, collect certain items and help J.J. traverse otherwise impossible terrain. Nadia M. (2018) reads this mechanic as a metaphorical expression for certain socio-culturally constructed trans experiences of self-loathing and perseverance. However, it is not this break from convention that creates the moment of revelation about the true nature of the game's narrative.

The game's actual narrative is finally revealed through another mechanic within the surreal dream-like setting, a series of text messages that J.J. has been receiving on her phone throughout the game. At set points during her journey, J.J. receives text-messages which are interactions between herself and her mother and also between herself and Emily. It quickly becomes clear that these messages occurred sometime before her experience on the island, and seem much more "real" and mundane than the surreal environment she currently finds herself in. She receives similar message interactions from friends and a teacher, but these are triggered when a certain number of donuts are collected and do not arrive at set points in the game (although there remains a timing aspect to them because the optional donuts can only be collected in certain amounts at set stages during the game). The only message-interactions that are happening concurrent to J.J.'s present experience come intermittently with the personification of her plush, F.K., who seems to be somehow aware of what is happening now.

At first, all of the texts from J.J.'s university acquaintances and Emily are relatively jovial, and her messages with her mom are mildly frustrating in a way that suggests that J.J.'s mom might be over-protective. As the game goes on and the messages continue, a clearer narrative begins to emerge. J.J. and Emily's relationship is unclear as they navigate whether they are just friends or something else and there is also a secret that one has told the other. In the latter parts of the game, J.J.'s mom becomes more clearly overbearing and suspicious of J.J. being "deviant" in some way and invades her privacy, until finally discovering that J.J. is trans. At around the same time, students at her university also discover that J.J. is trans, and she is both privately and publicly "outed."

It becomes obvious that this led to enormous distress as J.J. attempted to cope with what happened. Through the texts we learn that her relationship with Emily became strained and she felt isolated from her friends and demonised by her peers. In the end, the player learns that J.J.'s mom had forced her into "treatment" designed to "cure" her of her supposed "deviancy." The trauma of these experiences led J.J. to attempt to take her life, which is represented within the surreal gameplay world in an especially harrowing scene in which the player controls J.J. to hang herself. Soon after this it is revealed that the hairshrieker was a manifestation of J.J.'s own self-harm. The box-cutter was what J.J. had actually used to attempt suicide in the real-world, and in this surreal world she is attempting to comprehend what has happened and struggle with her own trauma.

J.J. does not die however. This goes against a common mainstream LGBTQ narrative trope that focuses on such trauma (Russo, 1987; Epstein & Friedman, 1995). J.J. manages instead to defeat the hairshrieker after realising her wish to live through another text message interaction that she has with F.K. What follows is a "final boss-battle" with the hairshrieker, and upon defeating this aspect of herself, J.J. is brought to consciousness in the school gymnasium, where she had attempted suicide. She finds doctors tending to her and Emily nearby. J.J. and Emily tearfully embrace which leads into the game's ending; which reads as positive and hopeful for the future, and once again emphasizes the sentiment "that nobody is wrong for being what they are."

CONCLUSION AND FURTHER RESEARCH

I have presented my close-reading of *The Missing* (2018) as an assemblage of its intersecting elements, specifically in how it deploys those elements to create a misdirect to initially obscure a trans narrative. This misdirect is set up at first in the opening messages. The boat, island and beach cut-scenes that follow, reinforce this. Within those scenes, the sound design and especially the sparsely used music cues make the message clearer, as does the imagery and lighting changes from scene to scene, as well as the dialogue. The achievement that pops onto the screen as soon as the cut-scenes are over reads, "Where's Emily?" which sets the apparent goal of the game. Later text messages and interactions reference a secret and what seems like the beginning or the exploration of a romantic relationship between the two. Once the player is given control of J.J., the 2-D puzzle-platform genre conventions also support this misdirect. Commentaries and playthroughs from others (Nadia M., 2018; Blondeau, 2019; Cox, 2019) demonstrate this narrative and game design functioning to misdirect the player in this way. The designer, SWERY, reports that his goal with the game and the misdirect, is to create an opportunity for a player to learn empathy for marginalised people (2019) – presumably trans people, like J.J.

But what is SWERY's reasoning behind attempting to create this pedagogical moment with such a misdirect and then revelation? Trans "gender revelation" is a media trope that has been well researched in other contexts, especially in how it is often problematically deployed in TV and film (Russo, 1987; Phillips, 2006; Stacey and Street, 2007). In these cases, the revelation of a character's "trans" subjectivity is often linked to problematic narrative tropes of "duplicity" or played for "comedic" effect at the expense of the trans character (Russo, 1987). In *The Missing* (2018), however, this is not how the moment of revelation plays out. Rather, J.J.'s outing shows why a trans person might not reveal their subjectivity openly and what impact it can have to exist in a society that would represent trans subjectivity as duplicitous or comic.

The misdirect and subsequent revelation also does not actually play out in the narrative per se, and is instead seemingly directed at the player, rather than the characters. SWERY has said that he felt the misdirect was necessary to get the player to accept the message of the game (2019). He does not mention explicitly how the misdirect functions to do so specifically, other than describing how the use of 2-D puzzle-platform conventions was to get the player to quickly accept the game's function and goal (2019). I have shown that SWERY and his game design demonstrates an awareness of, engagement with, and an attempt to impact on, a discriminatory socio-cultural context. For example, in response to a question at his GDC 2019 presentation, SWERY admits that he was worried there might be a negative (transphobic) response to the game but that he was pleased that the response had been generally positive. By inference, it seems clear that he understands the context in which his game exists and the design of the game was based upon that, which includes his stated reason to implement the misdirect in order to foment empathy in the player (2019). This also implies that he is only considering non-trans (cis) players' experience with his game, and his intention to educate them.

But what of trans players? Reactions to the game by trans commentators have been positive, some overwhelmingly so (Nadia M., 2018; Blondeau, 2019; Dale 2019), praising it not only for being a "good representation" of a trans character in the medium, but also for the opportunity to "play as" a character that evokes such recognition and opportunity for self-reflection. My own experience certainly evoked similar feelings and I started to wonder what opportunities games like these could offer us, as trans players. If the game was intended to teach empathy for trans people to cis people, but trans players also found something of value in the game, I wondered what experiences might games like these provide to trans people in relation to exploring our own subjectivity?

The lack of opportunity for trans people to explore and express identity and our sense of self – our trans subjectivity – in our socio-cultural context is well documented in the multiple publications of the Transgender Europe global research project, *Transrespect versus Transphobia* (annual publications and reports by global region, from 2009 to present) amongst many other such research initiatives. As a direct result of transphobia, it can be hard for trans people to fully embody ourselves; namely, inhabit our subjectivities experientially and physically in an open and safe way as members of society. This is well-represented in this game (Nadia M., 2018; Blondeau, 2019; Dale 2019). It may also be difficult to learn about our experiences in these often complex, dangerous, social circumstances. Griffiths et al. (2016) and Straus et al. (2017) suggest videogaming can be of direct benefit to queer people, specifically as a site to explore gender in a safer virtual environment. My further research will explore exactly *how* this might be possible from a videogame-based-learning perspective. I will investigate how trans people might explore, express and embody their subjectivities in videogames, and also, how videogames might provide opportunities for cis people to potentially learn and understand more about trans experience.

ACKNOWLEDGMENTS

I would like to thank Sabine Harrer, Doris Rusch and Katta Spiel for their insight and support which has not just helped me with the article I present here but has helped me achieve a position from which I can continue this research and my work in game design. I wish to also thank Fares Kayali, for the supervision for the PhD that I am currently writing based on this work. And I wish to thank Effie Baird, without whom I simply would not be here and doing this.

REFERENCES

Assassin's Creed. (2007) (XBOX360 version) [Video game]. Montreal, Canada: Ubisoft.

Assassin's Creed: Syndicate. (2015) (PC version) [Video game]. Montreal, Canada: Ubisoft.

Allen, S. (2016). Video Games as feminist pedagogy. The Journal of the Canadian Games Studies Association 8(13) 61-80.

Anthropy, A. (2012). Rise of the videogame zinesters: How freaks, normal, amateurs, artists, dreamers, drop-outs, queers, housewives, and people like you are taking back an art form. New York, NY: Seven Stories Press.

Bizzocchi, J. & Tanenbaum, T. J. (2011). Well read: Applying close reading techniques to gameplay experiences. In D. Davidson (ed.) Well played, 3.0: Videogames, value and meaning (289-316). Pittsburgh, PA: ETC Press

Blondeau, B. (2019, December 13). The Missing is the most vital queer game of the decade. The Gamer. Retrieved from https://www.thegamer.com/the-missing-is-the-most-vital-queer-game-of-the-decade/

Brulé, E. & Spiel, K. (2019) Negotiating gender and disability identities in participatory design. Proceedings of the 9th International Conference on Communities & Technologies – Transforming communities (C&T '19), 218-227. New York, NY: Association for Computing Machinery.

Cox, J. (2019, May 3). Scary game squad - The Missing: J.J. Macfield and the Island of Memories (Part 1) [Video, Video series]. Youtube. Retrieved from https://www.youtube.com/watch?v=uNmcMli0YwE&list=PLFx-KViPXIkHxcPOhHAiFcb-W4q74Q7t1&index=1

Dale, L. K. (2019, July 2). The Missing's playable trans protagonist is beautiful, with or without her wig. Syfy Wire. Retrieved from https://www.syfy.com/syfywire/the-missings-playable-trans-protagonist-is-beautiful-with-or-without-her-wig

Dale, L. K. (2020, July 17). A trans woman discusses SWERY, trans representation, and Deadly Premonition 2 - Access-ability [Video]. Youtube. Retrieved from https://www.youtube.com/watch?v=YtU2_P26Pj4&t=1128s

Dickerman, C., Christensen, J, Kerl-McCain, S. B. (2008) Big breasts and bad guys: Depictions of gender and race in video games. Journal of Creativity in Mental Health 3(1) 20-29.

Egan, J. K. (2019). Towards a critical game-based pedagogy in composition [Thesis]. Humboldt State University. Retrieved from https://digitalcommons.humboldt.edu/cgi/viewcontent.cgi?article=1333&context=etd

Epstein, R. et al. (Producers), Epstein, R. & Friedman, J. (Directors) (1995) The Celluloid Closet [Motion Picture]. United States; Sony Pictures Classics.

Flanagan, M. (2009). Critical play: Radical game design. Cambridge, MA: MIT Press.

Glennon, J, Johnston, D. & Francisco, E. (2020, May 14) The rest of us: The Last of Us Part II trans controversy, explained: Who are games for, anyway? Inverse. Retrieved from https://www.inverse.com/gaming/last-of-us-2-trans-controversy-explained-abby-tlou

Grant, B. K. (2010). Screams on screens: Paradigms of horror. Loading: The Journal of the Canadian Game Studies Association 4(6). Retrieved from https://journals.sfu.ca/loading/index.php/loading/article/view/85/82

Griffiths, M. D., Arcelus, J. & Bouman, W. P. (2016). Video gaming and gender dysphoria: Some case study evidence. Revista de Psicologia, Ciències de l'Educació i de l'Esport 32(2), 59 -66.

Hayward, P. (Ed.) (2009). Terror tracks: Music, sound and horror cinema. London, England & Oakwood, CT: Equinox Publishing.

Kazuma H (2018, November 5th). Interview: SWERY on The MISSING of J.J. Macfield and the Island of Memories. Rely on Horror. Retrieved from https://www.relyonhorror.com/in-depth/interviews/interview-swery-on-the-missing-of-j-j-macfield-and-the-island-of-memories/

Keogh, B. (2014). Across worlds and bodies: Criticism in the age of video games. Journal of Games Criticism, 1(1), 1-26.

Klein, R. (2020, March 8). Queer Teaching: How queer games can augment learning through play [Conference presentation]. QGCON: The queerness and games conference. Retrieved from https://qgcon.com/queer-teaching-how-queer-games-can-augment-learning-through-play/

Kostopolus, E. (2017). Using role-playing gamification to create safe spaces for LGBT Students in the composition classroom [Thesis]. Minnesota, MO: James C Kirkpatrick Library, University of Central Missouri. Retrieved from https://ucmo.primo.exlibrisgroup.com/view/delivery/01UCMO_INST/1284607370005571

M, N. (2018, December 17). The Missing: J. J. Macfield and the ownership of identity. Timber Owls. Retrieved from https://timber-owls.com/2018/12/17/the-missing-j-j-macfield-and-the-ownership-of-identity/

Marcotte, J. (2018). Queering control(lers) through reflective game design practices. Game Studies: The International Journal of Computer Game Research 18(3). Retrieved from http://gamestudies.org/1803/articles/marcotte

Murray, S. (2017). On video games: The visual politics of race, gender and space. London, England: Bloomsbury Academic.

Phillips, J (2006) Transgender on Screen. Hampshire, England & New York, NY: Palgrave Macmillan.

Philips, T. (2015, September 24). Assassin's Creed Syndicate takes a leap towards inclusivity with the series' first transgender character: Ubisoft also updates the series' standard opening text. Eurogamer. Retrieved from https://www.eurogamer.net/articles/2015-09-18-assassins-creed-syndicate-takes-a-leap-towards-inclusivity-with-the-series-first-transgender-character#:~:text=Ubisoft%20has%20also%20updated%20Assassin's,various%20religious%20faiths%20and%20beliefs.%22

Pozo, T. (2018). Queer games after empathy: Feminism and haptic game design aesthetics from consent to cuteness to the radically soft. Game Studies: The International Journal of Computer Game Research. 18(3). Retrieved from http://gamestudies.org/1803/articles/pozo

Rapp. A. (2018). Autoethnography in human-computer interaction: theory and practice. In M. Filimowicz & V. Tzankova (Eds.), New directions in third wave human-computer interaction: Volume 2 – Methodologies (25 -42). New York, NY: Springer.

Ruberg, B. (2019). Video games have always been queer. New York, NY: New York University Press.

Russo, V. (1987) The celluloid closet (Revised Ed.). New York, NY: Harpers & Row.

Sage, V. (Ed.) (1990). The gothick novel. London: Macmillan.

Shaw, A. (2009). Putting the gay in games: Cultural production and GLBT content in video games. Games and Culture, 4(3), 228–253.

Shaw A. (2011). Do You Identify as a Gamer? Gender, race, sexuality and gamer identity. New Media and Society 14, 25-41.

Shaw A. (2014) Gaming at the edge: Sexuality and gender at the margins of gamer culture. Minnesota, MO: University of Minnesota Press.

Shaw, A. (2015). The tyranny of realism: Historical accuracy and politics of representation in Assassin's Creed III. Loading…: The Journal of the Canadian Game Studies Association 9(14), 4-24.

Shaw, A. & Friesem, E. (2016). Where is the queerness in games? Types of lesbian, gay, bisexual, transgender, and queer content in digital games. International Journal of Communication 10, 3877-3889.

Smith, A. (2014). This essay was not built on an ancient Indian burial ground: Horror aesthetics within Indigenous cinema as pushback against colonial violence. Offscreen 18(8). Retrieved from https://offscreen.com/view/horror-indigenous-cinema

Smith, G. & Whitehead, J. (2008) A framework for analysis of 2D platformer levels. In Proceedings of the 2008 ACM SIGGRAPH Symposium on Video games (Sandbox '08) (75–80), New York, NY: Association for Computing Machinery https://doi.org/10.1145/1401843.1401858

Stacey, J. & Street, Sarah (Eds.) (2007) Queer Screen; A screen reader. London, England & New York, NY: Routledge.

Straus, P., Cook A., Winter, S. & Watson, V. (2017). Trans pathways: The mental health experiences and care pathways of trans young people. Perth, Australia: Telethon Kids Institute. Retrieved from https://nla.gov.au/nla.obj-539448700/view

Suehiro H (2018). The Missing: J.J. Macfield and the Island of Memories. (Switch version, PC version, PS4 version) [Video game]. Osaka, Japan: White Owls Inc.

Suehiro H (2019, March 18 – 22). 'The Missing': An attempt at complete cohesion of gameplay and narrative [Conference presentation]. Game Developers Conference (GDC), San Francisco. https://www.gdcvault.com/play/1025789/-The-Missing-An-Attempt

White Owls Inc. (2018) The Missing: J.J. Macfield and the Island of Memories press kit [Press Kit]. IGDB. Retrieved from https://www.igdb.com/games/the-missing-jj-macfield-and-the-island-of-memories/presskit

Taylor, T.L. (2009). The assemblage of play. Games and Culture, 4(4),331-339.

Zehnder, S.M. & Lipscomb, S. D. (2006) The role of music in video games. In P. Vorderer & B. Jennings (Eds.) Playing video games: Motives, responses, and consequences (241-247). New York, NY: Routledge

PLAYING AT KNOWING ANCIENT EGYPT. THE TOURIST GAZE IN *ASSASSIN'S CREED: ORIGINS*

Mona Khattab, Tanja Sihvonen, Sabine Harrer

*A*ssassin's Creed is a single-player, action-adventure stealth video game franchise by Canadian developer Ubisoft. The game series lets its player delve into history from Renaissance-era Florence to Victorian London. In this article, we perform a close reading of the 'discovery tour mode' function in *Assassin's Creed: Origins* (2018), a game taking place in ancient Egypt. Our goal is to understand the purpose of placemaking as a technology for virtual and identity tourism. We argue that due to its quasi-touristic staging of an ancient civilisation, the discovery tour mode is a particularly potent feature in exploring how games render history palatable for an implied white Western game audience.

Keywords: Game Studies, Ubisoft, postcolonial theory, placemaking, identity tourism

---

INTRODUCTION

Assassin's Creed (Ubisoft, 2007-) is a single-player, action-adventure stealth video game franchise that lets its player delve into history, from Renaissance-era Florence to Victorian London. In this article, we focus on one game in this series, *Assassin's Creed: Origins* (2018; henceforth ACO), and specifically its 'discovery tour mode' function that stages ancient Egypt as a site for virtual and identity tourism (Nakamura, 2002). Hence, we ask how ACO functions as a 'cultural technology' in the contexts of placemaking (Álvarez & Duarte, 2018) and storytelling. We argue that due to its quasi-touristic staging of an ancient civilisation, the discovery tour mode is a particularly potent feature in exploring how games render history palatable for an implied white Western audience (see Gilbert, 2019; Radošinská, 2018).

What makes the discovery tour mode specifically interesting is the focus that ACO has put on the othering and cultural appropriation of classical civilisations through gameplay (see Mukherjee, 2018). As a tool frequently employed in games to 'immerse' players into exotic, colonial landscapes and marginalised bodies, othering simultaneously mobilises a Western imperial cartographic memory while objectifying and mystifying people and places. This is actualised through the narrative strategy of the discovery tour mode that provides the assumed Western player with a particular cultural representation of an ancient non-Western civilisation.

Our article unpacks these dynamics of placemaking in ACO through three narrative viewpoints. The first of these, referred to as the cultural gaze, examines the transcultural depiction of Egypt as a particular kind of landscape, an exoticised 'Other' designed for white Western gamic consumption.

Secondly, we look at the protagonist, and how 'possessing' the bodies of available avatars becomes a sociocultural entry point into the discovery tour mode of the game. ACO's character creation tool raises questions pertaining to identity tourism (Nakamura, 2002) and digital blackfacing (Gray, 2012; Leonard, 2004). In this section we analyse how ACO's discovery tour mode facilitates such practices as part of the player's cultural gaze.

Thirdly, we look at the game's voice-over, exploring what might be termed *museumification*, a practice of knowledge making that is characteristic of Western 'ethnographic museums' rooted in colonialist presence in non-Western countries (Lidchi, 1997). In this part, we argue that ACO's discovery tour mode is a version of digital museumification in that it turns Egypt into a digitally rendered 3D space designed by and for a Western cultural gaze. This means that the local geographical space of Egypt is digitally reconfigured as a Westernised space, repurposed and refashioned to satisfy the expectations of non-local game audiences.

Methodologically, we conduct a textual analysis of the ACO discovery tour mode to shed light on the intersectional quality of these three narrative dimensions of cultural gaze, protagonists, and the voice-over. Our findings highlight the importance of postcolonial perspectives in understanding the use of non-Western locations in games. They suggest that the museumification of digital games continues traditions of Orientalist othering, making non-Western locations palatable for implied white Western game audiences under the guise of literacy and learning (Gilbert, 2019; Westin & Hedlund, 2016).

ASSASSIN'S CREED: ORIGINS AS A SITE FOR LUDIC HISTORY

The player of the *Assassin's Creed* series takes control of several narrative modes as they traverse through 3D-rendered historical periods. The games rely on temporal-spatial intersections as the central character relives their ancestors' experiences to fight an apocalyptic prophecy (see Seif El-Nasr et al., 2008; Veugen, 2016). The game franchise is loosely based on *Alamut*, a 1938 novel by Slovene author Vladimir Bartol set in 11th century Persia, displaying a fascination with a group of assassins known as *Ḥashāshīn* in Arabic. The group is said to have been pitted against the Knights Templar as the Arabic and Persian equivalent to the Christian military order. Bartol's novel, translated in English as late as 2004, is set in the castle of Alamut, where its master, Hasan ibn Sabbah, trains his soldiers through trickery to become fanatic assassins to repel foreign conquerors. The novel was published just before WWII and is often interpreted as a contemporary allegory of rising totalitarian regimes in Europe. It is also regarded to convey a nationalistic message, although this semantic level is subdued in it (Hladnik, 2004).

The novel, and in turn the games, offer a convoluted reinterpretation of Arabic and Persian mysticism through heavily convoluted narrativity and spatiality. The figure of assassin is, in Bartol's novel as well as in the game franchise, a tool with which to reshape the figure of modern Islamic fundamentalism and terrorism. Simultaneously as the assassin is conceived as an orientalised villain, they are transformed into an occidentalised hero, "thus enabling a self-othering of the Western subject and an identification, rather than disqualification, with this specific Arabo-Islamic Other" (Komel, 2014, p.525). In *Assassin's Creed: Origins*, the main character is a Medjay, a member of an ancient Egyptian police, called Bayek, who is responsible for protecting the Siwa Oasis during the troubled times of conflict in the Ptolemaic Egypt (305–30BCE).

The game dislocates the physicality of history through various narrative means; for instance, a voice-over narrator helps players make sense of the historical environment they are navigating. Players control aspects of the narrative by making selections that incur varying results, but they are also allowed to freely roam certain settings without specific game objectives to discover the distant time and place through the game character's embodied experience. This roaming experience, the discovery tour mode, is a particular feature of ACO, and its purpose is to bring a virtual, historical tourism aspect to the game. As the game studio Ubisoft promises in its promotional text (Ubisoft Support, n.d.), the discovery tour "will allow you to explore ancient Egypt without being interrupted by combat or quests," an activity which is supposed to be 'purely educational'.

NARRATIVITY IN THE ACO DISCOVERY TOUR MODE

In the context of this article, we explore how this promise of 'pure education' around ancient Egypt is constructed on the three ludo-narrative levels of spatiality, character design, and museumification through voice-over. In doing so, we are drawing on the theories of narratologists Mieke Bal (2009) and Gerald Prince (2003), especially their notion of 'sujet'. Defined by Bal (2009) as the quintessential story, or the basic plot, the sujet is the backbone of the narrative. The adaptation into various media, such as novels, films, or games, is the fleshing out of the backbone (Bal, 2009). In the case of ACO, the sujet is the journey in ancient Egypt and the form is a digital game. In order to adapt a sujet to a specific medium, narrative perspective is necessary. In ACO, such perspective is spatiotemporal, making the sujet, that basic storyline, oscillate between the parameters of the ancient landscape on the one hand, and the contemporaneity of the player on the other hand.

A core element of narrativity is the central character. As protagonist, their role is to anchor the sujet to an actantial model. A narratological term coined by Greimas (1986), the actantial model divides the acting participants in the narrative to protagonist, antagonist, object of quest, and helper. In the ACO discovery tour mode, the protagonist remains the same from the perspective of the player. The object of the quest is the unraveling of the sites. The 'antagonist' is a lack of educational knowledge, and the 'helper' is the educational information provided to players.

In addition to sujet and a character, setting is the third major element in narrativity. It is defined as "the spatiotemporal circumstances in which the events of a narrative occur" (Prince, 2003, p.88). Thus, discussing spatiality and temporality highlights the role the setting performs in the narrative structure of the ACO discovery tour mode.

In this mode, there are two time frames: The sujet can be seen in the historical in-game time, where the timeline of presented 'facts' occurs. The narrative plot time, however, is determined by the player, as they make choices around avatars and virtual tours. This sense of temporality is orchestrated by the game designers. The audience is assumed to be familiar with the convention of historical staging via Western museums. They are further assumed to be interested in 'learning' through a convenient breakdown of 'facts', structured in a particular quantifiable fashion. This aligns the narrativity of ACO's discovery tour with what game scholar Mukherjee (2015) describes as spatiality of empire building. While the discovery tour mode is not explicitly themed as an empire game, it still incorporates an imperial gaze which lays claim to knowing and controlling 'ancient' spaces and peoples.

SEEING THE OTHER: THE CULTURAL GAZE

This section examines the ACO narrative as othering of ancient Egyptian culture and proposes the concept of the cultural gaze as a tool for this process. 'Gaze' as a concept originates from feminist film criticism and most notably from Laura Mulvey's 'male gaze' (1975), which refers to the male objectification of women in film. From this, we derive the notion of a 'cultural gaze', whereby a viewer, in our case an (implied Western) player looks at an othered cultural object, in our case the historicised location of ancient Egypt in ACO.

The cultural gaze is related to the concept of Orientalism, famously introduced by Edward Said in *Orientalism* (1978). The term itself ranges from academic, cultural, and literary construction of the 'Orient' as opposed to the 'Occident', as part of the colonialist project to objectify and devalue non-Western, specifically African, Arab, and Islamic cultures, and to justify colonialist domination of those regions. Orientalism is thus a European invention, an imperialistic othering tool targeting African and Arabo-Islamic cultures, and sustaining white supremacy (Burney, 2012). This is evident in literary and artistic representations, ranging from translations, or mistranslations of works from that region into European languages to original Western media, including paintings, cinema, and more recently, gaming (Fickle, 2019).

ACO's discovery tour mode applies a cultural Orientalist gaze in its visual design of ancient Egypt in order to serve dominant stereotypical expectations potential Ubisoft audiences might already hold of ancient Egypt. Gamers traverse a detailed geographical map of 'ancient Egypt' rife with pyramids, temples, palm trees, and camels. Places of 'interest' are highlighted as starting points for virtual museum tours. Throughout each tour, a golden line is showing players the way, structuring the place for optimal consumption. Combined with this 'authentic' environment, the guided tour establishes its techniques of an Orientalist cultural gaze in Edward Said's sense (see also Burney, 2012), in that it frames 'ancient' space as controllable and knowable by the uninformed 'tourist'. In addition to the promotional material, the in-game text reinforces this message by addressing players directly: "Guided tours will take you through majestic landmarks, and acquaint you with ancient Egyptians and their culture" (Fig. 1).

Figure 1. Screenshot of the Assassin's Creed: Origins discovery tour mode. It promotes guided tours to the player featuring a bird's-eye view on Alexandria.

This narration, projected on a bird's-eye shot of Alexandria (Fig. 1), establishes the game's relationship of us/players versus them/ancient Egyptians which promises the player an Orientalist cultural intimacy ('acquaintance') with an ancient civilisation. Paradoxically, this offer is based on knowledge curated by a contemporary game company, whose educational materials are derived from colonial-ethnographic museum culture. This is best shown in the various instances where the game presents fact sheets and illustrations by French Orientalist artists like Jean-Léon Gérôme (Fig. 2).

Figure 2. Screenshot of Orientalist imagery used in Assassin's Creed: Origins discovery tour mode: The Sphinx of Giza by French painter Jean-Léon Gerôme.

BEING THE OTHER: ANCIENT AVATARS

The discovery tour mode allows players to choose one of many avatars from a selection of 'authentic' historical celebrities, including Cleopatra and Julius Caesar. Their selection happens freely and with no consequence to the game mechanics itself – it is mere decoration, and avatars can be changed at will at any time during a tour.

From a spatial perspective, it is interesting how the avatar literally becomes the player's entry point to the cultural landscape of ancient Egypt. Most of the avatars presented are bodies of colour. When moving around, the player takes possession of these bodies, and a third-person camera frames them as the player's bodies, therefore positioning the player into a 'subjective' experience, a way of walking in the Egyptians' shoes. When introducing historical facts, the game camera moves up, and we leave the avatar's body. This frames history as 'objective,' detached from bodily experiences and the lives of the historical characters we have just inhabited, and beyond our control.

Given the game studio Ubisoft's proximity to white supremacist gamer culture (e.g. Good, 2020), a core audience of ACO's discovery tour mode can be expected to be white gamers. This makes avatars susceptible to 'identity tourism' (Nakamura, 2002), a version of the cultural gaze whereby white users inhabit non-white bodies in an often unconsciously racist fashion. A similar term discussed in regard to racist game character performances is 'high-tech blackface' (Leonard, 2004; Marriott, 1999), which refers to players' desire to 'be black' for the duration of a game, for instance by inhabiting "stereotypical visions of strength, athleticism, power and sexual potency" in sports games (Leonard, 2004).

In the context of ACO's educational vision, these conversations can be extended to include the pleasures of becoming an ancient Other, thereby symbolically taking full control of ancient Egyptians' culture, traditions, and ways of living. Not only are players of ACO allowed to traverse and 'acquaint' themselves with the environments and people in the game; they can 'be ancient' for the duration of the game, inhabiting certain bodies of colour whose exciting personal histories are thereby exhibited as controllable and ready for appropriation.

Since ACO's list of period characters available in the discovery tour mode features persons of different races, white gamers have the hypothetical option to resist identity tourism and digital blackfacing; they might choose a character of their own skin tone. However, by providing game characters of different races as equally available to all players, the game positions white players as entitled to 'look through the eyes' of simultaneously historically famous and racially marginalised avatars. Thus, the game models a form of ludic white entitlement over brown bodies which – even if unintended by design – offers a platform for white supremacist fantasies.

Furthermore, as opposed to ACO's story mode where Bayek serves as a focus character in the player's 'discovery' of ancient Egypt, the discovery tour mode offers a long list of avatars to select from. The players can swap their avatars mid-tour, should they ever get bored by a character or desire to explore a location in a different 'skin'. This presents ancient Egyptian protagonists as ultimately replaceable and dehistoricized; they can be moved in and out of local contexts at the player's will. Overall, this treatment of protagonists as subjected to the cultural gaze presents ACO's education as a practice which does not necessitate decolonial deconstruction of colonial activity; it rather perpetuates such activity by molding it in the 'innovative' form of digital gaming.

KNOWING THE OTHER: VOICE-OVER

In narratology, omniscient and omnipresent narrators represent authority (Prince, 2003). Their ability to be present everywhere and to know more than the reader places them above readers and grants them unquestionable authority. The ACO discovery tour mode presents such an omniscient narrator in the form of a voice-over conveying information through a decidedly American English accent. Mimicking conventions of a classical museum tour, the voice-over is positioned to help players make sense of the game space by structuring and prioritising information to provide 'purely educational' knowledge. It thereby acts as an extension of institutionalised discourse on ancient Egypt, a voice of neutral reason and rationality which audiences are supposed to trust.

Paradoxically, by being American English, the perceived 'neutrality' of this voice is constructed through the use of an accent which does not reflect first-hand local experience of Egypt. The game thus uses a US-centric imperialist voice to mark the narrative contents as trustworthy, neutral, and suited to provide players with the necessary 'facts' of history.

CONCLUSION

Our goal in this article has been to understand the purpose of placemaking as a cultural technology for virtual and identity tourism in games. Examining the discovery tour mode in *Assassin's Creed: Origins* has been an exercise in unpacking how cultural representation works in a postcolonial theoretical setting. In this article, we have conducted a close reading of three related narrative elements of cultural gaze and space making, character design, and museumified voice-over in order to understand digital games as sites for contemporary colonial activity. What makes ACO's discovery tour mode a specifically interesting example in this regard has been the designerly effort it puts into the high-definition rendering of an ancient and mysterious version of Egypt, inviting an Orientalist player gaze on East vs. West (Burney, 2012; Said, 1978) while promoting this choice as 'purely educational' (see also Hammar, 2017; Karsenti et al., 2019). By dividing player and game landscape along the binary of Orient vs. Occident, us vs. them, the discovery mode mobilises a white imperialist cartographic memory for player edutainment (e.g. Gilbert, 2019: Mukherjee, 2018) under the guise of objective, 'pure' education.

Not only does this digital touristic 'museum package' stage ancient Egypt in a way which perpetuates a colonial cultural gaze, it also stages the bodies of 'authentic' historical celebrities of colour as freely accessible and 'possessable' by any consumer of the discovery tour. Furthermore, the neutral museumified voice-over frames the player's tour as an objective learning experience, delivering on Ubisoft's marketing promise to produce 'pure education' (Ubisoft Support, n.d.). This positions any player as entitled to inhabit the spaces and bodies of 'authentic' others, and to gain knowledge and 'education' via a Western quasi-omniscient narrator, the game designers. The three dimensions of a high-definition cultural gaze, the 'authentic' celebrity bodies, and the 'real' museum narration helps the game avoid questions around coloniality and accountability. Whose 'pure education' does the discovery tour facilitate in rendering its version of ancient Egypt? Who or what is actually being 'discovered' by players? Addressing such questions would expose the power dimensions baked into the Orientalist design decisions, including an analysis of who is entitled to 'discover' (White western audiences) and who is being discovered (the imagined 'Orient') in the game.

REFERENCES

Álvarez, R., & Duarte, F. (2018). Spatial design and placemaking: Learning from video games. Space and Culture, 21(3), 208–232. https://doi.org/10.1177/1206331217736746

Bal, M. (2009). Narratology: Introduction to the theory of narrative. Toronto, Ontario: University of Toronto Press.

Bartol, V. (2004). Alamut (M. Biggins, Tans.). Seattle, Washington: Scala House Press. (Original work published 1938)

Burney, S. (2012). Pedagogy of the Other: Edward Said, postcolonial theory, and strategies for critique. New York, NY: Peter Lang.

Fickle, T. (2019). The race card: From gaming technologies to model minorities. New York, NY: New York University Press.

Gilbert, L. (2019). 'Assassin's Creed reminds us that history is human experience': Students' senses of empathy while playing a narrative video game. Theory & Research in Social Education, 47(1), 108–137. https://doi.org/10.1080/00933104.2018.1560713

Good, O. (2020, September 10). Ubisoft CEO apologizes for controversies, including one that fired Assassin's Creed director. Polygon. https://www.polygon.com/2020/9/10/21430881/assassins-creed-director-fired-ubisoft-tom-clancys-elite-squad-blm-logo-controversy-yves-guillemot

Gray, K. L. (2012). Intersecting oppressions and online communities: Examining the experiences of women of color in Xbox Live. Information, Communication and Society, 15(3), 411–428. https://doi.org/10.1080/1369118X.2011.642401

Greimas, A. J. (1986). Sémantique structurale. Paris, France: Presse Universitaires de France.

Hammar, E. L. (2017). Counter-hegemonic commemorative play: Marginalized pasts and the politics of memory in the digital game Assassin's Creed: Freedom Cry. Rethinking History, 21(3), 372–395. https://doi.org/10.1080/13642529.2016.1256622

Hladnik, M. (2004). Nevertheless, is it also a Machiavellian Novel?: A review essay of Michael Biggins, 'Against Ideologies: Vladimir Bartol and Alamut' Pp. 383–390. In: Vladimir Bartol. Alamut, Seattle: Scala House Press, 2004. Slovene Studies Journal, 26(1/2), 107–116. https://dx.doi.org/10.7152/ssj.v26i1.4225

Karsenti, T., Bugmann, J., & Parent, S. (2019). Can students learn history by playing Assassin's Creed? An exploratory study of 329 high school students. Revue Chaire de Recherche sur le Numérique en Éducation, 1-22. http://hdl.handle.net/20.500.12162/3656

Komel, M. (2014). Re-orientalizing the Assassins in Western historical-fiction literature: Orientalism and self-Orientalism in Bartol's Alamut, Tarr's Alamut, Boschert's Assassins of Alamut and Oden's Lion of Cairo. European Journal of Cultural Studies, 17(5), 525–548. https://journals.sagepub.com/doi/pdf/10.1177/1367549413515253

Leonard, D. J. (2004). High tech blackface: Race, sports, video games and becoming the other. Intelligent Agent, 4(4.2), 1–5. http://www.intelligentagent.com/archive/IA4_4gamingleonard.pdf

Lidchi, H. (1997). Poetics and the politics of exhibiting other cultures. In S. Hall (Ed.), Representation: Cultural representations and signifying practices (pp. 151–222). London, UK: Sage.

Marriott, M. (1999, October 21). Blood, gore, sex and now: Race. The New York Times. https://www.nytimes.com/1999/10/21/technology/blood-gore-sex-and-now-race.html

Mukherjee, S. (2018). Playing subaltern: Video games and postcolonialism. Games and Culture, 13(5), 504–20. https://doi.org/10.1177/1555412015627258

Mukherjee, S. (2015). The playing fields of Empire: Empire and spatiality in video games. Journal of Gaming & Virtual Worlds, 7(3), 299–315(17). https://doi.org/10.1386/jgvw.7.3.299_1

Mulvey, L. (1975). Visual pleasure and narrative cinema. Screen, 16(3), 6–18. https://doi.org/10.1093/screen/16.3.6

Nakamura, L. (2002). Cybertypes: Race, ethnicity and identity on the internet. London, UK: Routledge.

Prince, G. (2003). A dictionary of narratology. Lincoln, Nebraska: University of Nebraska Press.

Radošinská, J. (2018). Portraying historical landmarks and events in the digital game series Assassin's Creed. Acta Ludologica, 1(2), 4–17. https://actaludologica.com/portraying-historical-landmarks-and-events-in-the-digital-game-series-assassins-creed/

Said, E. W. (1978). Orientalism. New York, NY: Pantheon Books.

Seif El-Nasr, M., Al-Saati, M., Niedenthal, S., & Milam, D. (2008). Assassin's Creed: A multi-cultural read. Loading . . ., 2(3), 1–32. https://journals.sfu.ca/loading/index.php/loading/article/view/51

Ubisoft Support. (n.d.). Discovery Tour Mode of Assassin's Creed: Origins. Retrieved March 30, 2021, from https://support.ubisoft.com/en-gb/Article/000062699/Discovery-Tour-Mode-of-Assassins-Creed-Origins Accessed 1 April 2021

Veugen, C. (2016). Assassin's Creed and transmedia storytelling. International Journal of Gaming and Computer-Mediated Simulations, 8(2), 1–19. https://doi.org/10.4018/IJGCMS.2016040101

Westin, J., & Hedlund, R. (2016). Polychronia – negotiating the popular representation of a common past in Assassin's Creed. Journal of Gaming and Virtual Worlds, 8(1), 3–20. http://doi.org/10.1386/jgvw.8.1.3_1

IN SEARCH OF IDENTITY THROUGH THE GAME "GRIS"

Steve Hilbert

This article deals with the question of whether and in what form the game Gris can contribute to the individual's search for identity. Using the scientific approach of symbolic interactionism, which is based on interaction and social processes, the investigation is always narrowed down into different categories by means of grounded theory until further investigation is no longer possible and the work thus arrives at a final hypothesis. Social interactivity is a crucial aspect in the search for identity. The coding process as well as the interviewers repeatedly point out its importance here, be it in terms of social influencing factors such as the emotions, the contact or the change or by means of the different phases of conflict within these. The ability to perceive and understand oneself and especially one's own emotions enables systemic processes within a peer group. The change of perspective is repeatedly demonstrated as the prerequisite in this regard. Since this aptitude is acquired through cognition, the educational aspect and equal access to education consequently represent the conclusion of this work.

Keywords: Identity, mental health, emotions, cognition, education

---

INTRODUCTION

How can the game scenario of *Gris* contribute to the individual search for identity?

The story of *Gris* is an emotional journey into the injured interior of the performer. The main protagonist tells an unusual story, supported by the colourful design of the game world as well as the supporting music by the Spanish band Berlinist. In keeping with the destroyed game world (e.g., in the form of ruins, which are a symbol for decay/destruction (Fuchs,2017)), *Gris* body language shows that the key character is in pain. The game's action thematises mental illness, especially depression. In many scenes, the vulnerability as well as the humanity of *Gris* is portrayed wordlessly A fitting example of this is a game scene in which *Gris* leans her face against the upper lip of the female statue. She begins to blink and a tear escapes from her left eye. *Gris* turns her head, puts her right hand on the statue and kisses it above the upper lip. When she releases her mouth from the statue (2:07:16), her mouth suggests a small smile. The right corner of her mouth turns slightly upwards, making her face look more relaxed.

In identity crises we must be aware of the many forms of identity crises, as social crises, individual crises, and individual crises of meaning. Differences between public and social identity imply a different view of perspectives, whether personal or public. The game offers the player the opportunity to make mistakes without experiencing a consequence in real life.

The social philosopher George Herbert Maed's symbolic interactionism considers human beings as social beings. He (Blumer, 2013) recognizes socialization as a process of identity formation. It involves the analysis of the human collective and human behaviour. This means that the perceived content is in interaction with future action. The meaning of this content be it physical, social, or abstract, influences further actions and thus one's own self at the same time. In communication, the person is both sender and receiver (in the sense of an expectation of feedback from the other person).

At the same time, Blumer (2013), a student of Mead, further developed symbolic interactionism, reckons that the common action is not newly created. Rather, it is based on events and knowledge already experienced and thus enables comparative conclusions. Social interaction is essential in the search for identity.

The game theme for its part promotes a more conscious exchange with the community and deals with the question to what extent *Gris* can be an identity-promoting game by illuminating the relationship between different variables and identity formation. Dealing with the topic in a reflective manner means exercising proactive behaviour. The person's own knowledge is broadened as he/she can recognize connections and assess the consequences of his own actions. He/she learns to communicate and to relate his/her own reflections to his/her own behaviour. Education is a permanent process of one's own self-reflection.

The ability to perceive and understand one's own inner being, and emotions enables system-oriented processes to be created within a peer group. Emotions, changes and (personal) contact are essential for identity, as they have a common/influencing effect on each of them. The prerequisite for this is the ability to change perspectives. This ability is acquired through cognition, which must be supported by the public community for people to benefit from this development. Accordingly, the educational aspect and equal access to education represent the conclusion of the search for identity.

RELATED WORK

A game can thus be a learning aid and contribute to the expansion of one's own identity. Games act as "possible catalysts for meta-learning" and help "to analyse educational processes" (Fromme et al., 2008, trans. by author). "Games ... are models of reality. As models, they do not (necessarily) imitate reality, but focus on action that are subject to a certain theology; in this respect, they are not completely different from our behaviour in reality" (Berg & von Sass, 2014, trans. by author).

"Non-formal education refers to a real-life practice in which children and young people become decisive co-producers of their own educational biography" (Ministère de l'Éducation nationale, 2018, trans. by author). Children as well as young people actively participate in their development. They influence it through aspects arising from discovery learning and intrinsic motivation.

In the game *Gris*, the possibility is opened to explore different topics without experiencing a direct impact. The player is in a protected space, which Huizinga (1949) defines as a magic circle. Experiences can be tried out here and then more easily adopted in the real world. "Reflections and appropriation are keys to thinking about the possibilities of game creation from a different perspective" (Sicart, 2013). Rengelshausen (2018) notes that "a reflective approach to media and one's own identity is required in order to be able to at least partially control the influence of digital technologies on our society." When shaping identity, it is thereby even more important to be aware of the fast pace of modernity, the large number of decision-making possibilities and the (emotional) uncertainties that go along with it. In digital games, a psychological identification of the person with the game character takes place. He/she looks for comparisons between the digital avatar and his/her own identity (Gast, 2017:9). Accordingly, a digital game can tie in here to bring a certain topic more strongly into the consciousness of the player. In this respect, it is essential to recognize that digital games can contribute to the formation of one's own identity.

Although at the beginning of the era of digital games the representation of gender in game characters did not play a significant role (see games such as Pong, Space Invaders or the like), this changed with technological progress. This enabled the development of more graphically sophisticated and detailed representations, such as the depiction of genders, which Kiel (2020) considered an important contribution to sexual education in digital games. The more accurate depictions of avatars allowed gender representations to thus gain relevance and become omnipresent in digital games at least since the 90s (Hellal, 2019:76). Iacovides (2012) asks: "Are people aware of what they are learning?" in the affirmative and refers to his own interviews with players who attribute a learning effect from digital games to themselves. "They just need to keep up enough of a pretence to be able to accept the projection of game elements in their daily activities" (Cassone, 2017).

METHODOLOGY

Gris is a non-violent game that presents a politically relevant issue in an artistic way. The game Gris deals with psychological darkness and the way out of depressive moods. The protagonist Gris tries to recover her initial weakness and the loss of her voice during the game. Fear and pain befall her in the form of a tacting flock of birds. By gaining new abilities, the young woman gains agility and inner strength, which is accompanied by an extraordinary soundtrack. Overcoming obstacles leads her to come to terms with her personal experiences and to inner acceptance.

Using Grounded Theory (Corbin & Strauss, 1990) is a systematic method of collecting and analysing data to create a construct of hypotheses, the difficulty of creating open data presents itself as a challenge. To better understand social interactions, the graph of conflict management forms by

Reinhardt & Tries (2008:99) is used to define the characteristics of interdependency or conflict forms.

When creating the first theoretical samplings, the following first aspects can be coded:

- conditions
- interactions, strategies, and the resulting consequences between actors
- degrees of expression found within a category.

Three different categories stand out in the observation of the scene:

1. emotions
2. change
3. contact

The orientation of the three categories (emotion, change, contact) develops regarding the perception of the significance of different communicative forms of expression. Here, one's own perception as well as the perception of other people or by other people is addressed.

The elaboration of further concepts makes it possible to pose new directional questions to analyse further contexts. The creation of one or more concepts is simplified. To obtain detailed data, a scaling is created to create a clear coding:

*Interpersonal

*Peer groups

Every interaction follows a conscious or unconscious intention. To better understand this interaction, the graph of conflict management forms by Reinhardt & Tries (2008:99, trans. by author) is used to subordinately define the characteristics of reciprocity or the forms of conflict in a grid.

Here, reference is made to:

1. preventive/prophylactic,
2. destructive and
3. constructive interaction,

focused.

Dealing with conflicts, whether physical or psychological, as well as with other people or with oneself, is a central point of analysis in the *Gris* game. The complexity of the story expects a certain empathy and understanding towards the social play theme. This attempt at explanation results in the decision to select social workers as participants. Conflict as a form of interaction is examined for its effect and influence by a semi-structured interview. All three forms of conflict were broken down into their individual components for the subsequent semi-structured interview and identified with possible potential solutions. These solutions were queried in a questionnaire with teams of two in an interview. The interviews took place via video conference. The interview structure starts with an explanation of the game by the developer studio of *Gris* as well as the illustration of a closing video scene of the game. The textual coding of the interview is based on the data collection transcription according to Bortz & Döring (2006:313, trans. by author).

AIM OF THE RESEARCH

The starting point of the beginning of the research work refers to my personal observation of a scene that was ground-breaking for me. More than ever, we are asking ourselves the question of our own identity. We humans also seek this identity every day in a different way. The search for one's inner self is in a permanent competition with the competing fellow citizens. One's own representation and positioning in society must deal with constant comparisons from the outside world (Rengelshausen, 2018:3, trans. by author).

Gris attempts, in a seeming beauty, to represent identity based on the clinical picture of depression. The visual representation of the game content causes players to consciously engage with a social, socially relevant issue.

Based on these considerations, the following research questions are investigated:

1. Accordingly, does *Gris* contribute to identity formation? The research attempts to explore the extent to which the figure of *Gris*, through his actions, can create an expanded awareness of other socially significant issues.

2. Can this theme, through the artfully designed game (including avatar), set new impulses in the movement for social acceptance and emancipation of this disorder?
3. To what extent can the social processes of the interactions be shown based on the symbolic interactionism?

FINDINGS

The computer game *Gris* takes up artistic impressions and transforms them into a playful work of art.

Mortensen, Linderoth and Brown (2015:100) add to this view by claiming that this means that the game design consciously follows an aesthetic objective and asks the players to rethink their perception of the game. The focus is not on the moral or social perspective, but on undermining the *game system* to promote a different form of interaction between the game (or the game designers) and the players and to draw attention to the situational moment in the game. The importance of the moment exceeds the importance of the single object (i.e., the actual game character) of the game.

The thematization of depression in *Gris* can, based on the author's own assumption, according to Caillois, be classified in the realm of mimicry and agôn. On the one hand, the player slips into an imaginary role, which is integrated into a theme or an illusion, as is the case with theatrical roles, for example. On the other hand, the avatar has to follow a few rules of the game (such as collecting shining stars, resisting windstorms, ...), which simulate a competition-like situation (Caillois, 2001:54).

In a peer group, a person finds him/herself in a similar situation. The ability to change perspectives with regard to de-escalation plays an essential role here, whether in the peer group or in one's own perception.

The interview analysis shows that recognizing the feelings of others and dealing with them within the peer group has a strong influence on the interaction of all the peer group members. It is the condition for being able or willing to solve conflicts within the group together.

Category emotions/preventive/prophylactic interaction: Identifying the feelings of others and dealing with them within the peer group exerts a strong influence on the interaction of all members.

Category change/ destructive interaction: At the same time, it is the precondition for being able to solve conflicts within the group together.

Category contact/destructive interaction: Successful conflict resolution guarantees the existence of the peer group and its norms. The positive experience here makes the emotions in interpersonal communication more visible.

The attempt to understand this emotion is connected to the individual personal experience and own experiences (with one's own feelings). This in turn raises the question of the degree to which an individual must have experienced a similar emotion in order to be able to absorb and understand another emotion.

An emotion can be felt to different degrees by each person. Since emotions are politically and socially relevant, the question is not whether, but how the subject of the game has a socially relevant impact.

Schopenhauer's observations indicate that in order to control a will, not only must the basic instinct be satisfied, but cognition is also required. Before the information one receives can be interpreted, it must be filtered. Strasser (2006:149, author's note) mentions that filtering is the prerequisite for greater cognitive abilities. This results from the process called apperception. If a person has the necessary cognitive abilities, he/she can adopt other perspectives. It therefore implies that players must also have the necessary cognition to be able to understand the game in all its complexity.

FUTURE RESEARCH

It was shown that cognition is needed to deal with the complexity of the game *Gris* in order to draw conclusions about socialization processes. The goal of further research work must therefore deal with the promotion of cognitive processes of an individual. From a social and educational perspective, this takes place through the path of education. The challenge in this consideration is to ensure the necessary condition for equal access to educational offers.

REFERENCES

Berg, S., & von Sass, H. (2014). Spielzüge: Zur Dialektik des Spiels und seinem metaphorischen Mehrwert. Freiburg/München: Karl Alber.

Blumer, H. (2013), Symbolischer Interaktionismus. Berlin: Suhrkamp Verlag

Bortz, J., & Döring, N. (2006). Forschungsmethoden und Evaluation. Heidelberg: Springer Medizin Verlag.

Caillois, R. (2001). Man, play and games. Urbana und Chicago: University of Illinois Press.

Cassone, V. (Vol.12(4) 2017). Mimicking Gamers: Understanding Gamification Through Roger Caillois. Games and Culture, S. 340-360.

Corbin, J., & Strauss, A. (1990). Grounded Theory Research: Procedures, Canons ans Evaluative Criteria. Zeitschrift für Soziologie, S. 418-427.

Fromme, J., Jörissen, B., & Unger, A. (2008). Bildungspotenziale digitaler Spiele und Spielkulturen. Zeitschrift für Theorie und Praxis der Medienbildung, Themenheft Nr. 15/16.

Fuchs, M. (2017). "Ruinensehnsucht": Longing for decay in computer games. Transactions of the Digital Games Research Association, 3(2).

Gast, A. (2017). Identification with Game Characters. Stockholm: Södertörn högskola.

Giesinger, J. (3 2010). Bildung als Selbstverständigung. Vierteljahrsschrift für wissenschaftliche Pädagogik 86, S. 363-375.

Harring, M., Rohlfs, C., & Palentien, C. (2007). Perspektiven der Bildung. Wiesbaden: VS Verlag für Sozialwissenschaften.

Hellal, K. (2019). Gamer Identity: Customized Avatar or struggling with Stereotypes. Linz: Johannes Kepler Universität.

Huizinga, J. (1949). Homo ludens: A study of the play-element in culture London. Citado en Nobert Elias & Eric Dunning: op. cit.

Iacovides, I. (2012). Digital Games: Motivation, Engagement and Informal Learning. York: The University of York.

Kiel, N. (2020). Using Digital Games for Sexual Education: Design Rules, Issues, and Applications. In

Games and Ethics (pp. 215-237). Springer VS, Wiesbaden.

Ministère de l'Éducation nationale, d. l. (2018). Nationaler Rahmenplan zur non-formalen Bildung im Kindes- und Jugendalter. Luxembourg: Ministère de l'Éducation nationale, de l'Enfance et de la Jeunesse & Service National de la Jeunesse.

Mortensen, E., Linderoth, J., & Brown, A. (2015). The Dark Side of Game Play. New York: Taylor & Francis.

Reinhardt, R., & Tries, J. (2008). Konflikt- und Verhandlungsmanagment. Berlin, Heidelberg: Springer-Verlag.

Rengelshausen, H. (2018). Die menschliche Identität in der digitalen Kultur. Hildesheim: Universität Hildesheim.

Sicart, M. (2013). Beyond Choices: The Design of Ethical Gameplay. Cambridge, Massachusetts: The MIT Press.

Strasser, A. (2006). Kognition künstlicher Systeme. Heusenstamm: Ontos Verlag

GAMES OF THE SOUL

Doris Rusch, Andrew Phelps

This paper explores a design framework for creating existential, transformative games - games that directly engage the player in the contemplation life - with the ostensible goals of reflection, awareness, empathy, and growth. Through this work, we seek to re-contextualize games as experiential, expressive works of art that can move us profoundly and evoke lasting inner shifts. These existential media experiences engage their players directly in the consideration of the human condition writ large, our position in the universe, the role and meaning of our lives and relationships in ways that are complex and at once both deeply personal and resonant across the human experience. In considering a design framework for creating games of this type, our work draws from the theory and practice of existential psychotherapy and its main themes and goals to inform the conception of game ideas and gameplay experiences. Through examination of existential, transformative games and their designs with the goal of theorizing a design framework, several elements emerge:

The first is that several of these games use myth to communicate existential ideas in a way that speaks to the unconscious and encourages self-reflection and environmental awareness. In this manner these games can be said to resonate with their players in ways that use shared culture, vocabulary, and societal backdrop to convey ideas well beyond and below the surface of the initial role of myth as a social, cultural, narrative, or aesthetic tool in game design. Such games use myth as a shortcut to the contemplation of the spiritual and to questions of existence.

The second is that, often in combination with myth, these games are deeply rooted in ritual (both in their play and, in a certain sense, their creation). The repeated patterns of gameplay speak to the way repeated patterns, symbols, and practices draw players into channels of thought and reflection in ways that feel deeply human. These games use myth and ritual as existential navigation and personal calibration tools, and in this manner exhibit similar characteristics to the practice of psychotherapy.

Third, these games can be said to be a form of experiential games in which the true narrative and purpose of the game is never overtly stated, and indeed is rarely an explicit narrative at all, but seeks to be felt rather than read. These games focus on the experiential nature of the game itself as it is played, seeking to convey their messages and resonances through this very act, to be evocative, and to invite emotional reflection and response via metaphor.

Given these elements and understanding their criticality, how do we go about creating new myths, in creating new resonances? What practices can help designers create more and better work in this area? This paper explores these questions in depth, presents early work on our theory of design for effective games of this type, and ponders the nudge that games can give us, when we listen, for meaningful, transformational change.

Keywords: existential design, mythology, ritual, game design, games for impact, games for change

———

INTRODUCTION AND BACKGROUND

This work builds upon several recent publications by the authors who are theorizing a design model for existential, transformative game design (Rusch & Phelps, 2020; Rusch, 2020; Phelps & Rusch, 2020; Rusch, 2018). In considering how to best design these games, and how to examine games for the purpose of meaningful transformation, we borrow approaches from existential psychotherapy. As designers, both of the authors have worked for significant periods in the areas of so-called "games for change" or "games for impact" and have become frustrated with certain tenets of this space; namely, that the kinds of change sought by these games must be measurable across different participants and effective at scale in ways that are predefined. The dominant theoretical model employed is that of the transformational framework (Culyba, 2018), which presupposes that the designed purpose of the game is to change the player, that said change is measurable in particular ways, and that said changes are both transferrable to the real world and last for a significant timescale. Furthermore, the transformation sought is posited by the designer of the game—the player is not asked or engaged on how they would like to change prior to playing, nor are they necessarily guided through reflective exercises or analysis afterward. The emphasis of design in this context is on small, discrete, measurable changes in behavior or attitudes. Paulo Pedercini noted in his critique that "the kinds of change we can clearly measure are not all that interesting" (Pedercini, 2014). We agree and further posit that—while this model is very useful e.g. for learning games of a specific kind—it does not capture all of the potential ways that games can aid in personal transformation. Humans are, in this sense, extremely complex. Design that seeks to focus on using games to help players feel more connected, and that they are living a meaningful life, needs to expand beyond these frameworks and evaluative foci.

In 2011, Sophia Ouellette played the video game *Journey* (Thatgamecompany, 2012) one last time with her father, who had been diagnosed with colon cancer and died shortly afterward (Comulada, 2016). The goal of the game is to reach the summit of a mountain, with significant but slow-paced obstacles. Journey is designed with a difficulty curve in which the game is at its most difficult just before reaching the peak, such that players are rewarded with the beauty of the view from the top. In an interview with the press, Sophia notes: "I think that that gave my dad some kind of peace because near the end of his life, he was playing a game that told him that in the end it would be all right" (Comulada, 2016, Takahashi, 2013). This material, Sophia's letter to *Journey* designer Jenova Chen, and the subsequent interview all illustrate how the profound nature of games as storytelling devices that can have lasting impact; indeed, this example further illustrates the unique and personal   nature that games can have on their individual players. While many players enjoyed *Journey*, not all of them had an experience similar to Sophia's. That does not, and should not, make her experience any less important. As Dr. Richard Senelick noted: "Storytelling, in its various forms, may be one way to connect more meaningfully with our patients, to both help us get to know them individually and help them understand their physical condition" (Senelick, 2012, para 5).

It is this focus on individuality that is critical. To account for this personal aspect of change, we draw here from existential psychotherapy as a source of inspiration to inform design. Existential psychotherapy holds at its root the so-called "existential givens", that 1) life is finite, 2) there is no external meaning but that we must find meaning for ourselves, 3) we must make choices, and 4) we are ultimately alone (Yalom, 1980). In that sense, existential psychotherapy takes the individuality and personal journey of people into account, while addressing them through the lens of universal concerns. Further, existential psychotherapy rejects the idea of "mental illness" and focuses on creating a deep connection with the true self, understanding our purpose, and our place within the world. It shifts the notion of patient to that of client—the goal of such therapy is not to "fix" a given individual but rather to help them achieve goals and objectives that they themselves are engaged in identifying. Thus, clients are met where they are currently positioned or self-identify, rather than via predefined analysis or assumptions of who they are supposed to be. Critically, we do not propose that these methods are all-encompassing, but rather that they imply an alternate line of inquiry that can inform design in ways that procedural rhetoric is unlikely to do.

In applying these ideas and themes to transformational game design, we begin by adapting several tenets into practical game creation techniques. First, this approach to transformational design does not seek to transform the player, but rather to recognize that transformation occurs in a complex process that involves the player being positioned for change and subsequently nudged through interaction with a game or media experience towards self-reflection, analysis, and enlightenment. In this sense, these games aim to provide a possibility-space to explore and contemplate existential themes and see what resonates with players out of their own accord.

As one example, consider the oft-used notion of teaching 'relationships with others' via games. This is often done by creating a small multi-user game requiring players to cooperate in order to complete some task, such as a typical multiplayer role-playing game with a fighter, a healer, and a damage-dealing class necessary for survival. But we could reinterpret the idea of collaboration in existential themes by recognizing that the issue is that we are all, eventually, alone, and we must understand the limits of intimacy. Thus, a starting point to improving our relationship to others is improving our relationship with ourselves—that "the ability to be alone is the condition for the ability to love" (Fromm, 1956, p. 88). This can be read as part of the design of Walden: A Game (n.d.), which explores solitude, transcendence, and a conversation with the self as a basis of conversation with others. "I had three chairs in my house, Thoreau wrote, one for solitude, two for friendship, three for society" (Heitman, 2012). It is these kinds of creative re-interpretations of human experiential themes through the lens of existential psychotherapy that can give rise to profoundly moving gameplay experiences that contribute to contemplations of life, death, purpose and relationship and thus to a meaningful life.

A related theme of existential psychotherapy is the idea that "viewed from an existential perspective, the good life is an authentic life, a life in which we are as fully in harmony as we can be. Inauthenticity is illness, is our living in a distorted relationship with our true being" (Bugental, 1990, p. 246). Games can contribute directly to authenticity and inner balance by recognizing the connectedness between authenticity and the unconscious. Put another way, it is in recognizing the alignment between the goals that we *think* we have, and the goals that we *really* have. This leads us in turn to seek out tools to connect with the unconscious, to invite the player to ponder and reflect, and to get in touch with the emotional and underlying mechanisms of what really drives and moves us towards change. Existential psychotherapy acknowledges that inner harmony and alignment of thinking and feeling self are key, but it does not offer its own psychological strategies or tools to foster this alignment. Instead, we have found that it turns to myth with its symbolism and imagery, as famously evidenced by Rollo May's work *The Cry for Myth* (1991).

DESIGN ELEMENTS AND COMPONENTS

Mythology and Symbolism

This desire to tap into the unconscious, and to deeper shared cultural understandings, leads to a focus on the use of myth and ritual. "A myth is a way of making sense in a senseless world. Myths are narrative patterns that give significance to our existence" (May, 1991, p.15). Myths purposefully speak the language of the unconscious through symbolism and imagery

(Rusch, 2017), yet their true meanings are internal processes (Kirmayer, 1999). They are based on a model of the human psyche (Bonnett, 2006), and through these models they can affect us deeply in ways that slide below the conscious and into the unconscious, they express something that is universal to all of us in a shared cultural context.

Myths work with us at this level through psychological resonance and reflection. "The power of myth is that it works without having to analyze it intellectually. If the recipient is in the right mindset—open to the themes the myth deals with—the symbolism and imagery within the story "resonate" and activate the recipient's imagination" (Rusch & Phelps, 2020, p. 3). Or, as Jung notes: "the auditor experiences some of the sensations but is not transformed. Their imaginations are stimulated, they go home and through personal fantasies begin the process of transformation for themselves" (Bonnett, 2006, p. 27). Thus, the goal of a game in this context is not directly as a transformative tool, but as a catalyst, a way to engage the player deeply in a myth that will stick with them and continue to resonate long after the game itself has ended.

How can designers harness the psychologically resonant power of myth (Goodwyn, 2016) to create existential, transformative games? There are certainly several ways, including working with existing mythical material, which has already been done and will continue to be one possible approach (e.g. *God of War* (Sony Interactive Entertainment, 2005), *Darksiders* (THQ Nordic, 2010), *Apotheon* (Alientrap, 2015), *Dante's Inferno* (Electronic Arts, 2010), etc.). There is nothing wrong with trying to mine existing myths for themes for games, but it is limiting because one is working with a limited pool of material. Our goal is to expand this pool and explore how new material can be birthed that carries the potential of myth to connect players with unconscious, resonant themes that can ignite individual change processes. We thus propose active imagination and dreamwork as two psycho-technologies that have been developed by Jung (1997, 2002) and further explored by more recent scholars (Feinstein & Krippner 1988, 1997; Goodwyn, 2018; Johnson, 1986; Moss 1998) to surface unconscious material. As Joseph Campbell stated: "Myths derive from the visions of people who have searched their own most inward world." (2004, p. 24). You do not have to be a poet to be allowed to create new myths or at least content that carries the psychological charge of myth to some degree (whether you really created a myth or not tends not to be visible until it has stood the test of time). Game designers as expressive artists can take this role just as well. By exploring the depths of their own minds — through listening and paying close attention to their "inner storyteller" (Goodwyn, 2018) as it speaks to them through the symbolism and imagery of emotionally charged dreams, or engaging in the focused and intentional practice of active imagination where one enters a conversation with the different parts of oneself while remaining fully awake (Jung, 1987; Johnson 1986; Rusch, 2020) — game designers can practice the generation of psychologically potent imagery and symbolism that can inform games that are infused with a more universally recognizable and resonant force of unconscious material.

Potent imagery has a way of appearing suddenly and with great emotional clarity. It "rings true" and you know it when you see it. It is not the result of hard, methodical labor or mental coercion but an openness to receive, and gentle coaxing through consistent, playful practice. An example of this process is the origin story of the game *Zombie Yoga* (Deep Games Lab, 2012), a Kinect game in which you do yoga poses to direct a "light ball'" — a representation of one's inner light — in order to heal emotional wounds from the past, diffuse metaphorical zombies that represent personal fears, and thus liberate one's inner child.

It all started in a Tai Chi class when one of the authors, Rusch, (and designer of *Zombie Yoga*) was engaged in an exercise called "push hands". This is a partner exercise in which both participants take turns defending their own and transgressing the other person's physical boundaries that guard their personal space. While trying to defend her space from the other person's attempts to break through the arm barrier, Rusch not only experienced physical but a great emotional distress. The exercise itself was understood by her unconscious as a perfect metaphor for the kind of psychological work she had been doing for a long time to defend her boundaries in regards to a taxing relationship. Key to this experience was the strength with which the mind-body connection had suddenly become tangible and obvious. The push-hands exercise was just one manifestation of that and had no place in the game that was inspired by it. It is important to not confuse the source of the inspiration with the finished design. The question that needs to be asked is: what is underneath the surface? What does this metaphor tell me? What does the image my unconscious responds to so strongly actually represent? In this case a concrete situation was understood as a metaphor and sparked the unconscious' further engagement with the psychological theme it had pointed towards. The night after this Tai Chi class, Rusch had a dream in which she used a force field from her hands to speed up and slow down objects around her. Again, the point was not the concrete action of impacting these objects, but the mind-body connection that allowed her to do so.

In the morning, the message was clear: her unconscious called her to get in the body to heal emotional wounds and deal with psychological challenges. Having engaged in some talk therapy before (with limited success), she suddenly had a strong feeling that her path forward was through embodied experience. "I need to feel, not think. I need to use my body as a guide. I need to let my body show me the way. I need to use my body and physical action as symbols to impact my mind." As a game designer, the desire to manifest these new insights in the form of a game was an obvious next step. Not only would this allow Rusch to further engage with these ideas over a longer period of time and in a highly intentional manner, but it could also help others strengthen their mind-body connection and harness the body's power to heal the mind. With this, *Zombie Yoga* was born—the love child of physical action, metaphorical comprehension, and dreamwork. Creating the game was not only an exercise in sourcing unconscious material. It manifested this material in two ways: the game creation can be understood as a ritual performed to honor the messages of the unconscious as well as providing an enactable myth in the form of ritualistic engagement to players.

Ritual is Myth Enacted

At this point, it is critical to note the transformative potential of symbolic action. Through action, players or participants get a handle on inner processes through the performance of these actions. As an example, consider the "poetic acts" of Jodorowski (2010, 2015), which have become famous as a form of theatre counselling. In one of his designed rituals to aid clients who feel they have failed in life and love and are contemplating suicide, he offers the conclusion that such a state cannot be cured and the only solution left is to simply "die to be reborn as a new person" (Jodorowski, year). This involves an elaborate set of rituals and actions, including being buried and dug up, washed and anointed in new clothes, and choosing a new name. As noted by Thompson et al. (2009), the emphasis here is on the action: "[w]e suggest that the efficacy of mental practice resides in its performativity — that is *doing* (even in the mind's eye) *makes it so*" (p. 134).

Games lend themselves directly to the idea of ritual action both at a content level and through their nature as interactive media. These acts are not merely pretend play — they are not simply acting out fantasies without consequence — but rather "various symbolic actions are *indexical* rather than *iconic*; the feigned death is not about pretending to be physically dead, but it refers to the more elusive concept of letting go of an old self" (Rusch & Phelps, 2020). By engaging with ritual in this fashion, players take direct action in the myth itself, they engage in its progression and determine their role in how it unfolds. Games, because of their very nature as interactive fiction, lend themselves to these kinds of interventions, as noted by Murray (1997) when she viewed them through the lens of symbolic dramas.

Experiential Game Design

It is also important to note that games, as interactive media, can move the action of symbolic enactment directly into the control of the player. This provides both a sense of agency and of direct effect, as noted by Phelps et al. (2020), in reference to expanded notions of "flow" (Killi, 2005), and numerous educational theories and ideas on media and interaction (Phelps et al., 2020). Ultimately the concept of experiential design can be thought of in procedural rhetoric terms of mechanics, coupled with complimentary aesthetic in a sort of "what you play is what you do" design space. It seeks to re-orient a game-for-learning approach to argue that no other act, other than playing the game and reflecting on the play, should be required for the learning to occur: the transformation happens inside the individual, and the game acts as a designed experience is the catalyst for this change rather than transfer of experiences to other external domains. This can produce a more ephemeral, resonant kind of game for learning, reflection, or change, focused less on rote knowledge retention or application and more on the experience of feeling and action.

There are several examples of games of this type that seek to use strong thematic elements coupled with mechanics that are less direct or obvious than some of their counterparts that focus more on a skill-and-drill approach. *Journey*, as previously mentioned, seeks to use atmospherics and a kind of literal mountain climb as a metaphorical engagement with challenge, reward, and reflection. *Splattershmup* (MAGIC Spell Studios 2016), engages players in exploring gestural abstract painting by slowly and deliberately creating a piece of art, and explores line, space, and motion through a fluid, spatially oriented interactive video game. Both *Elude* (GAMBIT, 2010) and *Fragile Equilibrium* (Phelps, 2018a) explore depression and anxiety by engaging players in repetitive, abstract mechanics that either seek to create a sense of peaceful and metaphorical engagement, or a deliberate sense of being squeezed and anxious, respectively. Indeed, as Phelps notes "[Fragile Equilibrium] is not a game that teaches someone about depression, it is not a game that aspires to educate someone or empower someone or God-forbid claim to cure someone. It is not a game that directly addresses its subject: it intends instead to evoke a feeling, a nostalgia, a sense of something..." (Phelps, 2018b, para. 19) This "sense of something" is meant to evoke a sense of depression through an experiential metaphor model (Rusch, 2017) grounded in the mechanics: players are forced to navigate a traditional "shmup" style video game while simultaneously dealing with an effect that simulates the breaking of the screen the game is being played on. The game further draws on a lilting, Wabi-Sabi aesthetic and a form of retro and ruinous decay that is nostalgic and somewhat haunting. The game seeks to simulate the notion that people with depression must deal with the world they inhabit, and yet simultaneously reflect on how they are interpreting the world and the degree to which their depression and anxiety is modifying their worldview.

These kinds of deeper experiential designs and metaphorical approach have been proven to be effective with individual players and streamers (Phelps et al., 2020), but again it is the individuality that is key. These experiences are purposefully left vague, left open, to allow players to bring their own histories, their own contexts, and their own lived experience to the forefront. Phelps notes this in his experiences with *Missile Command* (1980) in reference to thermonuclear safety drills common in elementary schools in the United States during the 1980s:

"I remember connecting the fact that a small elementary school desk was unlikely to have much effect given a nuclear detonation, and I remember connecting the fact that we lived within range of the [U.S. Air Force] base. If the Soviet Union fired, they were firing, quite literally, at me. I was terrified. I remember waking up in the middle of the night, wet with sweat, panicked. It wasn't a question of if, it was merely when. I remember thinking as a child it was less than even odds I would ever be an adult.

Then I went to an arcade, and I played *Missile Command*. By this time, I'd played lots of games, and I'd probably even played *Missile Command* before, it wasn't new by then, just still somewhat popular. But in that moment, playing *Missile Command* was transformative: it provided a way for me to process my frustration, my fear, and my anger. It offered an outlet for my grief, and it also, amazingly, provided a sense of agency and control over a situation in which I had none of either." (Phelps, 2020)

Note that these transformations are not always limited to the player but can actually occur for the developer through the course of making such games. In the case of Missile Command: in the original design of the game, the abstract "bases" that the player must protect are not abstract at all: originally, the "bases" in the game were named after six cities in California: Eureka, San Francisco, San Luis Obispo, Santa Barbara, Los Angeles, and San Diego. Designer David Theurer was haunted by visions of thermonuclear war, (Barkan, 2004, p. 140) and to keep the game on a morally neutral ground he insisted that it would not contain the ability for players to fire at other countries or cities, but rather only to defend their own. "The idea of defense was one that players could take pride in, while slowly realizing what the game was forcing them to do: choose between the death of the few or survival of the many" (Rubens, 2013). This was "one of the earliest instances of presenting a player-created narrative almost entirely through gameplay" (Rubens, 2013), which ties directly to experiential design, and yet also had a profound impact on the thinking and moral reflection of its creator.

MAPPING A DESIGN SPACE

Together these three elements: existential themes, the incorporation of myth and ritual in both aesthetics and mechanics, and a more open-ended experiential nature of play, work together to define a space wherein games have significant potential for transformation. While some designs may pull at any one of these elements more closely, games that incorporate two or even all three elements to a significant extent are incredibly powerful. This is visualized by Phelps and Rusch (2020) in the diagram seen in Figure 1:

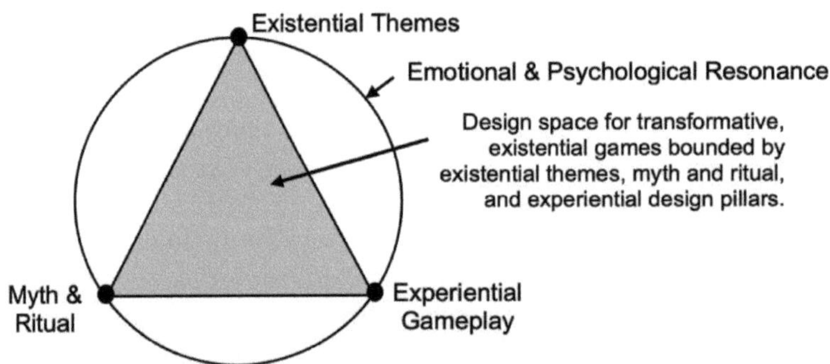

Figure 1. Mapping a design space for existential, transformative games.

Critically, this diagram notes that it is a bounded but continuous space: individual games when examined can map across it, but together these elements form a general notion of design that aspire to emotional and psychological resonance. They work together to create games that are primed to move us, that speak to us unconsciously, primally, even when they engage us intellectually and emotionally. We can then use this mapping in considering additional designs: does the design engage the player strongly on at least one, and ideally more than one, axis? Does the design use both mechanics and aesthetics to reinforce the relationship between these pillars? Does anything in the design threaten the interpretation and reflection of these pillars from the player perspective? This becomes a useful design framework to continue to iterate on a design proposal, and to advance the creation and analysis of more games of this type.

DESIGN SPACE IN PRACTICE

An example of a game that is situated in the midst of this design space is the hit indie-darling game of 2020, *Spiritfarer* (Thunder Lotus Games, 2020). In this game the player takes the role of a new "Spiritfarer" named Stella, whose job is to meet interesting souls throughout the world and guide them to a peaceful transition through the "Everdoor", where they will pass on to the next existence. While that may seem particularly dark (and it is), the game is a very loving and touching piece that first uses narrative and story engagements to get the player to feel connected to these characters, and then guides the player through the process of mourning their individual deaths as they occur throughout the game. At the end of the game, Stella herself realizes it is time to leave the world through the Everdoor, and so the last journey is her own.

The game can be read through a complex intersection of Eastern and Western mythology. "*Spiritfarer* is based on classical Greek mythology: the story of the river Styx, a waterway that's said to be a pathway between Earth and the afterlife. In the myth, spirits are transported by Charon, the ferryman, through the river and into the underworld, paying passage with a token" (Carpenter, 2020). In the game, each character you encounter gives you a token as they board the ship, and when they eventually cross through the Everdoor you receive a spirit flower in their memory. These flowers in turn are used to upgrade your boat and abilities in order to continue the quest. This is one example of the symbolism and metaphor that is rife throughout the game, ranging from a Satyr with regrets for past behavior to Stella herself. The designers of the game note that "[a]s *Spiritfarer's* creation evolved, it became clear to us that Stella had to have been a real person, and had to have lived a real life. The world of *Spiritfarer* then became a projection of Stella's life as an end-of-life care nurse, imagined by her while she, herself, was living through her final moments" (Thunder Lotus Games, 2020, p.12). The game is also clear, both in action and in visuals, of its many Japanese influences, from iconic cherry-blossoms to the altars   and shrines that are used throughout the game to unlock new skills and abilities. An example of such is shown in figure 2:

Figure 2. A shrine for ritual enactment in Spiritfarer. (Smith, 2019)

The actual action of the game involves several different aspects of asset management, sort of like tending a farm in *FarmVille* or an island in *Animal Crossing*. You grow crops and make food for your guests, you collect ore and other materials to build passengers houses on your boat, and each passenger has little side quests where you learn their story and you collect and craft different items for them. These all form a tapestry of indexical symbolic function. As Carpenter notes, "[t]he play may seem mundane and repetitive, but the role of the *Spiritfarer*, caring for these people in their deaths, gives each action depth." (2020)

In this fashion, the game functions at both a mythological and symbolic level, and it is of course also a deeply experiential game. Player-as-Stella advances the metaphor of the myth directly, literally sailing the ship to the Everdoor and rowing passengers, one by one, across it, even deciding in some instances when and in what order, engaging directly in the ritual of their passage. You provide comfort and care for your passengers by literally doing many of the things you would in life: cooking for them, providing them shelter, and engaging in some of their favorite activities. This caregiving is deeply moving (Smith, 2019). In addition, there are other gentle touches at this experiential, slow-paced invitation to reflection: fishing off the boat in the sunset, quiet moments at the top of a mountain peak, the violin that one of the guest spirits will occasionally play for you. Together these elements—existential themes, mythology and symbolism, and experiential play—are brought together to dramatic effect in ways that, as already discussed, rely not only on the game but also on the history of the player and their position in time and experiences and mood:

> "I'd prefer not to think about death. I'm afraid of it—so much that I often force myself not to talk about it, as if I might speak those fears into existence. But I played *Spiritfarer* at a time when I could not ignore death; in the middle of a pandemic that has killed (and continues to kill) numbers of people I can hardly comprehend. And days before receiving the review code, a dear family member died tragically and unexpectedly, the grief of which made starting the game almost too painful. Not necessarily because I didn't want to confront the things I was feeling, but because I was worried about seeing the experience flattened in a video game package—   reduced to a concept too simple to evoke any real feeling. I know that death is not neat, that there is no bow to tie up my own personal grief.

> At no point during *Spiritfarer* did I feel helpless, like there was not something I could do to bring some sort of relief to the spirits on board the ship. It's a stark contrast to how I'm experiencing the grief of death in real life, but it's comforting. […] It's that sort of help—in being able to help someone figure something out for themselves—that makes the repetition of management and care feel so rewarding." (Carpenter, 2020)

The game has won numerous awards and touched numerous players and reviewers, as evidenced in an outpouring of adoration for the title (Flynn & Flynn, 2020; Harrison, 2020; McKlusky, 2020). The themes of resonance, of peace, and of processing grief have reached a wide and varied audience and have done so, critically, in ways that individual players build on and reflect upon their own lived experience. In analyzing the design and effect, we find that this directly correlates with the theoretical design model, and what we can glean from existential psychotherapy in practice. Spiritfarer is a game that, because of its design, can be said to resonate, and for the right player at the right time, profoundly impact their views on mortality as an example of transformational game design.

> "[…] I don't want to fall into trite arguments of how games can make you feel, or soppy stories of how games can deal with serious themes. My goal is to call attention to how *Spiritfarer* gave me the room to find my refuge and work things

through through good narrative design. It is not only the writing, but how the goals of the game and the interactions guide you to do one thing or another. There's nothing preventing players from playing the game to find all the islands and collect all the items. The game does not force you to care about the story. You can keep playing the game and keeping a farm and cooking and building stuff for as long as you want. You can find the way to be happy with the game. There are many ways to play it. I found the game I needed at the right moment. I cared about the story of the characters, but it may be a different game for others, who may find it meaningful in their own way. And that's the beauty of it." (Fernandez-Vara, 2021)

FUTURE RESEARCH

This research is still in its infancy, as the authors continue to explore this model for design and to further theorize the area. They have been collaborating for some time and produced a series of four major journal articles in this space that have formed the core references for the major sections of this paper.   Each of these publications explores these themes and ideas in significantly more nuance and detail: these are complicated topics that require a lot of exposition to unpack. The authors are therefore in the process of producing a manuscript on the topic and engaging in additional studies and interviews with designers and players to better understand and document these ideas. At the same time, they are collaborating on a new video game, an interactive narrative titled *The Witch's Way*, which seeks to put these ideas and design methods into practice. Together, these materials will ideally help to inform other designers in their own work and inspire them to create new kinds of transformational games.

ACKNOWLEDGMENTS

The authors would like to thank the organizers and audience of the FROG 2020 conference for their support of this work and the discussion following its presentation. We would also like to thank Jocelyn Wagner and Drew Moger, both graduate assistants at the American University Game Center in Washington, D.C., for their help in the editing process. Finally, we would thank our respective universities, centers, and laboratories, and the students and colleagues therein, for giving us the space and support to explore these issues and to collaborate across great distances to further this work.

REFERENCES

Alientrap. (2015). *Apotheon* [Video Game]. Windows, Linux, OS X, PlayStation. Alientrap.

Atari. (1980). *Missile Command* [Video Game]. Atari.

Barkan, S. F. (2004). *Blue wizard is about to die!: Prose, poems, and emoto-versatronic expressionist pieces about video games, 1980–2003.* Rusty Immelman Press.

Bonnett, J. (2006). *Stealing fire from the gods.* Studio City, CA: Michael Wiese Productions.

Bugental, J. (1990). *Intimate journeys.* San Francisco, CA: Jossey-Bass.

Campbell, J. (2004). *Pathways to bliss.* Novato, CA: New World Library.

Carpenter, N. (2020, August 18). *Video games revel in death, Spiritfarer focuses on what happens next.* Polygon. https://www.polygon.com/reviews/2020/8/18/21373518/spiritfarer-review-release-date-switch-pc-ps4-xbox-one-thunder-lotus

Comulada, S. (2016, March 24). *A girl and her dad played one last video game before he died. She'll never forget it.* Upworthy. https://www.upworthy.com/a-girl-and-her-dad-played-one-last-video-game-before-he-died-shell-never-forget-it

Culyba, S. (2018). *The transformational framework.* Pittsburgh, PA: ETC Press.

Electronic Arts. (2010). *Dante's Inferno* [Video Game]. PlayStation, PlayStation Portable, X box. Electronic Arts.

Feinstein, D. & Krippner, S. (1988). *Personal Mythology. The Psychology of Your*

Evolving Self. Using Ritual, Dreams and Imagination to Discover Your Inner Story. Jeremy P. Tarcher, Inc.

Feinstein, D., & Krippner, S. (1997). *The mythic path.* New York, NY: Tarcher/ Putnam.

Fernande-Vara, C. (2021, February 16). Spiritfarer: Much more than the "feels." *Vagrant Cursor.* https://vagrantcursor.wordpress.com/2021/02/16/spiritfarer-much-more-than-the-feels/

Fromm, E. (1956). *The art of loving.* Manhattan, NY: Harper.

Flynn, D., & Flynn, D. (2020, August 18). *It's not something you get used to – Spiritfarer Review.* GAMING TREND. https://gamingtrend.com/feature/reviews/its-not-something-you-get-used-to-spiritfarer-review/

GAMBIT MIT. (2010). *Elude* [Video Game]. Adobe Flash (Windows / Macintosh). GAMBIT MIT.

Goodwyn, E. (2016). *Healing symbols in psychotherapy.* London and New York: Routledge.

Goodwyn, E. (2018). *Understanding dreams and other spontaneous images.* London and New York: Routledge.

Harrison, W. (2020, December 10). *'Spiritfarer' is The Blade's 2020 game of the year.* The Blade. https://www.toledoblade.com/a-e/culture/2020/12/10/spiritfarer-is-blade-2020-game-of-the-year/stories/20201210006

Heitman, D. (2012, October). Not exactly a hermit: Henry David Thoreau. *The National Endowment for the Humanities, 33*(5). https://www.neh.gov/humanities/2012/septemberoctober/feature/not-exactly-hermit

Jodorowski, A. (2010). *Psychomagic: The transformative power of shamanic psychotherapy.* Rochester, VT: Inner Traditions.

Jodorowski, A. (2015). *Manual of psychomagic: The practice of shamanic psychotherapy.* Rochester, VT: Inner Traditions.

Johnson, R. (1986). *Inner work.* New York, NY: HarperOne.

Jung, C. G. (1997). *Jung on active imagination.* New York, NY: Routledge.

Jung, C. G. (2002). *Dreams.* New York, NY: Routledge Classics.

Kiili, K. (2005). Digital game-based learning: Towards an experiential gaming model. *The Internet and Higher Education, 8*(1), 13-24. https://doi.org/10.1016/j.iheduc.2004.12.001

Kirmayer, L.J. (1999). Myth and Ritual in Psychotherapy. *Transcultural Psychiatry, 36*(4), 451-460.

May, R. (1991). *The cry for myth.* New York and London: W.W. Norton & Company.

McKlusky, K. (2020, August 20). *Review: Spiritfarer is a lavishly animated exploration of death.* Gamecrate. https://gamecrate.com/reviews/review-spiritfarer-lavishly-animated-exploration-death/26178

Moss, R. (1998). *Dreamgates.* Navato, CA: New World Library.

Murray, J. (1997). *Hamlet on the Holodeck.* Cambridge, MA: MIT Press.

Pedercini, P. (2014, May 1). *Making games in an f**** up world* [Video]. YouTube. https://www.youtube.com/watch?v=MflkwKt7tl4

Phelps, A. (2016) *Splattershmup* [Video Game]. MAGIC Spell Studios.

Phelps, A. (2018a). *Fragile equilibrium* [Video Game]. MAGIC Spell Studios.

Phelps, A. (2018b, December). Fragile equilibrium: extended artist's statement. *Medium.* Retrieved from https://medium.com/@andymphelps/fragile-equilibrium-extended-artists-statement-5f3d84548411

Phelps, A. (2020, September 12). Games of the soul: An anecdotal introduction by way of missile command. *Medium.* https://medium.com/@andymphelps/games-of-the-soul-mc-bbe67da8e27f

Phelps, A., & Rusch, D. (2020). Navigating existential, transformative game design. *DiGRA '20 – Proceedings of the 2020 DiGRA International Conference: Play Everywhere.* Digital Games Research Association 2020 Conference. http://www.digra.org/wp-content/uploads/digital-library/DiGRA_2020_paper_21.pdf

Phelps, A., Wagner, J., & Moger, A. (2020). Experiential depression and anxiety through proceduralized play: a case study of fragile equilibrium. *Journal of Games, Self, and Society, 2*(1), 104–149. doi: 10.1184/R1/12215417

Rubens, A. (2013, August 15). *The creation of Missile Command and the haunting of its creator, Dave Theurer*. Polygon. https://www.polygon.com/features/2013/8/15/4528228/missile-command-dave-theurer

Rusch, D. (Lead designer/game director). (2012). *Zombie Yoga* [Video Game]. PlayStation Kinect. Deep Games Lab.

Rusch, D. (2017). *Making Deep Games – Designing Games with Meaning and Purpose*. Boca Raton, FL: CRC Press Taylor & Francis Group.

Rusch, D. (2018). "21st Century Soul Guides – Leveraging Myth and Ritual for Game Design", DiGRA Nordic *Subversion, Transgression, and Controversy in Play*, University of Bergen, Norway.

Rusch, D. (2020). Existential, Transformative Game Design. *Journal of Games, Self & Society*, 2(1), 1-39.

Rusch, D., & Phelps, A. (2020). Existential Transformational Game Design: Harnessing the "Psychomagic" of Symbolic Enactment. *Frontiers in Psychology*. Special Issue on Digital Games and Mental Health. doi: 10.3389/fpsyg.2020.571522.

Senelick, R. (2012, April 9). *The healing power of storytelling*. HuffPost. https://www.huffpost.com/entry/patient-care_b_1410115

Smith, G. (2019, July 8). *Spiritfarer is about death, but it's how it treats life that makes it unusual*. Rock, Paper, Shotgun. https://www.rockpapershotgun.com/spiritfarer-is-about-death-but-its-how-it-treats-life-that-makes-it-unusual

Sony Interactive Entertainment. (2005). *God of War*. [Video Game] PlayStation. Sony Interactive Entertainment.

Takahashi, D. (2013), February 8th. An Interview with Jenova Chen: How Journey's creator went bankrupt and won game of the year. [web article] Retrieved from https://venturebeat.com/2013/02/08/an-interview-with- jenova-chen-how-journeys-creator-went-bankrupt-and-won-game-of-the- year/

Thatgamecompany. (2012). *Journey* [Video Game]. Thatgamecompany.

THQ Nordic. (2010). *Darksiders*. [Video Game] PlayStation, Xbox, Windows, Linux. THQ.

Thompson, J. J., Ritenbough, C., & Nichter, M. (2009). Reconsidering the placebo response from a broad anthropological perspective. *Cult. Med. Psychiatry* 33, 112–152. doi: 10.1007/s11013-008-9122-2

Thunder Lotus Games. (2020). *Spiritfarer Artbook 1*. https://thunderlotusgames.com/surprise-spiritfarer-is-out-now/artbookpreview_01/

Thunder Lotus Games. (2020). *Spiritfarer* [Video Game]. Thunder Lotus Games.

Walden, a game. (n.d.). Walden, a Game [Video Game]. https://www.waldengame.com

Yalom, I. (1980). *Existential psychotherapy*. New York, NY: Basic Books.

SYNCHRONICITY IN GETTING AN ITEM TO THE PLAYER

Frank Pourvoyeur

I t is a substantial part in game design to enable Players to obtain items. To make the process comprehensible, Players have to understand where to get the item or gain the information how this process works. This might also include knowledge from a previous experience how to obtain a specific object. The game can explain this through the narrative, a form of manifestation, a place where such items are typically stored or how this item was created. The occurrence of an object being available for the Player because the game requires it is then contra-posed with what Carl Gustav Jung called Synchronicity. With taking reference to the concept of "Suspension of Disbelief" it is still possible to overcome the appearance of implausibility when the world is perceived as highly abstract by the Player.

Keywords: synchronicity, verisimilitude, Game Studies

———

INTRODUCTION

In video games players often get to objects they just happen to find by a seemingly lucky incident. There is no causal connection between the object being available in the game other than being useful for the Player.

When psychoanalyst Carl Gustav Jung (1875-1961) noticed such occurrences in his life he speculated about a connection of events happening and a desire for them beyond statistical plausibility and Jung (1960) called this Synchronicity. While Jung could only speculate about an unknown force intervening, Players certainly know this was a decision in the game design. It is usually undesired to make Players aware they are playing a game as this would break the immersion.

When events that help to implement plans occur to an extent that does not correspond to the expected probability, this is referred to as Synchronicity. As a consequence, it is registered as an "unbelievable coincidence" which can lead to Players unintentionally becoming aware that they are playing a game.

Synchronicity by definition is also used in many games as a stylistic tool. Shape and placement of objects give visual clues that tell the Player which objects have significance or are mere decoration. This too has no explanation within the world beside serving the player.

This article summarizes possible ways to get an object to the player that do not stretch Synchronicity while they are within the borders of the established concept of Verisimilitude as discussed by Tichý (1974). Furthermore, it will refer to the "Suspension of Disbelief" as used by Brown (2012) to evaluate when Synchronicities are prominently noticeable by a Player.

RELATED WORK

The term Synchronicity was initially defined by Carl Gustav Jung in 1960 for applications in the field of psychology. In both performing arts and playing games, Synchronicity can be used as an indication of holes in the plot or imperfect world building. Verisimilitude was first used by Karl Popper (1976) for his work in Critical Rationalism. Albeit, this research focuses on the adaptation of the term for fictional storytelling which nowadays refers to the ancient Greek concept for dramatic storytelling.

AIM OF THE RESEARCH

A game is typically designed in such a way that the required items can be found in the game world. If attention to Synchronicity plays an aspect in the design, its occurrences can be identified to better elaborate them to fit naturally into the world.

The aim of the work is to enable game developers to more easily identify situations that work as Synchronicity to either consciously use it as a stylistic choice or to avoid unintentional use derived from not being aware of it. An overview of possible strategies not affecting Synchronicity can serve as a reference for both game developers and game researchers.

FINDINGS

The research produced a general flow graph Fig. 1 that can be applied to game design projects to detect the occurrence of Synchronicity.

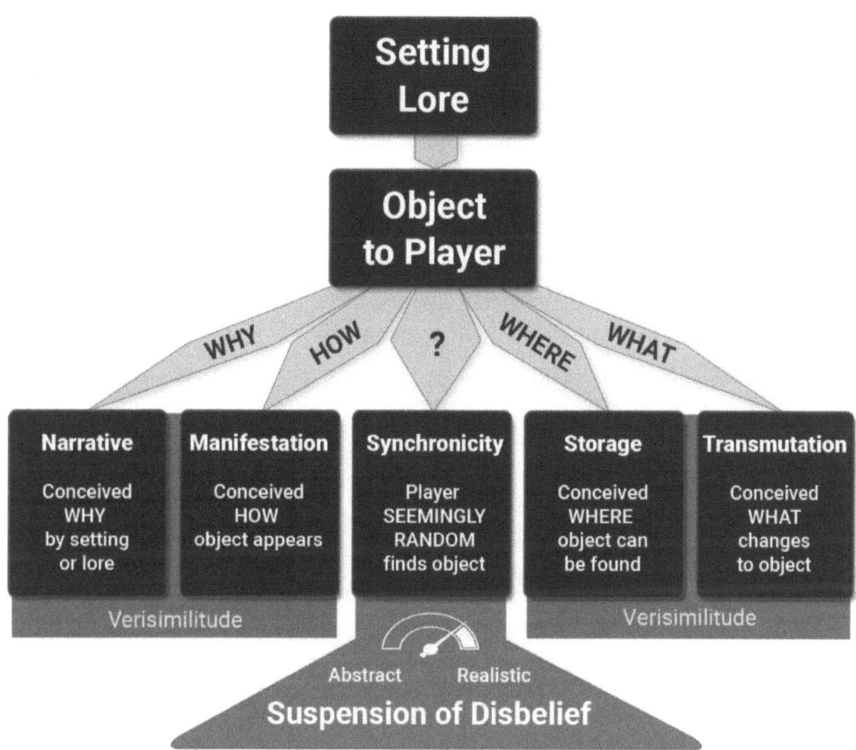

Figure 1. Getting an item to the Player general flow graph

The event of a player attaining an object has a certain believability in the context of the game world. This is high when it is explained and fitting the lore and the setting. The term Verisimilitude describes whether an event happening in the game world is regarded as plausible by the Player to happen inside this world. The setting and story established by world building and its lore allows the Player to make assumptions for the likelihood of events happening in this scenario.

There are 4 main explanations to the Player that support the Verisimilitude:

- **Narrative:** The Player gets the object for or in the end of a task. It is understood *why* the Player has received or will receive the object. Typical usage is a reward for a successful quest or an item that enables the Player to progress in the story. It needs to be taken into consideration that a vague relationship between the object and the task can also contribute to the occurrence of Synchronicity.

- **Manifestation:** Objects appear in the possession of the Player though with an explanation to comprehend *how* this had happened. This includes additional unlocked application possibilities for an item already in possession.

- **Storage:** The Player knows by understanding the game world *where* certain objects can deliberately be acquired. This can be stores dedicated to certain classes of items but also natural areas like a forest to obtain wood or water to catch fish. It is necessary for the player to have prior knowledge and experience where certain items can typically be found.

- **Transmutation:** When an object transforms it is intelligible to the Player *what* object changed. This action is known to the Players as Crafting. When the process is executed one or more items in possession are lost and Players receive something different in return when it is finished.

Note that being comprised in any of these categories does not necessarily dissolve Synchronicities. For example, having a very specific item availably in a tiny store does not give a satisfying justification for it being available.

CONCLUSION

When none of the conditions that satisfy Verisimilitude apply, the event appears as seemingly random to the Player and is best described with Jung´s term Synchronicity. However, the Player does not necessarily lose immersion when Synchronicity is noticed.

Due to "Suspension of Disbelief" consumers of fictitious work can willingly ignore certain flaws to enjoy the creation. Derived from cinematic experiences it can also be expected for games that the "Suspension of Disbelief" is more fragile in a realistic environment and tends to loosen up the more the setting and design is abstract. In stylized fictive worlds it gets harder for the Player to tell if the event is a Synchronicity when there are no references to compare probabilities of occurrences for this world. Synchronicity can be used to check for possible holes in the narrative and retaining believability in conjunction with established Verisimilitude and "Suspension of Disbelief".

FUTURE RESEARCH

Further research is needed to reveal a possible stylistic usage by intentionally using Synchronicity as a design choice. This potentially could lead to new experiences for Players when obtaining items. It is therefore proposed to conduct more research on Synchronicity in games to better understand this as a design device.

Since Synchronicities also occur when the focus of Players needs to be directed to any specific Game Object, further study can lead to an overall better game experience.

ACKNOWLEDGMENTS

The research presented in this paper was part of a master thesis at the Centre for Applied Game Studies at the Danube University Krems in 2020.

REFERENCES

Brown, D. (2012). *The suspension of Disbelief in Videogames*. Falmouth University. DOI:10.13140/RG.2.1.3175.8968

Jung, C G. (1960). *Synchronicity: An Acausal Connecting Principle*. Princeton University Press.

Popper, K. (1976). A note on verisimilitude. *The British Journal for the Philosophy of Science*, *27*(2), 147-159.

Tichý, P. (1974). *On Popper's Definitions of Verisimilitude*. Oxford University Press.

GAMES, PLAY
& HISTORY

GAMES, PLAY AND HISTORY

THE AUSTRIAN GAMES INDUSTRY AND THE FREE-MARKET ECONOMY 1991-2006. A POLITICAL HISTORY OF IDEAS.

Eugen Pfister

The Austrian games industry was particularly successful with business simulations and construction games. In these games we got to know the beautiful new economic world of the post-cold war period in a playful way: Capital had to be increased, production expanded, profits maximised and competition eliminated. During their heyday, the Austrian developer scene were honoured with state and federal awards, and Austrian politicians presented themselves to the press together with "their" shooting stars. After various bankruptcies, takeovers and company dissolutions, the young model entrepreneurs disappeared just as quickly from the collective memory.

It is remarkable that this peak phase of Austrian game production took place at the same time as a political transition phase of Austria, which has not yet been studied much. After the end of the Cold War, Austria joined the European Union. In addition to the paradigm shift in foreign policy, there were also far-reaching changes on the social and economic policy side. For example, the privatisation of Austria Tabakwerke, Telekom and Post took place between 1991 and 2006. In addition, Austrian federal governments have adopted several austerity packages since 1995. It can therefore be said that the development of the Austrian game industry took place in a time of political and social change. In this sense, it is necessary to examine whether the games that emerged can also be read as a sources of direct contemporary Austrian history.

Keywords: Video Game History, Austrian Video Games, History of Ideas, Political History

———

INTRODUCTION

In much-cited books like Steven Kent's "Ultimate History of Videogames" (2001), the history of digital games is most often presented in an almost teleological way as a continuous line of progress from *Tennis for Two* (William Higinbotham, 1958) over *Spacewar!* (Steve Russell 1962) and *Pac-Man* (Namco 1980) to *Call of Duty* (infinity Ward, 2003). Such chronicles accordingly focus mainly on the console platforms that were successful in the US and Japan. At the same time, they ignore the highly dynamic developer scene that was flourishing in nearly all European countries during the 1980s and the 1990s. The history of video games is not a simple line from A to B but a dynamic network of countless transnational transfer processes and the simultaneous emergence of extremely dynamic video game cultures occurring in many different places. For example, there is not the one first chess program to which all later ones can be traced back, but many different projects, which partly influenced each other and partly went their own ways. In the early 1970s, for example, a chess program was developed at the Technical University of Graz: *Frantz* (Fritz Königshofer & Gerhard Wolf, 1974)[1], which like so many other game programs has never made it into the official canon of game history. Computer platforms such as the IBM-PC, the C64, the Commodore Amiga, the Atari ST, the Sinclair ZX Spectrum and the Schneider CPC dominated Europe in the 1990s, but were often neglected by researchers in favour of the history of the console wars between Nintendo and Sega, later by Sony and Microsoft in the US. This omission has only recently and partially been remedied, for instance by the work of Graeme Kirkpatrick on British gaming culture (2015), Alexis Blanchet and Guillaume Montagnon on the French history of games (2020) and most recently the extremely well-researched book by Jaroslav Švelch on game development in the CSSR in the 1980s (2018) and Tom Lenting on the Netherlands (2019). In addition, Austria, Germany and Switzerland could also hardly be described as gaming deserts at the time (Pfister & Rochat, 2021). Rather, some of today's most popular game genres originated in Austria and Germany: building games and business simulations. In Austria, three studios dominated the scene until the 2000s:

---

[1] "Frantz" in Chess Programming Wiki, retrieved from URL:
https://www.chessprogramming.org/Frantz (01.03.2021). Many thanks to Tobias Winnerling for the information!

Max Design, Neo Software and JoWooD.[2] But hardly anyone apart from journalists and game developers now remembers the names of these developers, despite the fact that JoWooD's buidling game *Der Industriegigant* (1997) sold over 750,000 copies at the time (APA0231-5-WI-0234-WB/CI)[3] and *Anno 1602* (Max Design, 1998) over two million copies worldwide (OTS0015-5-WA-0271-EUN0001-CA). JoWooD was a public company listed in the ATX for more than ten years. Why and when did the history of a successful Austrian games industry fall into obscurity? Could the reason be that even at this time, Austrian game developers attracted little attention in the Austrian press?

The answer is no. The Austrian games industry was extensively reported upon in the 1990s, and, in particular, in the 2000s in the Austrian press. The success of the young game developers did not go unnoticed by politicians either. Around the turn of the millennium, both the press and politicians were happy to associate themselves with these successful Austrian products. In fact, I would argue that some game developers were even promoted to being figureheads of the so-called Austrian "New Economy". With the end of the Cold War - but actually even before that - there was a slow but profound change in economic policy in Austria from welfare state to a free market economy. Instead of welfare and stability, efficiency and competitiveness became desirable values. During this period, politicians and the media were looking for young successful entrepreneurs to serve as role models for the whole of society.

However, it is not only the press coverage that offers us sources for a history of political ideas in Austria in the 1990s and 2000s but the games themselves: *1869 – Hart am Wind!* (Max Design, 1992), *Burnout* (Max Design, 1993), *Whale's Voyage* (Neo Software, 1993), *Der Clou!* (Neo Software, 1994), *Oldtimer – Motor City* (Max Design, 1994), *Albert Lasser's Clearing House* (Max Design, 1995), *Der Industriegigant* (JoWood, 1997), *Anno 1602* (Max Design, 1998), *Die Völker* (Neo Software, 1999), *Anno 1503* (Max Design, 2002). All of these games can be understood as business simulations and/or building

---

[2] There is also an Austrian MUD called Holy Mission, created at the University of Linz in 1992 and still active today.

[3] The news items from the Austrian Press Agency (APA) and the press releases from the Original Text Service (OTS) were retrieved by the author via the APA-Online-Manager tool.

games. The fact that most of the games developed in Austria at this time were business simulations is exactly what makes them interesting to the historian: they are fascinating sources for a contemporary history of economic and political ideas in Austria. They translate dominant discursive statements not only in narratives and aesthetic but also in game mechanics. Not only do they thus reflect political and societal change, they also were in part the catalyst of these changes (Pfister, 2018). For the following primary overview of the Austrian game industry in the 1990s and 2000s, I will concentrate on this small corpus of games mentioned above and the press coverage – mainly articles of the Austrian Press Agency (APA) and press releases (OTS) – on the respective game developers from 1991 to 2010. In addition to this, I can also draw on an article by Konstantin Mitgutsch and Herbert Rosenstingl (2015) – until today the only published overview – as well as three thoroughly researched newspaper articles by Daniel Koller for Der Standard, and the interviews with Austrian developers that Sebastian Esberger (2016) conducted as part of his Bachelor's thesis at the Vienna University of Technology.

In a preliminary evaluation, I analyse the games and the press coverage based on the principles of a historical discourse analysis. This means that I will pay attention to the dissemination of specific terminology as well as narrative and ludic constructions of meaning to identify dominant political ideas. In this way I will reveal the ideological discourses underlying the games and show what could be said and thought through these games.

VIDEO GAMES AS A SOCIALIZATION INSTANCE

Games always function on the basis of common values and world views. These form the framework of the game experience. They help us to quickly orient ourselves when immersed in these games. Game rules and game mechanics are best learned when they feel intuitive. This means that developers fall back on dominant (widespread) thought patterns that have largely already been learned at school, at home and through the media. By unconsciously learning the mechanics of the game on the basis of existing knowledge, they potentially reinforce or naturalise these thought patterns. Thus, we quickly learn not only the rules, but also the moral parameters of the game (Pfister, 2018, p. 9-14; Pfister & Görgen, 2020). Whether first person shooter, racing game, platformer or economic simulation; they communicate political ideas about the idea of fairness (Zimmermann, 2020), the value of freedom and/or security, the question of what just wars are or how far one may go in the fight against organized crime and terrorism (Pfister, Winnerling & Zimmermann, 2020). They also communicate economic models and if we want to win these games, we have to learn to play them better, to expand our businesses, to make more profit, to reduce costs, to expand aggressively, to eliminate competitors in time etc. For a historian who wants to better understand the (hidden) ideological discourses of a contemporary society it can therefore be very rewarding to take a closer look at its popular culture and video games. By analysing games, we can research political ideas that might have reached people who otherwise may not have been exposed to them through consumption of other news media

When game developers design a business simulation, they rely on their own knowledge in addition to researching the topic. For Austrian developers at the end of the 1990s, it was therefore obvious to tap into the widespread and common ideas of economic growth and entrepreneurship of the time. We must assume that these game developers had no political agenda in doing so. Their goal was not to advertise a new economic policy, they simply reproduced the zeitgeist. These games offer the historian an outstanding and unique source with which to analyse dominant economic policy discourses in Austria in the 1990s, because they reflect the life-world of their developers.

FROM WELFARE STATE TO NEOLIBERALISM

It is noteworthy that this first peak phase of Austrian game production not only coincided with a phase of political tradition in Austria, but in a broader sense, also with a cultural one, something which has not been researched fully so far: After the end of the Cold War, Austria joined the European Union. The so called "acquis Communautaire" was accompanied by far-reaching political changes, which then became even tougher in the context of the common monetary union (Ederer, Stockhammer & Cetkovic, 2015, 54). In addition to the paradigm shift in foreign policy, there were also far-reaching changes on the social and economic policy side. Under the social-democrat chancellor Bruno Kreisky (1970-1983), Austria relied on public debt and state spending to achieve full employment and with the "Austrian Industrial Holding" was able to rely on a huge network of nationalized companies, an economic policy that was later called Austrokeynesianism (Rathkolb, 2005, p. 580). From 1977 on, Kreisky slowly turned away from deficit spending (Ederer, Stockhammer & Cetkovic, 2015, 34). This reorientation was reinforced especially under the Social Democrat Chancellor Franz Vranitzky (1986-1997), when "the SPÖ [Sozialistische Partei Österreich, Socialist Party Austria][4] followed a course of 'capitalist modernization' with the efficiency of the market taking precedence over the social democratic model" (Gehler, 2012, p. 166). The nationalized industry was criticized in public and not only by the Conservatives of the ÖVP (Österreichische Volkspartei / Austrian People's Party) as outdated, cumbersome, inefficient and, above all, as a loss-making business (Korom, 2012, p. 361). From 1987 and until 2006, public industries were partially or completely privatised one after the other: the Austrian oil and gas company OMV, the Austrian machine and engine factor SGP (Simmering Graz Pauker), the Austrian steel production, a production company for hospitals VAMED, the Austrian salt production, the Telekom Austria, the Austrian tobacco production, the Bank Austria, the Postal Service etc. (Jandratsits, 2015). The privatization of the nationalized industries was accompanied by some drastic austerity measures – amounting to the near final rejection of state Keynesianism (Hemerijck, Unger & Visser, 2000, 204).

The Austrian federal governments of the 1990s and early 2000s – first a coalition of the Social Democrat Party SPÖ and the Conservative Party ÖVP (1986-2000), later a coalition of the ÖVP and the far-right FPÖ (2000-2007) – adopted several austerity packages. The role of the state was redefined. The intention behind privatisation and deregulation policies were to reduce the state's responsibility to provide a regulatory framework for the economy (Ederer, Stockhammer & Cetkovic, 2015, p.37). "With a new law, they transformed the state holding agency tasked with managing state participation in industrial sectors (Österreichischen Industrieholding Aktiengesellschaft) into a privatization agency" (Röth & Spies 2018).

The simultaneous transition from a manufacturing-based economy to a service-based economy (Angelo/Marterbauer, 2015, p. 61) offered many

---

[4] The „Sozialistische Partei Österreich" was renamed under Franz Vranitzky in 1991 to „Sozialdemokratische Partei Österreich".

young Austrians new opportunities, the so-called new economy.[5] This era saw the first success of the Austrian games industry. In this way, a group of computer hobbyists was able to grow into successful entrepreneurs within a short period of time. A stock corporation was created and expanded at a breath-taking pace far beyond the borders of Austria.

MAX DESIGN (1991-2004)

Max Design was founded in Schladming, a small former mining town in Styria by Wilfried Reiter and brothers Albert and Martin Lasser in 1991. In an APA-news item from 1992 they declare that they have made a job out of their hobby (APA0002, 09.09.1992). In an interview conducted by Daniel Koller, we hear from Reiter about how the three developers came from a hobbyist-culture. According to Reiter, their passion was sparked by the C64 in 1985. Directly after unpacking their home computer they started to program their first animations: "Due to a lack of available documentation it was endless trial and error in the beginning, but we got better and better and we spurred each other on to better and better programming skills and better and better looking pixel graphics"[6] (Koller, 07.08.2018). Their first commercial game was the appropriately titled *Cash,* a shipowner business simulation game for the Commodore Amiga 1991 (Nettelbeck, 1991).

Notwithstanding the mixed reviews of the first game (Nettelbeck, 1991) the company achieved recognition in Austria and Germany for its challenging business simulations. (APA0013, 06.05.1998) In 1992 the historical trading simulation *1869 - Hart am Wind* followed, in which the players took on the role of a shipowner at the end of the 19th century. The 120 pages thick manual testifies that the developers had invested a lot of time in the research for the game. The acknowledgements show that colleagues from the University of Salzburg, among others, had helped with the research (manual, title page).

---

[5] Today, we inevitably associate the term 'new economy' with the dot-com bubble, and in Austria the biographies of Werner Böhm, founder of Y-Line, which filed for insolvency in 2001 and André Rhettberg, who led Libro AG into bankruptcy in 2002 and was sentenced to three years in prison for fraudulent bankruptcy. „Neue Anzeige gegen Andre Rettberg" in: derstandard.at, 17.08.2010, URL: https://www.derstandard.at/story/1281829382457/ex-libro-chef-neue-anzeige-gegen-andre-rettberg (01.03.2021)
6 All quotations were translated into English by the author.

This led to an almost forty-page historical overview, written by one Johann Schil[ch]er with the help of "John Wood"[7] describing the topics of imperialism (manual, p. 16-19), colonialism (manual p. 20-33) and migration (manual, p. 34-41). The (German) manual was explicitly praised in a separate APA news item (APA0001, 09.09.1992). The game itself is a lovingly designed trading simulation (ibidem). Players have to buy goods cheaply and sell them expensively and at the same time keep an eye on the running costs of their team and possibly also the coal consumption. At the same time, it is important to keep track of the political developments in the 'world', because unrest and wars mean additional threats to profit The game principle is similar to *Sid Meier's Pirates* (MicroProse, 1987) and *Ports of Call* (Ageis Interactive Software, 1987) – or for that matter *Cash*. Aesthetically, I suspect that the television series *The Onedin Line* (Cyril Abraham, BBC, 1971-1980), which was shown on Austrian television in the 1980s, inspired the shipowner narrative. The game was extremely well received in the German gaming press. The German editors were particularly taken with the economic mechanics (Nettelbeck, 1992). In ASM, the developers at Max Design are referred to as "business experts" (Hink, 1992).

Max Design's subsequent game *Burntime*, released in 1993 was basically another trade simulation game, but this time was in the guise of a roleplaying game and set in the aftermath of a devastating ecological disaster. There was again a manual, which reported in detail about the accident at Chernobyl (Burntime manual, p. 24-27) but also about global warming and the greenhouse effect (Burntime manual, p. 32-57) across almost 30 pages. Besides the trade mechanics, the game also contains attempts at political simulation. The players have to find followers to become the emperor of the Wasteland, here reminiscent of the game *Kaiser* (1984). In 1994 and 1995 *Oldtimer* (in English Motor City) and *Albert Lasser's Clearing House* followed. The first was a simulation of the automobile industry and the second a stock trading simulation. While the former was still praised for the Authentic historical replication of the automobile industry in Europe at the beginning of the 20th century (Duy 1994, p. 10-11), the latter received only mixed reviews (Borovskis 1995, p. 76). According to the developers themselves, the mid-

---

[7] Johann Schilcher later founded JoWooD together with his colleagues Dieter Bernauer, Johann Reitinger and Andreas Tobler [most probably John Wood].

1990s saw a low point in the gaming studio. In an interview with Koller for Der Standard, Wilfried Reiter explains that the wrong partner was brought on board and it was difficult to secure the rights to one's own games and thus they barely scraped past insolvency (Koller, 07.08.2018). "But we pulled ourselves together, tightened our belts and kept going. And we succeeded incredibly well. The first game after this low phase was to become the most successful game in German gaming history at that time and for many years afterwards," Reiter explains. He was talking about *Anno 1602*, the strategy game and origin of a game series that is still successful today (now without Austrian input). In several respects, this game marks the high point in the company's history. It was described as being "cutting edge software" in a news item from the APA (APA0013, 06.05.1998), and was lauded for its intricate economic simulation n the German video game press (Duy, 1998, p.128-130). The game mechanics focus on trade and profit margins, but are supplemented by rudimentary aspects of diplomacy and most importantly the construction of settlements, here comparable to the *Sim City* (Maxis, 1989) or *Die Siedler* (Blue Byte, 1993) series. The manual was again – despite a tutorial – comprehensive, but this time a historical contextualization was omitted. Nevertheless, it is an interesting read. The following sentence is marked bold: "Keeping your citizenry satisfied should always be your main goal." (Anno 1602 manual, p.3). The majority of the buildings can be technologically upgraded throughout the game to please the colony's citizens, which produces more cash for the colony, with which the player can in turn continue upgrading his colony and expand to other islands. This also means that the actual goal of the game, despite the aforementioned sentence in the manual, is not to satisfy the settlers' needs but rather to maximise expansion. *Anno 1602* (1998) and its successor *Anno 1503* (2002) were developed together with the German game developer Sunflowers. *Anno 1602* sold more than 2.5 million copies and became the best-selling game ever developed in the German-speaking world (OTS0015, 06.09.2002). It won the Austrian Staastpreis for Multimedia in 1999 in the category "Unterhaltung, Spiele, Freizeit" (OTS0183, 24.06.1999). *Anno 1503* was also a success and sold more than 200,000 copies in Germany in less than two weeks (OTS0015, 06.09.2002) - both games sold more than 4.4 million copies worldwide (Mitgutsch & Rosenstingl, 2015, 72). After the success of *Anno 1503*, the team decided to dissolve the studio. Talking to Daniel Koller, Reiter explains in the interview that in order to be able to continue to exist on the world market, a lot of money and time would have had to be invested. Also, it would be necessary to leave their home town of Schladming, something they were unwilling to do (Deppe 2019). In 2004 Reiter and the Lasser brothers decided to dissolve Max Design (Koller, 2018).

NEO SOFTWARE (1993-2006)

In January 1993 – two years after Max Design – Neo Software was founded by Hannes Seifert, Niki Laber and Peter Baustädter, then located in Seifert's house in the small town Hirtenberg in Lower Austria. They too emerged out of a hobbyist culture. Seifert had at this point already sold his first game for the Commodore 64 in 1987, the text adventure *Der Verlassene Planet* then published by the journal Markt & Technik.[8] According to Seifert this was the moment that his father stopped complaining about his hobby, when he saw that he could make money out of it (Esberger, 2016). In contrast to Max Design they decided however to relocate their studio to Vienna in 1994 after the initial success of *Whale's Voyage* (1993). In this part roleplaying game, part "trade simulation" (Stiller 1993, p. 45) the player takes the role of a crew of four space travellers stranded in space after making a bad deal purchasing the spacecraft "Whale". Left orbiting a remote planet, they must use trade and skill to upgrade the ship enough to escape. In fact, the players quickly become involved in a political revolution against a character called General Noth and the "Federation". Pirates also feature in the game (Nettelbeck, 1993, p.44). Matt Barton later described it as being "a combination of Firebird's epic space-trading game *Elite* and SSI's *Eye of the Beholder*, and vaguely reminiscent of Binary System's earlier and much more successful *Starflight* series (1986, 1989) and Electronic Art's *Sentinel Worlds* (1989)" (Barton, 2007). In 1994, Neo Software published *Der Clou!* (in English *The Heist*) first for the Amiga then for the PC, an adventure game heavily inspired by heist movies like *The Sting* (George Roy Hill, US, 1973), *Topkapi* (Jules Dassin, US 1964) and *Ocean's Eleven* (Lewis Milestone, US, 1960). Players assume the role of Matt Stuvysant in London after World War II. As a burglar, the player is tasked with finding accomplices, scouting potential targets, and plotting burglaries, down to the finest detail. The loot from successful heist must then be sold to one of the

---

[8] In: C64-Wiki, URL: https://www.c64-wiki.de/wiki/Der_verlassene_Planet (01.03.2021).

three available dealers in order to earn money, which in turn can be invested in better tools. In the game, it is possible to steal the remains of Karl Marx out of the cemetery. Neo Software's biggest success, however, was the building game *Die Völker* (in English published as Alien Nations), which is very similar to *Anno 1602* – or for that matter the German game *Die Siedler* (The Settlers: Blue Byte, 1993), but with a Science Fiction setting. In contrast to Max Design's games which are characterized by serious themes and extensive research, Neo Software's games and especially *Die Völker* are characterized by a humorous approach. By 2001, *Die Völker* had sold over 1 million copies worldwide, most of them in Austria, the Czech Republic, Slovenia, Hungary, Germany and 80,000 in Russia.[9] Shortly after that, the German company Computec Media AG bought 51 per cent of Neo Software (APA0250 01.06.1999), and sold it in 2000 to the British Gameplay.com PLC group (OTS0190, 23.02.2000). In 2001, Gameplay, which was on the verge of bankruptcy, had to sell Neo Software to its shareholder Take Two for £1 (Fletcher, 2001). Since then, Neo Software have taken on commissioned work, and for example developing the extremely successful Xbox ports of *Max Payne* (2001), *GTA 3* (2003) and *GTA: Vice City* (2003). Neo Software was rebranded by Take Two as Rockstar Vienna (Mitgutsch & Rosenstingl, 2015, 74). The games they developed have achieved a turnover of 300 million dollars - with only 110 employees according Daniel Koller (14.10.2018). In 2006, however, the Vienna studio was closed overnight by its parent company without warning and despite good results. A situation that most Austrians only know from the movies: The locks on the doors had been changed, security men dressed in black denied them access to their desks and led them in small groups to the meeting room and each employee was invited individually to a severance interview. Hundreds of employees were dismissed without notice. The KPÖ alone publicly criticized the procedure (APA0504, 12.05.2006; OTS0278, 12.05.2006). Some of the employees then founded the company Games That Matter. The studio was bought by Koch Media and ran for a short time under the name Deep Silver Vienna (Mitgutsch & Rosenstingl, 2015, 74).

---

[9]   „JoWood in Russland", in: gamestar.de 20.04.2001, via archive.org, URL: https://web.archive.org/web/20181202032438/https://www.gamestar.de/artikel/jowood-in-russland,1336864.html (01.03.2021)

JOWOOD (1995-2011)

JoWooD was the last bigger Austrian developer studio to be established. It was founded in 1995 in Ebensee, a market town with 7100 inhabitants in Upper Austria by Dieter Bernauer (later Bernauer-Schilcher), Johann Reitinger, Johann Schilcher, and Andreas Tobler with its headquarters later moved to Liezen, also in Sytria. In an interview from 2018, Dieter Bernauer-Schilcher explains that the studio emerged out of a "hobby project" (Koller, 30.09.2018). According to Johann Schilcher, who already had some experience having programmed a game for the Atari ST, they began working on their first game Der Industriegigant in 1994 (Nettelbeck, 1998, p. 68). The game was published in 1997. It is interesting to note that the game was closely followed in the Austrian press from the very beginning. As early as June 1997, a detailed news item was published by the Austrian Press Agency (APA0023, 16.06.1997). It was not short on superlatives. At the beginning of the interview, Johann Schilcher refuted comparisons with Bill Gates put to him by the interviewer. It is worthwhile to take a discourse-analytical perspective here to see which values are highlighted in the news item as positive: Since February 1996, at least 80 hours of work have been on the timetable week after week and when it comes to family, 'quality instead of quantity' is the motto (Ibidem). Der Industriegigant received the Austrian state prize "Prix MultiMediaArt '98" in March 1998 in the category "best computer game" (APA0132, 04.03.1998). The business simulation for Windows 95 sold a total of 800,000 copies and was translated into twelve languages. It has a similar game-play to Anno 1602 and Die Völker is a building game, heavily inspired by the success of Transport Tycoon (Nettelbeck, 1998, p. 64), with the difference that it clarifies from the beginning that the game focuses on industrial production and distribution of goods. Already in the tutorial you will learn that production and sales strongly influence the growth of cities. For example, department stores are extremely conducive to growth.

In the year 2000, the GmbH (limited company) went public. "The decision to lead 'JoWooD' to the capital market is the logical consequence of the outstanding corporate success of the last few years and the excellent future prospects" according to Managing Director Andreas Tobler (APA0231, 31.01.2000). In the same year, the traffic simulation game, Der Verkehrsgigant, was published. Politicians, like the Viennese City Councillor for traffic Fritz

Svihalek were now also happy to show their support for the game developers (APA0288, 03.03.2000). The game was followed in 2002 by Der Industriegigant 2 and in 2004 by Der Transportgigant. Both aesthetically and in terms of game mechanics, the four economic simulations were very similar. The rendered intro videos set the tone: growth and progress. The game mechanics educate the players to plan production processes as efficiently as possible. Wages should be kept low if feasible. But what should not happen under any circumstances - as you can read in the tutorial of Der Industriegigant 2 for example - is that the workers have nothing to do.

The IPO of JoWooD was accompanied by an internal reorientation: "The company intends to use the proceeds from the [...] to pursue an active acquisition policy, expand its online product range and increasingly use the Internet for online gaming, e-commerce, merchandising, promotion and community building." (APA0403, 30.05.2000). This was followed by an almost breathtakingly rapid, if not purposeful expansion: The company changed successively from a developer to a publisher and went on a Europe-wide shopping tour of smaller developers and publishers (APA0610, 18.05. 2001; APA0099, 01.09.2000; APA0078 5, 12.12.2000; APA0127, 03.05.2001). At first everything went well. Sales increased by 36 per cent in 2000 (APA0202, 25.08.2000). The turnover doubled in comparison to the previous year (APA0258, 18.10.2000). International distribution partnerships and branches in Germany, Great Britain and Japan followed in 2001 (APA0580, 10.07.2000; APA0111, 01.02.2001; APA0607, 21.03.2001 and APA0092, 11.05.2001). In May 2001, Wilhelm Hamrozi, former CEO of Sunflowers, was added as Chief Operating Officer COO to the Executive Board (APA0084, 01.05.2001). The turnover continued to increase in 2001 (APA0151, 23.08.2001). There were however signs of internal problems: Johann Schilcher resigned as technical director and officially wanted to concentrate on developing games (APA0268, 05.06.2002). In September 2002, there were further changes in the supervisory board "due to differing views on the corporate strategy" (APA0565, 30.09.2002). Shortly afterwards, a sales and profit warning for the whole of 2002 was issued (APA0040, 08.10.2002). Six days later, the share price plunged by 80 per cent (APA0194, 14.10.2002). The workforce of 280 employees was to be halved (APA0169, 18.12.2002). Shareholders – relieved by these "measures" – then gave the "green light for capital increase" (APA0531,

18.12.2002). The company's future seemed to be secure once again. In the first quarter of 2003, a positive consolidated result was again achieved (APA0099, 14.05.2003). While supervisory board remuneration in Austria rose by an average of 2.9 per cent from 2002 to 2003, the average for the JoWooD supervisory board was 33 per cent (APA0145, 15.10.2004). In December 2004, however, the JoWooD supervisory board resigned as one: shareholder groups made realignment impossible, the APA reported (APA0488, 16.12.2004). Andreas Rudas, former SPÖ federal managing director, became the new chairman of the supervisory board and stayed until 2006 (APA0131, 23.12.2004). In January 2005, the share price plunged again by 16 per cent after a sales warning (APA0258, 19.01.2005). As a reaction to this, among other things, JoWooD closed the original site in Ebensee (APA0124, 26.01.2005). In March, insolvency rumours made the rounds (APA0619, 04.03.2005). Four days later it became public, JoWood had made a loss of 21.3 million euros in 2004 (APA0096, 08.03.2005). A package of measures was put together. Again, there were job losses (APA0133, 03.05.2005). The share recovered once again at the beginning of 2006 (APA0160, 27.04.2006). After changes in the executive board, JoWood boss Alfred Seidl then examined claims for damages against his two predecessors, company founder Andreas Tobler and Michael Pistauer (APA0404, 27.04.2006). The company expanded again and the turnover increased again in the next years. In October 2009, JoWooD went on another shopping spree. In 2010, disillusionment followed. The annual result for 2009 was revised downwards (APA0454, 22.09.2010). In January 2011, a final APA-EILT announcement: "Listed JoWooD insolvent - reorganisation proceedings applied for" (APA0319, 07.01.2011).

Looking back, Schilcher later explained in an interview that the early IPO (initial public offering) had been a big mistake; "An Economically healthy development was practically no longer possible." Excessive expectations, claims to shareholders that were out of touch with reality, and a resulting increase in excessive demands led to a toxic working climate (Fröhlich, 2011). Starting with an internationally successful game – Der Industriegigant – the company had – here following the zeitgeist – rapidly expanded. Since its games themselves did not generate enough profit for further expansion despite their success, the company went public and shifted the business to sales, to accumulate enough capital to rapidly expand. The company's own game production had apparently not been enough for the growth promised to shareholders. An aggressive expansion policy kept the share price high. The press and politicians were enthusiastic. The management board was expanded. The company expanded rapidly and without a real goal – except for expansion as an end in itself. But when crisis hit, it became apparent that there was no real substance to fall back on. In the process, quality control soon fell by the wayside. A famous example of this was Gothic 3 (Piranha Bytes, 2006), which was unfinished at the time of release, and received terrible reviews and flopped.

CONCLUDING REMARKS

These are preliminary observations of a turbulent and fascinating period of Austrian contemporary history. For a truly meaningful analysis, however, extensive research of all games and the entire news coverage would be required. I am convinced, that it is urgently necessary to finally scientifically reappraise the history of the Austrian games industry, also in the sense of a thorough global history of digital games (Köstlbauer, Pfister, Winnerling & Zimmermann, 2018). Hopefully in the near future I will have the opportunity to deal with it in detail. Nevertheless, even these first observations allow some interesting and illuminating conclusions about the history of ideas of the games themselves as well as the history of the developers. There is for one a mesmerizing resonance between the narratives and mechanics of the games and the story of their developers – expansion as a goal of its own.

The story of Max Design, Neo Software and JoWooD reads like a didactic play on the New Economy. One company – Max Design – became successful, realized that despite its success it could no longer keep up with the world market and stopped at the peak of its success. One company – Neo Software – became successful, looked for investors, was bought by different bigger companies and then closed down overnight without official reasons. One company finally – JoWooD – became successful, wanted to grow faster and therefore went public and shifted its business towards sales. It grew and grew until it suddenly collapsed into itself. The developers at the origin of these companies presented themselves as former hobbyists (mostly computer science students) that made a fortune out of their hobby. Most of them knew each other from their student days and, especially in the beginning, they worked together across company boundaries. Seifert, for example, made the music for *1869* and *Burntime*. Johann Schilcher wrote the manual for both games.

So, while the biographies of the developers clearly followed a neoliberal expansionist narrative, the situation in the games themselves is a little bit more complex. Here we can for instance find some faint remnants of a welfare state way of thinking. We could for instance argue that in *Anno 1602*, the aim is to keep an eye on the contentment of the population and to serve them centrally by satisfying their needs. The same is true for *Die Völker* and *Der Industriegigant*, where the satisfaction of the workers has to be regulated. This is all the more remarkable, when we take into account that the satisfaction of those workers of JoWood who were dismissed over the years, was not mentioned in the related news items. Especially in the games of Max Design, the detailed additional materials and information in the manuals showed a strong interest of the developers in the solution of overarching societal problems, such as global warming in *Burntime*. At the same time, we encounter in these games - whether consciously or unconsciously - central thought patterns of neoliberalism: There is - and this is perhaps the most important moment here - the crushing supremacy of economic thinking. Whether colonizing the New World, building up society after the ecological collapse, a career as a thief or a SF adventure: trade and sale of goods are always core game mechanics (Baur, 1999). Regardless of the setting, we always learn to maximize our profits quickly and are encouraged by the games to continue to expand. In this sense, we can assume - with all caution - that these games also functioned as learning instances in Austria. They propagated core statements of a neoliberal economic system, which were "naturalized" through constant repetition. Above all, however, they are exciting sources of a popular cultural discourse on economic policy, which is particularly evident in the fascinating resonance between game content, or rather game mechanics, and the history of game developers.

REFERENCES

Angelo, S. & Marterbauer, M. (2015). Ist alles besser in der Realwirtschaft? Zum Revival der Industriepolitik. Beigewum (Ed.), Politische Ökonomie Österreichs. Kontinuitäten und Veränderungen seit dem EU-Beitritt, Wien: Mandelbaum. 59-71.

Barton, M. (2007). The History of Computer Role-Playing Games Part III: The Platinum and Modern Ages (1994-2004). gamasutra.com. Retrieved from: https://www.gamasutra.com/view/feature/129994/the_history_of_computer_.php?page=2

Baur, S. (1999). Historie in Computerspielen: „Anno 1602 – Erschaffung einer neuen Welt", WerkstattGeschichte 23, pp. 83–91.

Blanchet, A. & Montagnon, G. (2020). Une Histoire du Jeu Vidéo en France. 1960-1991: Des Labos aux Chambres d'Ados, Paris(?): Pix'n Love.

Borovskis, T. (1995). Review of Albert Lasser's Clearing House. PC Games 12/95. 76. Retrieved from URL: https://www.kultboy.com/index.php?site=t&id=18492.

Deppe, M. (2019). Jahre Inselraffen - Der große Anno-Report. Gamestar Black Edition 02/2019. 135.

Duy, M. (1994). Review of Oldtimer. PC Joker 12/94. 10-11. Retrieved from: https://www.kultboy.com/index.php?site=t&id=12597.

Duy, M. (1998). Review of Anno 1602. PC Player 5/98. 128-130. Retrieved from URL: https://www.kultboy.com/index.php?site=t&id=8559.

Ederer, S., Stockhammer, E. & Cetkovic, P. (2015). 20 Jahre Österreich in der EU – Neoliberale Regulationsweise und exportgetriebenes Akumulationsregime. In Beigewum (Ed.), Politische Ökonomie Österreichs. Kontinuitäten und Veränderungen seit dem EU-Beitritt, Wien: Mandelbaum 34-58.

Esberger. S. (2016). Austriangames.at. [most parts of the webpage are now no longer online] Retrieved from URL: https://web.archive.org/web/20180811190648/https://austriangames.at/.

Fletcher, L. (31.01.2001). Gameplay sells subsidiary for £1. citywire.co.uk. Retrieved from: https://citywire.co.uk/funds-insider/news/gameplay-sells-subsidiary-for-1/a218731.

Gehler, M. (2012). Paving Austria's Way to Brussels:Chancellor Franz Vranitzky (1986-1997) – A Banker, Social Democrat, and Pragmatic European Leader, JEIH 18 (2012), 159-182. 166

Hink, A. (1992). Review of 1869. ASM 8/92. 118. Retrieved from: https://www.kultboy.com/index.php?site=t&id=6469.

Jandratsits, F. (2015). Privatisierungen: Tops und Flops, in kurier.at, 30.12.2015, Retrieved from URL: https://kurier.at/wirtschaft/privatisierungen-tops-und-flops/172.192.075.

Javet, D. & Rochat, Y. (2017). Jeux vidéo suisses : état des lieux (2012-2017), Culture en Jeu 54, 30.04.2017, pp. 6-7.

Kent, S. L. (2001). The Ultimate History of Video Games, New York: Three Rivers Press.

Kirkpatrick, G. (2015). The formation of gaming culture: UK gaming magazines, 1981-1995. Basingstoke / New York, NY: Palgrave Macmillan.

Koller, D. (07.08.2018). Max Design: Als die erfolgreichsten Games aus Österreich kamen. derstandard.at. Retrieved from URL: URL:

https://www.derstandard.at/story/2000087657379/rueckblick-auf-max-design-als-die-erfolgreichsten-games-noch-aus

Koller, D. (30.09.2018). Rückblick auf Österreichs erfolgreichsten Spielehersteller Jowood. derstandard.at. Retrieved from URL: https://www.derstandard.at/story/2000087592969/vom-aufstieg-bis-zur-insolvenz-rueckblick-auf-oesterreichs-erfolgreichsten-spielehersteller

Koller, D. (14.10.2018). Rockstar Vienna: Als Wien die Heimat von "GTA" war, derstandard.at. Retrieved from URL: https://www.derstandard.at/story/2000088999416/rockstar-vienna-als-wien-die-heimat-von-gta-und-max.

Korom, P. (2013). Austria Inc. Forever? On the Stability of a Coordinated Corporate Network in Times of Privatization and Internationalization, World Political Science Review 2013; 9(1). 357-383. 361

Lenting, T. (2019). Gamegeschiedenis van Nederland, 1978-2018, Karel van Mander Academy.

Mitgutsch, K. & Rosenstingl, H. (2015). Austria. In Wolf, M.J.P. (Ed.). Video Games Around the World. Cambridge (MA): MIT Press. 71-86.

Nettelbeck, J. (1991). Review of Cash. Amiga Joker 9/91. Retrieved from URL: https://www.kultboy.com/index.php?site=t&id=713.

Nettelbeck, J. (1992). Review of 1869. Amiga Joker 9/92, 28-29. Retrieved from URL: https://www.kultboy.com/index.php?site=t&id=9937

Nettelbeck, (1993) J. Review of Whale's Voyage. Amiga Joker 3/93. 44. Retrieved from URL: https://www.kultboy.com/index.php?site=t&id=791.

Nettelbeck, J. (1998). Review of Der Industriegigant. PC Joker 8/98. 64-68. Retrieved from URL: https://www.kultboy.com/index.php?site=t&id=15410.

Pfister, E. (2018). Der Politische Mythos als diskursive Aussage im digitalen Spiel. Ein Beitrag aus der Perspektive der Politikgeschichte. In: Thorsten J. & Schumacher, C. (Ed.). Digitale Spiele im Diskurs. Fernuni Hagen: Hagen 2018. Retrieved from URL: http://www.medien-im-diskurs.de

Köstlbauer, J., Pfister, E., Winnerling, T. & Zimmermann, F. (Eds.). (2018). Weltmaschinen: Digitale Spiele als globalgeschichtliches Phänomen. Wien: Mandelbaum.

Pfister, E. & Görgen, A. (2020). Politische Transferprozesse in digitalen Spielen. Eine Begriffsgeschichte. In Görgen, A. & Simond, S. (Ed.) Krankheit in Digitalen Spielen. Bielefeld: Transcript, 51-74.

Pfister, E. & Winnerling, T. (2020). Digitale Spiele und Geschichte. Ein kurzer Leitfaden für Student*innen, Forscher*innen und Geschichtsinteressierte. Glückstadt: Werner Hülsbusch.

Pfister, E., Winnerling, T. & Zimmermann, F. (2020). „Democracy Dies playfully. Three Questions –Introductory Thoughts on the Papers Assembled and Beyond". Gamevironments

13 (2020). 1-34. Retrieved from URL: https://media.suub.uni-bremen.de/bitstream/elib/4603/1/01_gamevironments_Pfister_Winnerling_Zimmermann.pdf.

Pfister, E. & Rochat, Y. (2021) Des débuts de l'histoire des jeux vidéo en Autriche et en Suisse: une étude comparative des industries. Retrieved from URL: http://spielkult.hypotheses.org/2880.

Rathkolb, O. (2015). Die Zweite Republik. In Winkelbauer, T. Geschichte Österreichs, Reclam: Stuttgart. 525-594.

Röth, L., Afonso, A., & Spies, D. (2018). The impact of Populist Radical Right Parties on socio-economic policies. European Political Science Review, 10(3), 325-350.

Stiller, H. (1993). Review of Whale's Voyage. ASM 4/93. 45. Retrieved from: https://www.kultboy.com/index.php?site=t&id=15410.

Švelch, J. (2018). Gaming the Iron Curtain. How Teenagers and Amateurs in Communist Czechoslovakia Claimed the Medium of Computer Games, MIT Press.

Zimmermann, F. (2020). Mein blauer Freund: Gerechtigkeitsmechaniken in Digitalen Spielen, Spiel-Kultur-Wissenschaften, Retrieved from URL: http://spielkult.hypotheses.org/2787.

SOURCES

APA0001, 09.09.1992 = APA0001 5 CI 0352, „In meinem Computer bin ich Kapitän", 09.09.1992.

APA0002, 09.09.1992 = APA0002 5 CI 0340, „Steirische Softwarefirma wagt den Sprung über den großen Teich", 09.09.1992.

APA0023, 16.06.1997 = APA0023 5 CI 0417, „,JoWooD' – ,Die Garage ist keine Garage'", 16.06.1997.

APA0132, 04.03.1998 = APA0132 5 CI 0198, „,Der IndustrieGigant [sic] gewann ,Prix MultiMediaArt 98'" 04.03.1998.

APA0013, 06.05.1998 = APA0013 5 CI 0313, "Handel und Strategie Anno 1602", 06. 05. 1998.

APA0250 01.06.1999 = APA0250 5 WI 0167, „Computec beteiligt sich an österreichischem Softwareentwickler", 01.06.1999.

APA0231 5 WI 0234 WB/CI, Oö. Spiele-Entwickler "JoWooD" plant Börsegang für 2. Quartal, 31.01.2000.

APA0288, 03.03.2000 = APA0288 5 CI 0424 „,Der Verkehrsgigant' freut Stadrat Fritz Svihalek", 03.03.2000.

APA0403, 30.05.2000 = APA0403 5 WI 0352 WB, „JoWooD will massiv in den Online-Spielemarkt einsteigen", 30.05.2000.

APA0580, 10.07.2000 = APA0580 5 WI 0130 WA/WB, JoWooD schließt Vertriebsvertrag mit führendem japanischen Publisher, 10.07.2000.

APA0202, 25.08.2000 = APA0202 5 WI 0137 WB, 41 JoWooD steigert Umsatz im Halbjahr um 36 Prozent, 25.08.2000.

APA0099, 01.09.2000 = APA0099 5 WI 0233 WB/WA, JoWooD übernimmt deutschen Softwareentwickler Neon, 01.09.2000.

APA0258, 18.10.2000 = APA0258 5 WI 0189 WB, JoWooD-Umsatz bis September mehr als verdoppelt, 18.10.2000.

APA0078 5, 12.12.2000 = APA0078 5 WI 0083 WB, PC-Spiele-Erfinder JoWooD übernimmt deutsches Entwicklungsstudio, 12.12.2000.

APA0111, 01.02.2001 = APA0111 5 WI 0238 WA/WB, Computerspiele-Entwickler JoWood lässt sich in Deutschland nieder, 01.02.2001.

APA0607, 21.03.2001 = APA0607 5 WI 0107 WB, JoWooD eröffnet Niederlassung in Großbritannien, 21.03.2001.

APA0084, 01.05.2001 = APA0084 5 WI 0165 WB, JoWooD erweitert Vorstand, 01.05.2001.

APA0127, 03.05.2001 = APA0127 5 WI 0215 WB/WA, JoWooD übernimmt deutschen Spielevertreiber Leisuresoft, 03.05.2001.

APA0092, 11.05.2001 = APA0092 5 WI 0183 WA/WB, JoWooD eröffnet Niederlassung in Japan, 11.05.2001.

APA0610, 18.05. 2001 = APA0610 5 WI 0135, 27 Börsekandidat Jowood übernimmt Spielesoftware-Entwickler, 18.05. 2001.

APA0151, 23.08.2001 = APA0151 5 WI 0287 WB, JoWooD mit starkem Umsatzplus und besseren Ergebnissen im Halbjahr, 23.08.2001.

APA0268, 05.06.2002 = APA0268 5 WI 0456 WB, JoWooD finanziert 2 - Kapitalerhöhung soll 20 Mill. Euro bringen, 05.06.2002.

APA0565, 30.09.2002 = APA0565 5 WI 0132, Wechsel im Vorstand von JoWooD, 30.09.2002.

APA0040, 08.10.2002 = APA0040 5 WI 0144 WB, JoWooD mit Umsatz- und Gewinnwarnung für Gesamtjahr 2002, 08.10.2002.

APA0194, 14.10.2002 = APA0194 5 WI 0182 WB, JoWooD-Aktie bricht nach Gewinnwarnung um 80 % auf 0,65 Euro ein, 14.10.2002.

APA0169, 18.12.2002 = APA0169 5 WI 0247 WB, JoWooD bis Herbst mit Megaverlust - Belegschaft soll halbiert werden, 18.12.2002.

APA0531, 18.12.2002 = APA0531 5 WI 0162 WB, JoWooD 2 - Aktionäre geben grünes Licht für Kapitalerhöhung, 18.12.2002.

APA0099, 14.05.2003 = APA0099 5 WI 0370 WB, JowooD schaffte im ersten Quartal 2003 positives Konzernergebnis, 14.05.2003.

APA0145, 15.10.2004 = APA0145 5 WI 0278 WB, Österreichs teuerster Vorstand werkt in der Bank Austria, 15.10.2004.

APA0488, 16.12.2004 = APA0488 5 WI 0215 WB, JoWooD-Aufsichtsrat geschlossen zurückgetreten, 16.12.2004.

APA0131, 23.12.2004 = APA0131 5 WI 0151 WB, Rudas soll in JoWooD-Aufsichtsrat einziehen, 23.12.2004.

APA0258, 19.01.2005 = APA0258 5 WI 0276 WB, JoWooD weiter auf Talfahrt - Aktie stürzt um gut 16 Prozent ab, 19.01.2005.

APA0124, 26.01.2005 = APA0124 5 WI 0170 WB, JoWooD schließt Standorte Ebensee und Offenbach, 26.01.2005.

APA0619, 04.03.2005 = APA0619 5 WI 0459 WB, JoWooD-Chef über Insolvenzgerüchte "verwundert", 04.03.2005.

APA0096, 08.03.2005 = APA0096 5 WI 0095 WB, JoWooD schrieb 2004 tiefrote Zahlen - 21,3 Mio. Euro Jahresverlust, 08.03.2005.

APA0133, 03.05.2005 = APA0133 5 WI 0441 WB, JoWooD 2 - Weiterer Mitarbeiterabbau bis Juni geplant, 03.05.2005.

APA0160, 27.04.2006 = APA0160 5 WI 0179 WB, JoWooD drehte Konzernergebnis im ersten Quartal 2006 ins Plus, 27.04.2006.

APA0404, 27.04.2006 = APA0404 5 WI 0234 WB, JoWooD lässt Klagen gegen Ex-Vorstände prüfen, 27.04.2006.

APA0504, 12.05.2006 = APA0504 5 WI 0201 WA, 12.05.2006 "Wiener Spiele-Entwickler Rockstar Vienna offenbar zugesperrt".

APA0454, 22.09.2010 = APA0454 5 WI 0217 WB/IT, JoWooD - Halbjahresverlust zwischen 10 und 15 Mio. Euro erwartet, 22.09.2010.

APA0319, 07.01.2011 = APA0319 3 WI 0075 WB/IT, Börsenotierte JoWooD insolvent - Sanierungsverfahren beantragt, 07.01.2011.

OTS0183, 24.06.1999 = OTS0183 5 II 0345 MWA004 WI, press release by the Austrian Wirtschaftsministerium, 24.06.1999.

OTS0190, 23.02.2000 = OTS0190 5 WA 0737 HER026, „Computec-media: 150-Millionen-Mark Deal ebnet Weg für Ausbau der Internet Aktivitäten", 23.02.2000.

OTS0015, 06.09.2002 = OTS0015 5 WA 0271 EUN0001 CA, press release by Sunflowers, 06.09.2002.

OTS0015, 06.11.2002 = OTS0015 5 WA 0271 EUN0001 CA, Sensation! Österreichisches Computerspiel ANNO 1503 verkauft sich in Deutschland über 160.000 mal in nur sieben Tagen, 06.11.2002.

OTS0278, 12.05.2006 = OTS0278 5 WI 0213 NKP0002, Wien: Computerspiel-Riese liquidiert Niederlassung - 100 MitarbeiterInnen gekündigt!, 12.05.2006.

GLOBAL HISTORY, FACTS AND FICTION IN EARLY COMPUTER GAMES: *HANSE, SEVEN CITIES OF GOLD, SID MEIER'S PIRATES!*

Wilfried Elmenreich, Martin Gabriel

For many years, historiography has ignored the importance of computer games for the general perception of past events, focusing instead on "conventional" receptions in film or literature. Mainstream historiography propagated the thought that computer games could not meet academic standards. On the other hand, there had been early computer games using historical events as background. Meanwhile, it is logical to also keep in mind the motives of designers for producing games set against a (quasi-)historical background: Historic settings are attractive for designers because they provide an already existing logical framework for a game, while making costly license fees obsolete. In this work, we analyse three games set against the background of – what Europeans call – the Late Middle Ages and Early Modern Period. The game *Hanse* is a game featuring elements of the 14th century Baltic trade. *Seven Cities of Gold* deals with European conquest in the Americas (during the 1500s). It is a real-time strategy game focusing on exploration. Sid Meier's *Pirates!* can be seen as a microcosm of the power struggle between European countries in the Early Modern Caribbean. It gives the player a sketch of complex economic or strategic issues where pirates (buccaneers) were operating under different circumstances to support the ambitions of colonial powers. The games were released between 1984 and 1987 for various platforms. Among the systems that have seen releases of all three games, the Commodore 64 versions have been used to analyse the games because of high market share and successful preservation of games. The games were in general very successful and well-received. From a technical perspective, those games did not max out the computer's capabilities, but rather attracted the players via the interesting setting and the historical connection.

Keywords: Video games, History, Commodore 64, Late Middle Ages, Early Modern Times

INTRODUCTION

For many years, mainstream historiography propagated the thought that computer games had little to do with history and could even less meet academic standards of historiography. However, historians have begun to intensify their research during the course of the last two decades, using computer games as a lens to look at how millions of people perceive (fictional?) histories. History is attractive for designers in providing an existing logical framework for a game, while, at the same time, making license fees obsolete (Ahammer, 2014: 35). Video games are no longer perceived as leisure for "socially incompetent teenagers", but as being of interest for the general public (Buse, Schröter & Stock, 2014: 61). However, computer games set against a historic background are not new. The text-based game *The Oregon Trail* has been developed as early as 1971 and is addressing 19th-century pioneer life on the Oregon Trail, a large-wheeled wagon route and emigrant trail in the United States. Although most early computer games leaned towards arcade games, there have been many games with historical settings since then.

In this article, we analyse three games set against a historical background that have been published in the 1980ies for the Commodore 64, which had been the prevailing platform at that time. Switching from arcade games to games run on a home computer opened up the possibility for game designers to also make for story-rich, slow paced games, since there was no need to let the player die and throw in another quarter for the next game.

The three games selected for this paper are set against the background of what could be called (from a European perspective) the Late Middle Ages and Early Modern Times. *Hanse* features elements of the Eastern and Northern European Trade of the 14th century, *Seven Cities of Gold* deals with the topic of European exploration in the Americas (around 1500), while *Sid Meier's Pirates!* can be seen as a microcosm of the conflicts between European powers in the Caribbean during the 16th and 17th century.

The three games are limited to a single platform (Commodore 64) and have been released between 1984 and 1987, which is a comparable short period compared to the overall time where games for the C64 platform have been produced. The aim of this paper is thus not to provide a comprehensive statistic over history-driven games but to analyse the three games from the perspective of historical accuracy and from the perspective of game design and technical realization.

The main research question addressed by this paper is to investigate what level of historic storytelling and what level of technical implementation were necessary to create those games.

RELATED WORK

Schwarz argues for four characteristics of a virtual past inherent to most computer games: (predetermined) linearity vs open endings, the mighty position of the player, authenticity as marketing feature, and a "contemporaneousness of the uncontemporary" (Schwarz, 2009: 15 f.). Most games featuring historical contents are not designed to meet the expectations of modern historical education and are sometimes diametrically opposed to such intentions (Grosch, 2002: 79 f.). Game narratives have to be counterfactual, while most players want them to seem authentic at least at the starting point – this correlates with the definition that a "game has to begin at a clear point in real-world history and that history has to have a manifest effect on the nature of the game experience." (MacCallum-Stewart & Parsler, 2007: 204). Video games can function as transmitters for ideologies, and many are based on the idea that "the ultimate answer of how to win is to spread your cultural, national and military influence all over the world" (Valdres, 2001: 70). Important features of life are missing from many games: social and cultural diversity, racial conflicts, or everyday problems (Grosch, 2002: 80). As Perry has stated with regard to the Middle Ages, homogeneity is problematic: "(…) it is certainly possible in any historical period to find a remote spot where everyone there looks completely homogeneous. But that is not (…) what medieval history looked like" (Chu, 2015).

Computer games also tend to marginalize the importance of women in history. Surely, in many cultures, politics, war, and economy – which are central for success in most historical games – were a male domain. Computer games often also relay images of a world in which there are no ethical limitations or consequences for violent behaviour (Knoll-Jung, 2009: 175).

Quite often, women are missing entirely from the gameplay, and if there are any, they are either unimportant or feature the same characteristics as men. Even if a game features female characters, their agency in a male-dominated general context is usually limited (Reinecke, Trepte & Behr, 2007: 12). While well-known female political or military leaders are relatively few, there are possibilities of introducing more female roles and countering a male-centric view of history (Knoll-Jung, 2009: 191). The topic of marginalization can be gainfully used by historians. Research in this regard should not end with asking the question of whether games are academically "accurate". Games feature the dichotomy of being historicized (by being put into a historical context), while at the same time historicizing themselves – by providing certain images of the past (Kerschbaumer & Winnerling, 2014: 14). Therefore, the central question for historians should be which ideologies and images of the past are being popularized and why.

Well into the 16th century, the most important maritime trade routes for European economies ran through the Eastern Mediterranean and the Baltic (see Ashtor, 1983; Harreld, 2015). One of the most crucial global developments between around 1350 and 1700, however, was Europe's overseas expansion.[1] Portugal as well as Castile and Aragon were in the vanguard of this development.[2] Explorations were financed by banks and merchants from Seville, Italy, or Upper Germany (see Otte, 1996; Verlinden, 1986). Transported goods often originated in Western Europe. Sailors, soldiers, and settlers represented a microcosm of diversity, even though a majority hailed from the Iberian Peninsula.[3]

In the 16th century, after the Spanish had conquered (at least on paper) large parts of the Americas, the importance of the Atlantic trade[4] began to increase

---

[1] Seven Cities of Gold (1984) can be seen as archetype of a video game focusing on the issue of Europe's (overseas) expansionism and as potential inspiration for Sid Meier's Pirates! (1987).

[2] The conquests of Ceuta, Madeira, the Azores, and the Canary Islands are the most important examples of early Iberian expansion.

[3] Avellaneda identified the birthplaces of 45 men taking part in the "Federmann Expedition" of 1537; 39 were Spaniards, the others were Dutch, German, Italian, and Portuguese (Avellaneda, 1987: 388).

[4] Most of the European trade with China or Southeast Asia ran through the Southern Atlantic, but Atlantic commerce was also closely connected to the Spanish trade between China and Mexico.

at the expanse of the Baltic and the Levant trade (see Acemoglu, Johnson & Robinson, 2005). Spanish merchant fleets carried silver, dye, or sugar to Europe, while the Acapulco-Manila trade, starting in 1571, signalled the advent of a global economy (see Flynn & Giráldez, 1995). The most important product of Spanish America's economy was silver. Spaniards were not able to secure large gold deposits, but at least 150 000 tons of silver were legally exported from Spanish America until the early 19th century. Mexico, Peru, and Bolivia were the world's most important silver producers (see (Burkholder & Johnson, 2001: 134 f.; Garner, 1988). Inhumane conditions caused the death of tens of thousands of indigenous workers, social structures were destroyed by forced labour, and epidemics like measles or smallpox depopulated large areas of the Americas (see Austin Alchon, 2003; Cole, 1985).

When Spain, from the early 17th century on, was challenged by European rivals, the Caribbean and Spanish Main became an international battlefield. This period was also the heyday of a phenomenon best known as "piracy". However, this term does not fully describe the complexity of an issue characterized by far-reaching economic, social, and strategic implications. Pirates, buccaneers, and privateers were operating under a set of different legal and social circumstances, even though, in many cases, they were *doing* the same thing (Latimer, 2009: 4). Nations that gave so-called letters of marque to "privateer" captains while at war, ordered their navies to hunt down these "pirates" after a peace agreement. In the South China Sea, pirate-merchants sold protection for "water money" (*baoshui*) as part of a larger economic cycle (Andrade, 2004: 427). Popular culture often differentiates between "bad pirates" and "honest buccaneers". This dichotomy is often set against the background of a disciplining process (Pfister, 2014: 201). In many cases, however, becoming a pirate was a matter of survival. Even though life on a pirate or buccaneer ship had to deal with violence on a daily basis, it was not *per se* more brutal than service on a navy or merchant vessel (Rediker, 1981: 206 f.). Some pirate leaders treated the masters of captured ships well if their crews described them as fine captains; others were killed to avenge their mistreatment of sailors.

METHODOLOGY

Each game was played for several hours. If available, also handbooks have been studied. Visuals, gameplay and texts have been analysed with respect to its historic setting. One focus was on the general dichotomy of "historical" (outdated as well as up-to-date) versus "counterfactual" narratives in the *overall* game design, the other on the presentation of characters, cultures or technology in specific situations *within* the gameplay. Historical analysis was based on edited sources and scientific findings published in modern scholarly literature on the fields of colonialism, piracy, and early modern global history as laid out previously. Central questions in this survey were, among others: Did the games try to "rebuild" the world of the past or is the past simply the setting for adventure? Were the games designed with modern or outdated knowledge of global history in mind? How did the games deal with the complexity and diversity of early modern societies?

The games were run on VICE (Versatile Commodore Emulator) 3.2 set to emulate a Commodore 64C model with the European PAL system and an emulated 8580 SID sound chip. As a disk drive, a VC 1541-II drive was emulated within VICE. For reference to the used game versions, see the respective entries at the Gamebase64 database.[5]

The built-in machine language monitor of VICE was used to identify the used programming language, the used graphic modes and sprites and the use of raster interrupt techniques for switching sprite positions or graphic modes during a single frame.

Commercial games on the Commodore 64 have been implemented mostly in Assembly language for two reasons: first, the overhead in size and execution time could be kept to minimum with Assembly language and second, there was a lack of compiled languages that would come with a fast and optimizing compiler. Many programs, among them also hobbyist games, have been implemented in the built-in Commodore BASIC 2.0. Commodore BASIC 2.0 is based on Microsoft Basic for 6502 and comes with a more

---

[5] Game descriptions at Gamebase64:
 Pirates! http://www.gamebase64.com/game.php?id=5727
 Hanse http://www.gamebase64.com/game.php?id=3403
 Seven Cities of Gold http://www.gamebase64.com/game.php?id=6722

convenient handling than Assembly language, but at the cost of execution speed. Besides being an interpreted language, the standard arithmetic system was built on 5-Byte floating point numbers, which makes even simple incrementation of a variable taking hundreds of processor cycles (for comparison, a similar operation takes six cycles in assembled machine language). One cycle corresponds to roughly 1 microsecond. An option to achieve a speed-up of a factor 3 to 5 was to compile the BASIC program with a specific BASIC compiler, however the achievable performance was still insufficient for most kind of games.

The graphics system of the Commodore 64 is able to display up to 8 movable objects, typically called sprites, within an area of 320 times 200 pixels. The video output is generated from top to bottom, where a single frame on a PAL system takes 1/50 of a second. Many games used the raster technique interrupt to redraw sprites with different coordinates further down on the screen, if the switch happens exactly between the two sprite positions, the sprite appears twice on the screen without flickering. Since the look of a sprite is defined by a single-byte pointer, it is also possible to change the look of the sprite on the fly so that the sprite below has an entirely different position and look.

FINDINGS

Hanse

Hanse was written by the German game designer Ralf Glau. The first version of the game has been developed by Glau for the Schneider CPC system. The C64 version was written by Bernd Westphal based on the CPC version. The game is in German language and is using *Mark* as currency, which was the official currency in Germany at the time the game was made. The game's name is derived from the Hanseatic League, a commercial federation that grew from a few North German towns across northern Europe.

Hanse is set in an earlier era than both *Seven Cities of Gold* and *Pirates!*, as dictated by historical logic. Players can trade in resources like fur, linen, or honey; a historically correct depiction, even though one of the most important commodities - fish - is missing. Activities in stock exchanges seem to be overrated, since Europe's stock exchanges were located far to the west (e.g. in

Antwerp), and so these kinds of financial enterprises were of little importance to the *Hanse* traders' everyday business. The use of a currency standard (*Mark*), while not entirely ahistorical, suggests the existence of a financial system based on a single coinage, which was not the case. Some features like the random loss of stored goods in a fire or the need to keep ships seaworthy are definitely in line with economic realities of the time as are the regionally diversified commodities. Notably for a 1980s game, *Hanse* can also be played using a female character, which hints at the possibilities for middle-class women to participate in trade; however, there is no specific female agency, since the choice of character does not influence the gameplay. While the game does not depict the "realm" of the Hanse as a whole (and also includes ports that were not a *Hansestadt*, like Novgorod), it does give an impression of the alliance as a trans- or even supranational enterprise par excellence.

The Commodore 64 version of *Hanse* uses the textmode with a modified character set and a hires bitmap mode for intermediate sequences. The major part of gameplay takes place on a screen with a drawing of north eastern Europe and overview information about number and quality of ships, available money, stocked trading items and a menu where the player can select trading, stock market, ships, kontore (branching offices) and chronicles. The screen is built up using textmode, the miniature drawing of north eastern Europe has been realized with custom characters. The textmode has the limitation of only a single colour in addition to a common background colour, which is visible on the game screen. The game code is mostly written in BASIC, in the analysed version it appeared to be compiled with the Data Becker Basic 64 compiler. The sprite functions were not used in any of the observed game screens, since the game play does not need moving elements. The interrupt vector was changed to a different ROM address in order to disable the RUN/STOP functionality, but there is no custom interrupt service routine.

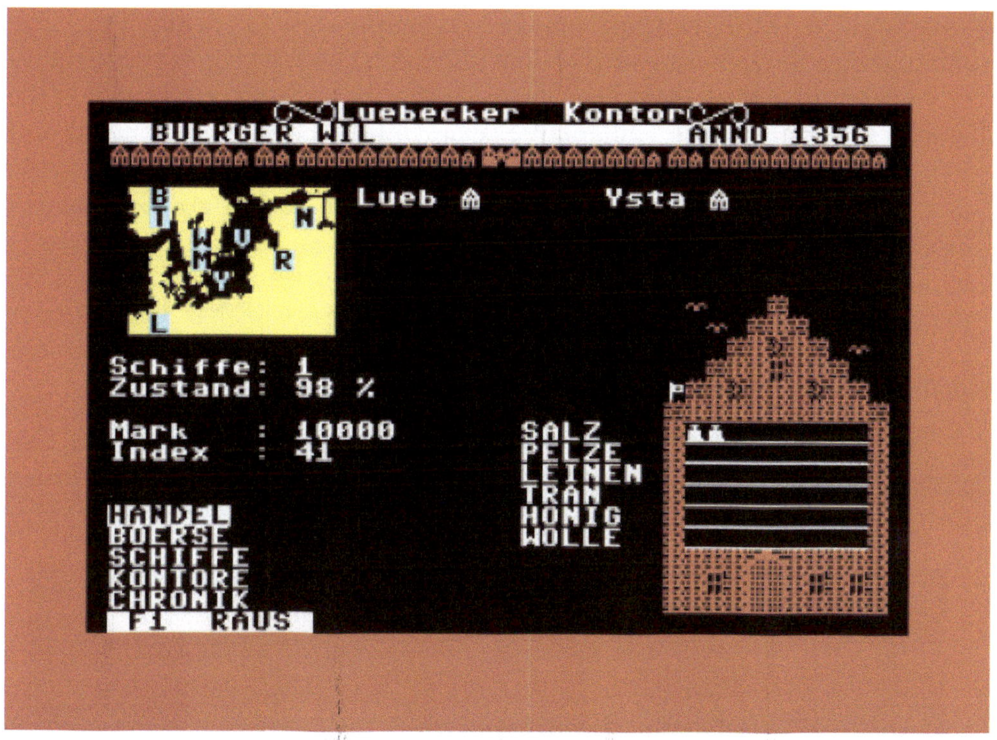

Figure 1. Screenshot of the main interaction screen of Hanse

Seven Cities of Gold

Seven Cities of Gold is an open-world strategy video game by Ozark Software, created by main developer Danielle Bunten (credited as Dan Bunten) together with Bill Bunten, Jim Rushing and Alan Watson. The topic of the game is the discovery and exploration of the Americas. The game's name is derived from the 16th century myth of the "Seven Cities of Cibola".

Some of the game's features correspond with historical authenticity, for example, the player can move faster when using rivers and the voyage back to Europe depends on sailing along a specific latitude. Resources central to successful gameplay are men, food, (unspecified) goods, and gold, while silver is of no importance, which is problematic, since the game - according to its name - is set in the Americas, where silver was much more important. The myth of "golden cities" alone cannot explain the pertinence of gold. Quite stunningly, the European garrisons of forts simply starve to death, if they run out of food and there is no artificial intelligence at least trying to prevent this. Horses, whose introduction could rapidly change societies, are shown in the graphics, but not in the gameplay. Social contacts between Europeans and indigenous groups run through one "chief", which correlates to the early modern European idea of a central ruler; however, such a position did not exist in most pre-Columbian societies. The player can "amaze the natives" - which relegates them to an irrational sphere of superstition - and also convert them (as a randomized feature). Overall, it becomes clear that *Seven Cities of Gold* is much more historically accurate on a technological than on a social or political level.

The version for the Commodore 64 was released briefly after the first version was built for the Atari 800 computer. Likely some game design decisions have been done with the capabilities of the Atari 800 in mind.

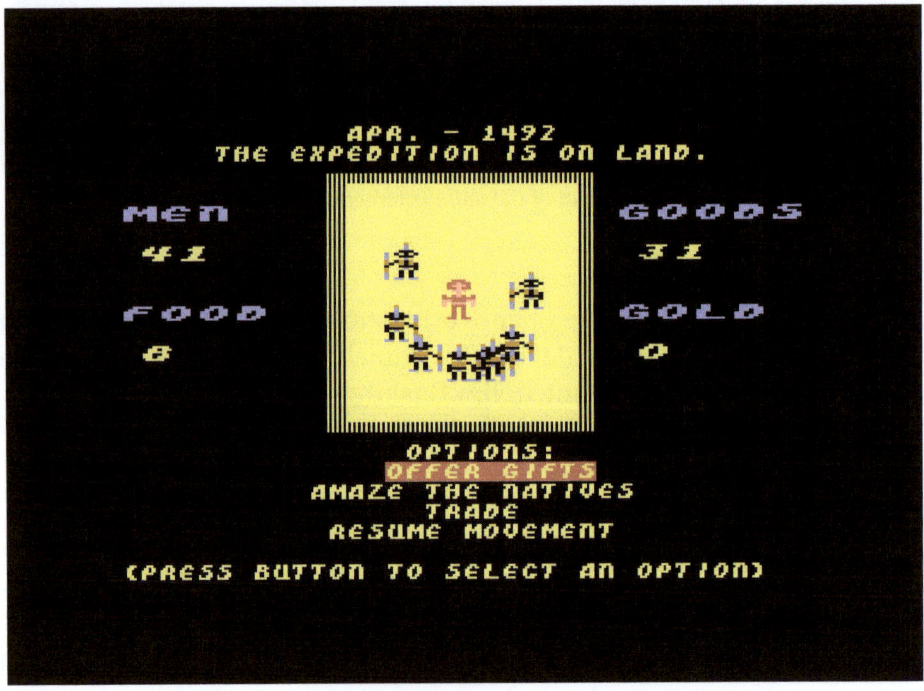

Figure 2. Interacting with villagers in Seven Cities of Gold

Except for a multicolor bitmap screen in the opening sequence, the game is using the multicolor text mode throughout the game. The beginning sequence which takes place in Europe is featuring a side-view scrolling screen, however only using one third of the screen height. After boarding, the game's characteristic top view is shown, where a compass is shown as an avatar of the player in the middle of a small game screen of 12 x 12 characters. For comparison, the full screen excluding borders would be 40 x 25 characters in size. The avatar is controlled by the player via joystick and always stays centered in the screen while the map moves according to the direction that was chosen. The avatar is realized with two overlaid sprites, while the map is drawn using an array of 12 x 12 custom characters. The characters also stay in place, but their content is dynamically written, thus emulating a small bit-mapped screen that is scrolled at pixel granularity. The same graphics set up is used when entering a village or viewing the map. The game is written in assembly language, which was expected, since scrolling the game screen needs to be done on a byte-per-byte basis, which, even in pure Assembly code, requires several thousand cycles. At a rate of 50 (PAL) or 60 (NTSC) frames per second, this leaves less than 20000 cycles per frame, making the timing for an on-the-fly update critical.

In general, graphics in the side-scrolling and in the exploration part are a bit coarse in comparison to other side scrolling or top view games using the same colour mode. The bitmapped graphics mode using an array of 12 x 12 characters offers many interesting possibilities for drawing the world or showing a lot of villagers at once, however it comes with the limit of a very small game screen. Nevertheless, the small game screen in combination with a huge game map gives the feeling of exploring a very large unknown continent, which contributes positively to the gameplay.

Sid Meier's Pirates!

Pirates! is another open-world strategy video game by Microprose. The game was created by Sid Meier, who is credited for game code and sound. Graphics were made by Michael Haire, documentation and scenario design by Arnold Hendrick. In the game, the player takes over a ship as captain and has the possibility to become a trader or pirate with different allegiances.

Pirates! is set in a later era than the other two games - the "golden age of piracy" (mainly the 17th century). Like *Seven Cities of Gold*, the game features accurate details such as the need to consider specific wind conditions, governors providing letters of marque - thus turning "pirates" into "privateers" - or the importance of certain resources such as food or tobacco. However, silver is again missing from the equation and replaced by gold. A player can be rewarded by receiving land grants, which shows similarities to colonial policies, but has no implications whatsoever as the game continues. On the other hand, duels between the player's character and the masters of ships he has attacked are very important; duels are also featured in the quest for brides - while colonial societies were patriarchal, the objectification of women in these scenes is obviously very problematic. *Pirates!* is best understood as an adventure game set against the backdrop of the Caribbean as a nexus of different European imperial interests, but without a deeper understanding of its historical complexities.

The game's title screen uses the multicolor textmode in a sophisticated way showing a nice depiction of a sailing ship and the game logo. In some text-based game screens, an additional picture is shown using custom characters.

Parts of the game featuring story parts of visiting a tavern or the governor of a city are implemented in BASIC, with other parts involving the traveling sequence are implemented in Assembly language. Other than in *Seven Cities of Gold*, the screen where the player's ship is moved uses the full visible screen area (without borders). In this mode, the environment is draw with only four colours: light blue background for the water, green for islands and a combination of black and yellow for vegetation and towns. Waves are also drawn in green colour. In addition, two overlaid sprites are used for the player's ship and some additional sprites for clouds, which are used to indicate wind direction and strength. Other ships are not shown but appear as a sudden notification when sailing.

Figure 3. Sailing ship and coastal areas in Pirates!

Pirates! features a very nice gameplay with suitable graphics and, among the three presented games is the most feature and story-rich game. Being released in 1987, *Pirates!* also depicts a continuous improvement in game productions on the Commodore 64 platform. Still, in comparison to contemporary games on the same platform, the game mechanics, graphics and sound of *Pirates!* are not overly demanding for the platform.

CONCLUSION

All three games were in general very successful and well-received by contemporary game magazines. From a technical perspective, the analysis shows that these games did not max out the computer's capabilities, but rather attracted the players via the interesting setting and (the promise of) historical authenticity. Our survey shows that the three games did not break with the long-standing master narrative that focused on European males being able to make their fortune in an international and/or colonial environment, while non-Europeans or also women do either not appear or are relegated to the status of "objects". The notable exception is the possibility to play a female character in *Hanse* who, nonetheless, embodies no specific female variation of political, social or economic agency. *Hanse*, *Seven Cities of Gold*, and *Pirates!* are materialist, not culturalist in their approach - an orientation they share with mainstream world or global history (Strasser 2020: 19). There are elements of economic history corresponding with scholarly historical knowledge; in general, however, economic content remains at an idealized, structuralist or undifferentiated level, with natural resources being there for the taking. Historical facticity and authenticity are of importance in setting the stage for the games; what follows are economic and/or adventure simulations that are counterfactual in nature.

FUTURE RESEARCH

Although today's technical specification of gaming platforms and today's game development tools differ significantly from the systems used in the 1980ies, there are a couple of lessons that can be learned from the analysed examples. Setting your game against a historical background immediately creates some resonance with the player's world and mind. The necessary material is available for free in real and digital libraries and, even more important, provides a license-free but well-known background. In addition, history is rich of different settings and stories, so that there is a vast set of materials available at the game makers' disposal. Furthermore, for the development of short games (Elmenreich, 2019), a historical background can raise interest in a game despite a comparably simple game play. Especially for low-budget independent game productions, the historic setting can thus be a helpful means for a successful game.

REFERENCES

Acemoglu, D., Johnson, S., & Robinson, J. (2005). The Rise of Europe. Atlantic Trade, Institutional Change, and Economic Growth. The American Economic Review, 95, 546-579.

Ahammer, A. (2014). "Nichts ist wahr, alles ist erlaubt". Das Mittelalter im Computerspiel. Master Thesis, University of Klagenfurt.

Andrade, T. (2004). The Company's Chinese Pirates. How the Dutch East India Company Tried to Lead a Coalition of Pirates to War against China, 1621-1662. Journal of World History, 15, 415-444.

Ashtor, E. (1983). Levant Trade in the later Middle Ages. Princeton, NJ: Princeton Univ. Press.

Austin Alchon, S. A. (2003). A Pest in the Land. New World Epidemics in a Global Perspective. Albuquerque, NM: Univ. of New Mexico Press.

Avellaneda, J. I. (1987). The Men of Nikolaus Federmann. Conquerors of the New Kingdom of Granada. The Americas, 43, 385-394.

Burkholder, M. A. & Johnson, L. L. (2001). Colonial Latin America. New York: Oxford Univ. Press.

Buse, U., Schröter, F., & Stock, J. (2014). "Du sollst spielen!". Der Spiegel, 3, 60-67.

Chu, A. (2015). The Inaccuracy of "Historical Accuracy" in Gaming and Media. Retrieved from https://www.techcrunch.com/2015/09/15/missing-the-target.

Cole, J. A. (1985). The Potosí Mita, 1573-1700. Compulsory Indian Labor in the Andes. Stanford, CA: Stanford Univ. Press.

Elmenreich, W. (2019). Short Games. In Wilfried Elmenreich, René Reinhold Schallegger, Felix Schniz, Sonja Gabriel, Gerhard Pölsterl, and Wolfgang B. Ruge, editors, Savegame, Perspektiven der Game Studies, pages 41–53, Berlin und London, December 2019. Springer Nature.

Flynn, D. O. & Giráldez, A. (1995). Born with a "Silver Spoon". The Origin of World Trade in 1571. Journal of World History, 6, 201-221.

Garner, R. (1988). Long-Term Silver Mining Trends in Spanish America. A Comparative Analysis of Peru and Mexico. The American Historical Review, 93, 898-935.

Grosch, W. (2002). Computerspiele im Geschichtsunterricht. Schwalbach/Taunus: Wochenschau.

Harreld, D. J. (Ed.). (2015). A Companion to the Hanseatic League. Leiden: Brill.

Kerschbaumer, F. & Winnerling, T. (2014). Postmoderne Visionen des Vor-Modernen. Des 19. Jahrhunderts geisterhaftes Echo. In F. Kerschbaumer & T. Winnerling (Eds.). Frühe Neuzeit im Videospiel. Geschichtswissenschaftliche Perspektiven. Bielefeld: Transcript, 11-26.

Knoll-Jung, S. (2009). Geschlecht, Geschichte und Computerspiele. Die Kategorie „Geschlecht" und die Darstellung von Frauen in Historienspielen. In A. Schwarz (Ed.). Wollten Sie auch immer schon einmal pestverseuchte Kühe auf Ihre Gegner werfen? Eine fachwissenschaftliche Annäherung an Geschichte im Computerspiel. Münster: Lit, 171-197.

Latimer, J. (2009). Buccaneers of the Caribbean. How Piracy forged an Empire. Cambridge, MA: Harvard Univ. Press.

MacCallum-Stewart, E. & Parsler, J. (2007). Controversies. Historicising the Computer Game. In A. Baba (Ed.). Situated Play. Proceedings of the Third International Conference of the Digital Games Research Association (DiGRA), The University of Tokyo, Tokyo, Japan, September 24th-28th, 2007. Tokyo: University of Tokyo, 203-210.

Otte, E. (1996). Sevilla y sus mercaderes a fines de la Edad Media. Sevilla: Fundación El Monte.

Pfister, E. (2014). "Don't eat me I'm a mighty pirate." Das Piratenbild in Videospielen. In F. Kerschbaumer & T. Winnerling (Eds.). Frühe Neuzeit im Videospiel. Geschichtswissenschaftliche Perspektiven. Bielefeld: Transcript, 195-210.

Rediker, M. (1981). „Under the Banner of King Death". The Social World of Anglo-American Pirates, 1716 to 1726. The William and Mary Quarterly, 38, 203-227.

Reinecke, L., Trepte, S. & Behr, K.-M. (2007). Why Girls Play. Results of a qualitative Interview Study with female Video Game Players. Hamburg: University of Hamburg.

Schwarz, A. (2009). Computerspiele. Ein Thema für die Geschichtswissenschaft? In A. Schwarz (Ed.). Wollten Sie auch immer schon einmal pestverseuchte Kühe auf Ihre Gegner werfen? Eine fachwissenschaftliche Annäherung an Geschichte im Computerspiel. Münster: Lit 2009, 7-28.

Strasser, U. (2020). Missionary Men in the Early Modern World. German Jesuits and Pacific Journeys. Amsterdam: Amsterdam Univ. Press.

Valdres, M. (2001). Dominance and Security in the Power Discourses. Sid Meier's Civilization as an Example of Discourses in Power Games. CEU Political Science Journal, 1, 58-71.

Verlinden, C. (1986). Wirtschaftliche und soziale Kräfte der frühen Europäischen Expansion im Mittelmeerraum. In C. Verlinden & E. Schmitt (Eds.). Die mittelalterlichen Ursprünge der europäischen Expansion. Munich: Beck, 135-148.

"THAT'S NOT HOW IT WAS!": THROUGH THE DARKEST OF TIMES IN THE CONTEXT OF A CULTURE WAR

Benjamin Kirchengast

The historical strategy and survival game Through the Darkest of Times (TtDoT) challenged the way in which history was previously represented in digital games. That is especially true for its representation of the National Socialist Germany and the resistance occurring in parts of the society at that time. However, this representation does not always spark huge excitement among the players of TtDoT. While the game was generally well received by the public, there has been some – quite loud – criticism in the community of digital games. I looked at the criticism that was voiced by some players in a total of 86 negative reviews on the gaming-platforms GOG and Steam. Based on Phillip Mayrings research method of the qualitative content analysis I formed different categories of criticism on TtDoT. First of all, some reviewers simply call the game mechanics boring. Other critics focus on the interaction of game mechanics and narrative elements, stating that the players actions do not have any impact on the outcome of the game, which is only partially true. The way in which you can or cannot interact with Nazis did not feel authentic for many users. One of the greatest points of criticism of the negative reviews, though, is the games representation of history. Many reviewers claim that the game transmits a modern perspective on history asserting that TtDoT is historically inaccurate and/or that it feels like the game is forcedly teaching history. Finally, there is a group of reviewers who call out the narrative of the game for supposedly transmitting a leftist political agenda, using the holocaust to discredit contemporary right-wing politics focusing on Donald J. Trump. The different categories mostly do not manifest themselves in isolation but in combination with one another giving way to deeper analysis. One major output of the study was that the resistance against TtDoT is to be understood in a broader context of an ongoing culture war between the left and the right. In connection with the culture war, the erosion of the masculine space of video games and their often highly masculine content is harshly opposed, which manifests in a negative reception of games that break with established narrative customs.

Keywords: Authenticity, Gender, Player Reviews, WWII, Video Game Culture

———

INTRODUCTION

Especially in the German speaking sphere, *Through the Darkest of Times* (TtDoT) received widespread media attention (not only in video game journalism). The long-lasting debate about the depiction of the Holocaust in digital games and the game's rather innovative approach to it, seemed to be an important catalyst for this media attention. The game being one of the first to be able to depict swastikas and other Nazi imagery in the German speaking space was another reason for the attention it attracted. In Germany the use of swastikas and other Nazi imagery is forbidden by §86a of the German criminal code. However, there is an exception, the "Sozialadäquanzklausel". It allows the use of Nazi imagery in art, education, and research. Since 2018 digital games that were reviewed and approved by the USK (*Unterhaltungssoftware Selbstkontrolle*) are allowed to use such depictions (Horcher, 2018). TtDoT was generally received very well by video game journalists. The narrative was praised as captivating and a learning experience; however, the gameplay was also being criticised for repetitiveness and lack of innovation. (Cramer, 2020/Kemp, 2020/Sigl, 2020).

However, the reception and critique of digital games is not a privilege of established media channels anymore. Through game purchasing platforms such as *Steam*, players themselves can rate games and publish reviews about them, helping other potential players decide whether or not to buy a game. I will call those reviews, written by users of game purchasing platforms, player reviews throughout this paper. Player reviews seem to be especially subjective in their approach to looking at the games they criticise. While no review can offer an objective approach, video game journalism usually follows certain guidelines of how to approach the process of reviewing and rating games. This does not apply to most player reviews, with the exception of some players who create their own templates for writing reviews or copy them from other reviewers.

Approaching TtDoT the player reviews seem to focus a lot on the historical setting of the game, which is not at all surprising. However, while the game is generally quite well received on platforms such as *Steam* and *GOG*, there is also some controversial discussion about its historical narrative. Quite a few reviews criticise the game for its supposedly faulty depiction of the Nazi regime, the Holocaust and World War II. The game is therefore criticised for missing historical accuracy. Yet the critique goes further than that. The developers are accused of trying to send a political message and instrumentalizing the Holocaust for that purpose. While the former is arguably true, as one of the developers himself mentions in an interview (Deutschlandfunk Nova, 2020), the latter seems to be itself motivated by historical and political beliefs, as I will show in this paper.

Negative player reviews that are based on a supposedly missing historical accuracy were therefore my focus of study. As I will show below, many of the negative player reviews are not only to be seen in context of TtDoT, but as part of a larger conflict and an erosion of a masculine video game culture.

As games become an ever-expanding part of Western-society and lose their subcultural "underground" stigma, there is also an erosion of a mainly masculine space taking place. Not only is the player base diversifying, but also the definition of what constitutes a digital game is in flux. This ongoing change ignites major resistance by players that identify with what the medium used to stand for and still does in a lot of minds. We have seen outbursts of such resistance before. The most prominent example is the hashtag-movement #*GamerGate* in 2014. The movement started off with a game-developer, Zoe Quinn, being openly accused (via a blogpost) of cheating on her ex-boyfriend, Eron Gjoni, with a video game journalist, Nathan Grayson, in exchange for positive reviews on her game *Depression Quest*. The blogpost got attention by some YouTubers, namely *Internet Aristocrat* and *MundaneMatt*, who each posted videos about the alleged scandal. On Twitter, the actor Adam Baldwin posted a link to the video by Internet Aristocrat with the hashtag #*GamerGate*. In subsequence the story blew up fast. What started as a supposed scandal of video game journalism quickly showed its true face: misogyny and antifeminism. Zoe Quinn and other female public figures of video game culture became the target of rape, murder, and bomb threats in combination with doxing, in other words the leaking of personal information online including home addresses (Kidd/Turner, 2016). The academic sphere got into the crosshair of the movement as well, as a fishbowl at the *DiGRA* was spun into a conspiratory meeting, trying to overthrow video game culture or gaming as whole (Chess/Shaw, 2013).

#GamerGate was not the first outburst of resistance against the erosion of video game culture as a masculine space and surely it was not the last one. In this paper I try to answer the question of how certain negative reviews on TtDoT can be understood as part of that culture war, and what implications about the historical beliefs of the authors those reviews carry with them. For this purpose, I am analysing negative player reviews on TtDoT on the platforms *Steam* and *GOG*.

In what follows I will start with going into related works concerning the representation of World War II and the Holocaust and the term of authenticity in this context. I will then briefly explain the methodology I used to analyse the postings, namely the qualitative content analysis by Phillip Mayring. After that I will go into the results by first presenting some important context about player reviews in general and second by discussing and interpreting the different categories of negative criticism that I have established. Finally, I will summarize the results in context of the research question.

RELATED WORK

World War II as a narrative setting

Using World War II as a setting for digital games is by no means a new phenomenon. As Eugen Pfister lays bare a short history of World War II and the Holocaust in digital games, starting in the 1980s the war between the Allied forces and the Axis powers was used as a narrative for all sorts of games. However, as Pfister shows there is a significant difference in the way World War II was depicted during that time and today. In the 1980s World War II was framed as a fight between even opponents with mutual respect for each other's military power. Throughout the years this changed, while in the

1990s Nazi symbolism disappeared from advertisement and the cover of games, the general trend going into the 2000s was showing the war but was not showing the atrocities of the Holocaust. Pfister points out: The attempt to avoid controversy generally leads to the Nazi regime basically being white-washed in these games. The reason for the depoliticization of World War II is not a latent antisemitism in the games industry; in fact, it may have been the opposite. It was feared that a staging of the Holocaust in the supposedly "fun medium" of digital games, would be barbaric. (Pfister, 2018b) Only slowly there is change to be seen.

Finally, Pfister notes two different ways in which digital games deal with World War II and the Holocaust. Firstly, the narrative of equal and fair enemies already mentioned above, which dispenses with any thematization of ideological backgrounds. And secondly, the tradition of the heroic epic, which has been adopted from the feature film and focuses on individual fates. (Pfister, 2018b)

Jonathan M. Bullinger and Andrew J. Salvati also address the latter. Borrowing from Roland Barthes' theory of myth and Aleida and Jan Assman's theory on cultural memory, they describe the phenomenon of *BrandWW2* as a concentrated form of myths that are anchored in the collective memory of readers of various media, which are used to emotionally bind consumers to the narrative of the medium. Historical accuracy hardly plays a role in this. The two authors refer to American media in their analysis, describing the prevailing narrative as follows: "Even though America had been hurting in the Depression, when Hitler began invading Europe and torturing the Jews, the American workforce mobilized and single-handedly stopped the ultimate evil. In order to save even more lives, we had to bomb Japan. Much of the success we enjoyed afterwards can be attributed to all those remarkable young men who stood up and fought for what's right. That's something that is missing today in America." (Bullinger & Salvati, 2011, p. 29)

The narrative is repeated and reproduced over and over again in a wide variety of media and thus solidified in the cultural memory of consumers. In order to sell *BrandWW2* it is necessary to create a *selective authenticity*. An experience, in other words, that is not based on a representation of detailed historical facts, but on a general sense of authenticity. (Bullinger & Salvati,

2013, p. 154) Tim Raupach makes a similar point about authenticity when he writes: Games with a fictional frame of reference consequently always measure themselves in the representation of their content against the representational conventions of their genre with which fictional as well as non-fictional models of reality are associated. (Raupach 2014, p. 115)

Bullinger and Salvati go into further detail, however, noting three components that are constitutive for producing *selective authenticity*: *technological fetishism, cinematic conventions*, and *documentary authority*. By *technological fetishism,* the authors understand a representation of contemporary technical devices that is as close to reality as possible. This form of authenticity formation is particularly constitutive in first-person shooter (FPS) games. The details sometimes go as far as accurate serial numbers on weapons. *Cinematic conventions* refer to the resorting to cinematic elements. Cutscenes can be used to advance the narrative, deepening characters, and focusing on details of the game world. In addition, a sense of authenticity can be created through scenes such as news reports, black and white depictions, and old photographs. Sometimes it is even sufficient to reference old war films that have already become entrenched in the cultural memory of the consumers and are related to the depicted past. The definition of *documentary authority* appears somewhat more blurred. In addition to the use of quotes from famous historical figures and historical maps, the representation of historical information systems also falls into this category. (Bullinger & Salvati, 2013, pp. 158-160)

Due to the widespread use of such media, an equally widespread and continuously reproducing idea of historical processes can be identified. The results are stereotypical ideas and narratives that are negotiated repeatedly in similar form. What must not be ignored, however, is that not every individual was/is exposed to these stereotypes to the same extent and in the same way. Thus, the concept of authenticity always remains a subjective concept that is perceived differently by different people.

METHODOLOGY

Qualitative content analysis

As a methodological approach to analyse the player reviews I chose to use the qualitative content analysis by Phillip Mayring. While one would need another or an additional method to analyse the player review sections of other digital games because of the sheer number of postings that are to be examined for AAA-titles, for TtDoT a simple qualitative method seemed to be sufficient. One of the appealing aspects of the qualitative content analysis is the option to form different categories of criticism that are visible in the negative player reviews. Mayrings model is separated into eight steps. 1.) The researchers must decide on the material of interest. It should be representative of a larger volume of data. 2.) The researchers define the context of the data's origin. 3.) There needs to be a formal characterisation of the data. It needs to be in an operable form. In interviews for example Mayring refers to the transcription process. 4.) The researchers need to define what aspects of the data they want to analyse. This is usually defined through the research question. 5.) The material should not be analysed in a vacuum but in context of a theoretical framework and in the context of related work. This is necessary to ensure that the findings are intersubjectively plausible and retraceable. 6.) The researchers need to decide on a method for analysing the content. While there are different ways of analysing data, Mayring suggests three different approaches that can be used in isolation from or in combination with each other: summarization, explication and structuring. 7.) The researchers need to define a section of analysis. Depending on the research material there are different standards that come to pass. 8.) The last of the eight steps is the interpretation of the gathered and analysed material. (Mayring, 2010, pp. 54-86)

The data of interest as well as the research question of this study were already defined in the introduction of this paper. In what follows I want to give an overview over the data, namely the player reviews on the platforms *Steam* and *GOG*, why I chose this material and what needs to be kept in mind when working with this kind of data.

Data

Through the Darkest of Times

TtDoT was produced as the first game of the Berlin-based indie developer studio *Paintbucket Games* and published by *THQ Nordic*'s subsidiary *Handy Games*. The game was released on January 30, 2020, the day the Nazis came to power in Germany. It is to be roughly classified in the genres of survival, strategy, and serious games.

The plot begins on February 1st, 1933 two days after Hitler came to power. The players take on the role of a critic of the NSDAP (National Socialist German Workers' Party). Together with two other critics, the character meets in a safe house and discusses how to stop the Nazis from seizing power completely. The plot is divided into four chapters in which the resistance group tries to help people, sabotage the Nazis, get the attention of the media etc. With some minor changes, each chapter works similarly, with the final chapter ending with the end of the war in Europe. It is important to note that players do not have the agency to change the outcome of the war or end it early, they can only influence minor events.

The game, basically offers players two ways to interact with the game world, which alternate. Players have the possibility to send the members of the resistance group on missions once per round (in the narration this is equivalent to once per week). These can be simple missions like recruiting supporters or collecting donations, or more difficult tasks from spying on the *Sturmabteilung* (SA) to breaking into a radio tower and broadcasting messages over the radio. However, many missions require that certain (usually easier) missions have been completed beforehand.

Secondly, there are narrative sequences in which the players can make decisions for the main character, like in text-based adventure games. There are longer narrative sequences after each chapter, but also shorter narrative sequences are inserted between the individual rounds. Before a round starts, there are short conversations between the main character and other members of the resistance group, in which the player can decide the statements of the main character, sometimes settling conflicts between other members or even making one of the members leave.

Furthermore, players have the possibility to choose between two game modes: "Story Mode" and "Resistance Mode". The "Story Mode" allows players to play through the game without the risk of losing the game. In the "Resistance Mode", the resistance group disbands when the group's morale drops to zero or the main character gets arrested three times. The players then has no way to continue the game and is forced to start over.

Player reviews on Steam and GOG

Player reviews provide a good insight into the opinions of players about the games they have played. Although the analysis of player reviews provides only a relatively small fraction of the opinions of the people who have actually played a game, the material seems to be richer than material gathered with questionnaires and broader than material gathered by qualitative interviews or participant observation. For answering the research question of this study, player reviews to me seemed to be the appropriate material, since the research question is embedded in a broad discourse about the change of video game culture, in which players who publish reviews actively participate.

The platforms *Steam* and *GOG* seem to be fitting sites for data collection. *Steam* is the largest computer game distribution platform (in terms of number of users) on the internet and thus offers the most material. *GOG* is a much smaller platform that presents itself as an alternative to *Steam*. It primarily advertises that games are sold without digital rights management (DRM) (as they used to be in the past). This is also the reason for the platform's name, which was originally "Good Old Games" and is now only used in an abbreviated form. The name suggests a connection to the original form of digital games, whatever that may imply. That is another reason why the platform offers itself for analysis in the given context. Other platforms like *Epic Games* or *Origin* are out of question because they do not distribute TtDoT and do not have review sections. At the time of writing this paper TtDoT was also available on *iOS* and *Android*, where players can also leave reviews in the *App Store* and the *Play Store* respectively. Those sources were not available during the study but could prove to be an interesting addition to the data I used here, as mobile gaming is perceived as a more casual sphere and as part of the erosion of gaming as a masculine space (Chess, 2017).

On *Steam*, players only have the option to rate a game either positively or negatively, whereby the game can only be rated by people who own the game on their respective *Steam* account and have started it at least once. At the time of my study, there were 480 reviews of TtDoT on *Steam*. 407 of the reviews on the platform were positive, 73 were negative. In percentage, the game received 87% positive reviews. The game was thus rated "Very Positive" on *Steam*. For language reasons, however, only reviews in German and English were used for this study. A total of 291 reviews were available in German and/or English, with 43 of the reviews being negative. This corresponds to a positive rating in 85.3% for German and English language reviews.

On *GOG*, the system is a bit more complicated. Reviewers have the option to give a game stars, with a minimum of one star and maximum of five stars per game. One star being the worst possible rating and five stars being the best. Unlike on *Steam*, on *GOG* people can also review a game they do not own on their *GOG* account. Accordingly, even people who have not played the game can theoretically publish a review. However, for reviewers who have linked the game to their *GOG* account, the tag "Verified owner" is displayed next to the date of publication. In total, there were only 13 reviews on *GOG* at the time this study was conducted. All 13 reviews were written in either German or English. Eight of the 13 reviewers awarded the game the maximum number of five stars each, one reviewer awarded four stars, one awarded three stars, and three reviewers awarded only one star. On average TtDoT thus received 3.3 stars from the reviewers. It must be mentioned that from the three player reviews that rated the game with one star none were written by "Verified owners".

It should also be noted that both *Steam* and *GOG* provide reviewers with guidelines they have to follow. On *GOG*, these guidelines are kept quite vague. Users are asked to refer to the game in question in their reviews and to be constructive in they critique as well as: "Please refrain from using language that may be considered offensive, inappropriate, or abusive."([Rules and Guidelines GOG], 2020) What is meant by "offensive, inappropriate or abusive [language]" is not clarified anywhere, however. Other users have the option of reporting reviews to moderators via a separate function, who reserve the right to delete reviews that do not comply with the guidelines. The *Steam* moderators also have this right, although the guidelines on this platform are more precisely defined. For example, reviewers are not allowed to insult other users, use profanity, post personal information, post links to phishing sites, threaten violence, post unlicensed material, post advertising, post spam, post racist or discriminatory comments, post pornographic content or content about drugs and alcohol, and post tips to cheat in games or bypass copyright protection. Furthermore, it is not allowed to discuss with moderators or to try to intervene like moderators. Most interestingly, however, it is not allowed to post "Religious, political, and other 'prone to enormous arguments' threads." ([Rules and Guidelines Steam], n.d.) What is to be defined as political, religious, or "prone to huge arguments" remains unclear. More importantly this rule explicitly places the medium of digital games outside of a political and societal discourse. However, as was shown time after time in Game and Media Studies, the content of popular culture has political implications no matter if intended or not (Murray, 2018; Pfister, 2018a). The rule seems most absurd in context of games like TtDoT that aim to make players reflect and discuss about serious topics.

For the study on hand this means that all player reviews are especially interesting in the context of the research question, as they could be deleted in accordance with the rules of conduct and are therefore at the mercy of individual moderators.

FINDINGS

Categories of criticism

In this section, I will discuss the results of my analysis of negative reviews on TtDoT on the platforms *Steam* and *GOG*. In an inductive process, different categories which exemplify the reviewers' points of criticism of TtDoT were formed during the course of the study. In the following, the characteristics of the individual categories are presented and illustrated with exemplary quotes.

Insignificance of Gameplay

"This briefly seems interesting, but very quickly you realize that absolutely nothing you do matters." (Mongoloid Mike, 2020)

The criticism refers to the fact that the player has hardly any influence on the course or the ending of the game, apart from the possibility to save non-player characters (NPCs). For example, user StevenWolfe writes: "Quite literally, at no point in time did I feel like I'd done anything useful..." (StevenWolfe, 2020) The reviewers thus deny TtDoT to be a game: "...at its core, it's a kind of graphic novel without any gameplay content." (FynniTheCat, 2020)

In terms of the historical content of the game, this means that the players wished to have more influence on the historical events that are narrated in the game. The insignificance of the players action in contrast to the game's political message is therefore problematized. The user graefinS for example suggests the developers' decision to keep the gameplay close to the documented historical events, would convey to players that resisting would be pointless. (graefinS, 2020)

Unauthentic fight against National Socialism

"The game doesn't give you the feeling that you're fighting one of the greatest evils of all time from the inside." (StevenWolfe, 2020)

While suggesting the game would not give the player enough agency, there is also a reference to preferring to fight Nazis - presumably meaning mainly Wehrmacht soldiers - in a violent way: "Yes, I know Nazis are bad, one of my past times is shooting the grenades on their belts in Sniper Elite, because they thought finding new and creative ways to murder roughly 6 million humans was a 'good idea'." (StevenWolfe, 2020) This is a reference to the game series *Sniper Elite*, in which players, in the role of a soldier, fight their way through levels to hunt down the perpetrators of fascist crimes in classic heroic fashion as is custom in *BrandWWII*. It is also pointed out that TtDoT does not convey the right feeling of fighting against the Nazi regime, that it does not feel authentic in this portrayal.

Boring Gameplay

"...the gameplay and its loop are terrible." (Seedy, 2020)

Like "irrelevance of gameplay", this category references the ludic elements of TtDoT. The two forms of critique often occur in combination with each other. In the criticism of the gameplay itself, the historical narrative elements of the game are not addressed. However, there are many player reviews in which criticisms of gameplay and narrative elements occur in combination. There are those player reviews which rate the game negatively because of the gameplay in spite of a positive evaluation of the historical-narrative content and those which combine negative criticism of the gameplay and negative criticism of the historical-narrative content in order to devalue the game: „At no point was I invested in the characters, the limited *extremely limited* gameplay, or the nonsensical narrative being force fed to me." (ZerO, 2020)

Modern Perspective on History

"The game makes the critical mistake of perceiving the past through a modern lens, and that critically hamstrings the game." (StevenWolfe, 2020)

The biggest criticism that affects the historical narrative of TtDoT is the accusation that the game views the past from a contemporary perspective and thus ignores the historical reality of the 1930s and 1940s. The criticism happens in different ways and is therefore once again divided into three subcategories (historically inaccurate, political agenda and forcedly teaching history), which, however, should not necessarily be understood in isolation from each other, but intertwined.

Historically inaccurate

"No care for historical accuracy." (Derek, 2020)

Reviewers claim that the description of events in TtDoT is not accurate, and the newspapers in the game do not reproduce original headlines of the time: "and not even original statements at that. Imagine if the papers read 'Hitler promises hope and change'." (Bloodrunsclear, 2020) These statements seem interesting especially since the game was generally perceived positively by historians. In criticizing historical accuracy, reviewers usually refer to something specific that they perceive to be inaccurate. However, this is not refuted by historical sources, but - and this brings me to the second subcategory - by comparison with current events.

"It is almost ironic how a game about Nazis is filled with so much propaganda." (Wayne, 2020)

The developers are accused of instrumentalizing the topic of National Socialism to take a stand in current political debates and devalue contemporary right-wing politics: "Using a tragedy of many nations to push a contemporary controversial agenda using a single nation's representatives as protagonists is insulting and racist." (Mannelig1, 2020) This reference to current political debates and actors becomes even sharper in several other reviews. For example, user Wildcard writes, not alone in making an explicit reference to Donald J. Trump: "If you have a brain and like challenging games that make you think, this one probably isn't for you. If you're a liberal who loves fairy tales about how Trump is Hitler, you will absolutely love this game." (Wildcard, 2020) The game is further portrayed as a "hyper-left, SJW [Social Justice Warrior, a devaluing term for vocal leftists on the internet] game [that] destroys any real sense of history or politics." (Slayer_Gauge, 2020) The review by Slayer_Gauge just cited is a good example of linking criticism of the lack of historical accuracy and the supposed political agenda of the developers of TtDoT. At the end of the review, she*he writes, obviously referring to positive reviews: "The irony seeing some of these comments about 'knowing history' are from people with only the thinnest grasp of history themselves. Perhaps you should think about the adage 'history is written by the victor'." (Slayer_Gauge, 2020) Not only does this implicitly deny the Holocaust, but it also calls into question the integrity of all historical scholarship, using a widely known phrase.

The user Wayne also offers a particularly blatant example. She*he accuses the developers of changing history in a politically motivated way and then drifts into a statement about how Nazis were actually left-wing because they were socialists and right-wing politics actually stand for freedom, whereby the most extreme form of right-wing politics is anarchy. Furthermore, she*he continues: the world is in danger of sinking back into systems similar to Nazism, as seen in the success of political figures like Bernie Sanders. Wayne concludes: "So if you are going to make a game focused around history BE PRECISE. Teaching a false history is worse than not teaching it at all." (Wayne, 2020)

It is also interesting to note that reviews related to political agenda received the most attention in review ratings, and while they received lower overall ratings than other player reviews, they also received significantly more "helpful" ratings than any other player reviews. It is also notable that most of the player reviews criticising the political agenda were written just one day after the release of TtDoT.

Forcedly teaching history

"...the game very quickly ducks into safe avenues, waves around atrocities almost for shock value." (StevenWolfe, 2020)

Another point of criticism, which is also discussed in connection with the accusation of a political agenda of the developers, is that the game wants to force players to learn something about the cruelty of the National Socialist system and relies on shock factors. The user derblaueClaus feels like the game developers think of the players as stupid and uneducated. He points out that the supposedly intrusive way the game uses to convey historical content is detrimental to immersion (derblaueClaus, 2020). Accordingly, the game loses authenticity in its depiction of Nazi Germany. This type of criticism, however, is quite different from the accusation of political propaganda as it does not go into avenues of denying the holocaust. Yet the two forms can also go hand in hand.

Contextualising the critique

The negative reviews of TtDoT do not show a clear picture, as can be seen above. Basically, however, only two aspects of the game are criticized. On the one hand, the gameplay, that is, the ludic elements of the game, and on the other hand, the historically saturated narrative. The criticism of the two aspects is usually brought separately from each other, even if narrative and gameplay are criticized in the same review. The notion that the gameplay intentionally lacks the options of agency that are accessible in other games in order to support the narrative is never discussed.

The critique on the insignificance of gameplay, takes place on the axis of private and public. While TtDoT only allows players to influence fates on a private level within Nazism (e.g., saving the lives of NPCs), players are accustomed to having an impact on simulated political-public events within digital games. The meaning of actions is located on the axis between private and public, with the private realm being historically assigned to women and the public realm assigned to men (Frevert, 1995). TtDoT, however, also breaks with the classic heroic epic of *BrandWW2* and the male fantasy of omnipotence to overthrow the Nazi regime unaided and thus being able to single-handedly change the political-public sphere. Instead, the game only gives players the opportunity to interact on a private level, which apparently feels trivial to some players.

The criticism of gameplay addresses a certain idea of what a digital game must have to offer to legitimately call itself a game. TtDoT is denied the character of a game since it cannot be won in the true sense and the players have no tangible influence on the games narrative. The game thus joins a debate about what defines a digital game. In the Game Studies, the definition of what constitutes a game has long been debated, even more so in recent years due to the widespread success of innovative indie games. Older definitions, such as that of cultural anthropologist and Game Studies forebear Johan Huizinga, who defined games as: a voluntary action performed within certain fixed limits of time and space according to voluntarily accepted but unconditionally binding rules, having its goal in itself and accompanied by a feeling of excitement and pleasure and a consciousness of being 'different' from 'ordinary life' (Huizinga, 1956, p. 34), still refer to the emotional level of the players. While older definitions still consider fun as a constitutive element of play, new definitions open up further. Jesper Juul, for example, describes games as: "1. a rule based formal system; 2. with variable and quantifiable outcomes; 3. where different outcomes are assigned different values; 4. where the player exerts effort in order to influence the outcome; 5. the player feels emotionally attached to the outcome; 6. and the consequences of the activity are optional and negotiable." (Juul, 2005, p. 6)

Even Juul's definition, however, leaves open the possibility of not defining TtDoT as a game, because his definition includes a very subjective variable with the emotional attachment to the outcome of the game. The debate, which is conducted in Game Studies for reasons of scientific acuity, is negotiated within the gaming community with (even more) emotional attachment. Feminist criticism of the representation of women in digital games results in mediatized waves of outrage within the gaming community. A private matter of a female game developer is turned into a conspiracy within video game journalism, leading to antifeminist and misogynistic attacks. On the other side the identity of the "gamer" as such is being declared dead, with gamers being understood as the stereotypical image of a young, white, socially incompetent, poorly fed male. (Alexander, 2014) Which in turn ignites even more anger by those who identify as "gamers". The criticism of games that, according to some "gamers", are not games, is part of a discourse about the future of digital games. About what is allowed to be in this definitory space and what is not.

In this understanding games with a World War II setting must have heroic epics to tell or experience. They have to give players the opportunity to single-handedly overthrow the Nazi regime and change the past, in order to fit into the internalized structure of such narratives. They have to fit into *BrandWW2*. The use of violence against Nazis is a constitutive element. A game that takes place in Nazi Germany and reveals the precarious situation and powerlessness of people who did not support the Nazis and the dilemma of choosing between resistance and being a bystander is not perceived as a game.

Considering this, the criticism on the developers' supposed political agenda and instrumentalization of Nazi crimes for contemporary left-wing politics is hardly surprising. Here, too, the erosion of a space for "gamers" that is separated from the rest of society becomes apparent. The desire to maintain games as an apolitical medium (not that they ever were, but they were and are often perceived as such) is vehemently defended. Depicting Nazism aside from the battlefields of World War II becomes a statement about contemporary political conflicts. Parallels between Nazi propaganda and the populist appearance of Donald J. Trump are seen as an underhanded attack against Trump, even as evidence of the political intentions of the developers. The idea that there might be actual parallels is not even negotiated. In fact, in this framework politics seems to be seen as something negative that tries to change the beloved status quo.

Of course, these observations pertain to only a fraction of the players of TtDoT. Overall, the game has received very positive reviews from players on *Steam* and *GOG*, and many of the negative reviews do not indicate a general rejection of any explicitly political theming or change in the gaming landscape. Still, the vocality of the minority should not be underestimated. The reviews resonated by far the most in their ratings of other users and were most often rated as "helpful" or "funny" by readers, though at the same time they received the most "unhelpful" votes. Moreover, similar to Hinrich Rosenbrock's observation of the antifeminist men's rights movement, the approach of this small group of critics seems to be coordinated (Rosenbrock, 2012, p. 142). This is especially evident when one looks at the date of the contributions that accuse the developers of having a leftist political agenda. The vast majority of relevant posts appeared on 31.1.2020 or 1.2.2020, i.e., one or two days after the release of the game. Another indication is the user Bloodrunsclear posting a review under the same name on both *Steam* and *GOG*. (Bloodrunsclear, 2020a, 2020b) This suggests that the devaluation of TtDoT happened in the first days after the release in a coordinated way to prevent other potential consumers from buying the game.

Authenticity matters here as players use it as an argument to devalue the game. It is not mentioned as such, but implicitly addressed in the form of questioning the historical accuracy of the game. Supposed historical facts, which the game reproduces, are automatically indexed as false due to the subjectively perceived lack of authenticity. This is not followed by any reference to historical sources or works of historical research. The perceived political agenda and/or crowbar with which history is conveyed seems to lead to the feeling of lacking authenticity, whereby the entire representation of historical content can be declared erroneous.

CONCLUSION

In an inductive procedure, I formed four main categories and three subcategories of negative criticism of TtDoT in this paper:

(1) Insignificance of gameplay
(2) Unauthentic fight against National Socialism
(3) Boring Gameplay
(4) Modern perspective of history
    a. Historically inaccurate
    b. Political agenda
    c. Forcedly teaching history

The categories exemplify the various forms of criticisms of the game that can be observed in player reviews and, in some cases, can be placed within a broad discourse about the erosion of the male space of digital gaming. For example, the accusation of the irrelevance of gameplay points to a negation of the private space (associated with femininity) as a meaningful space. The fight against Nazism has long been framed in digital games either as a battle of equal and fair opponents or, as in many films, as a heroic epic. TtDoT's break with this masculine fantasy, reflected in *BrandWW2*, leads not only to criticism of a lack of authenticity, but also to accusations of a political agenda. The developers are accused of projecting current political debates onto the topic of the Holocaust and instrumentalizing it in order to equate Donald J. Trump with Adolf Hitler and thus muzzle him. The harsh criticism of the game's supposed political agenda by a small minority of reviewers appears to have been coordinated, given the controversial but massive volume of ratings of the relevant player reviews and the short time frame (shortly after the game's release) in which they were written.

In part, then, the negative reviews can therefore be placed within a broad discourse of the erosion of the masculine realm of digital gaming. Be it through the intrusion of explicitly historical-political content or the disruption of established narratives.

REFERENCES

Alexander, L. (2014, August 28) "Gamers" don't have to be your audience: "Gamers" are over. Gamasutra. Retrieved from URL
https://www.gamasutra.com/view/news/224400/Gamers_dont_have_to_be_your_audienc e_Gamers_are_over.php

Bloodrunsclear. (2020, January 1). [Player review]. Retrieved from URL
https://www.gog.com/game/through_the_darkest_of_times

Bloodrunsclear. (2020, January 1). [Player review]. Retrieved from URL
https://steamcommunity.com/profiles/76561198093350917/recommended/1003090/

Bullinger, J.M., Salvati, A.J. (2011). A theory of brand WW2. Reconstruction: Studies in Contemporary Culture, 11, p. 19. Retrieved from URL
http://web.b.ebscohost.com/ehost/detail/detail?vid=3&sid=e8b9b545-0066-4e5d-ad7b-c07388c8287f%40pdc-v-sessmgr01&bdata=JnNpdGU9ZWhvc3QtbGl2ZQ%3d%3d#AN=79434613&db=hus

Bullinger, J.M., Salvati, A.J. (2013). Selective authenticity and the playable past. In: A.B.R. Elliott, M.W. Kapell (Eds.), Playing with the past: Digital games and the simulation of history (pp. 153-168). Bloomsbury.

Chess, S., & Shaw, A. (2015). A conspiracy of fishes, or, how we learned to stop worrying about #GamerGate and embrace hegemonic masculinity. Journal of Broadcasting & Electronic Media, 59(1), 208–220. DOI https://doi.org/10.1080/08838151.2014.999917

Chess, S. (2017). Ready player two: Women gamers and designed identity. University of Minnesota Press.

Cramer, E. (2020, January 29). Test: Through the Darkest of Times. 4Players. Retrieved from URL https://www.4players.de/4players.php/dispbericht_fazit/PC-CDROM/Test/Fazit_Wertung/PC-CDROM/39649/84244/Through_the_Darkest_of_Times.html

derblaueClaus. (2020, February 29). [Player review]. Retrieved from URL https://steamcommunity.com/id/derblaueClaus/recommended/1003090/

Derek. (2020, July 6). [Player review]. Retrieved from URL https://steamcommunity.com/profiles/76561198035769525/recommended/1003090/

Frevert, U. (1995). „Mann und Weib und Weib und Mann": Geschlechter Differenzen in der Moderne. ["Man and woman and woman and man": Gender differences in modern times.] C.H. Beck.

FynniTheCat. (2020, February 13). [Player review]. Retrieved from URL https://steamcommunity.com/profiles/76561198030396995/recommended/1003090/

Game: Anleitung zum Widerstand gegen Hitler. [Game: Manual for resistance vs. Hitler.] (2020, February 2). Deutschlandfunk Nova. Retrieved from URL https://www.deutschlandfunknova.de/beitrag/computerspiel-ueber-nazi-deutschland-widerstand-organisieren-mit-through-the-darkest-of-times

graefinS. (2020, February 3). [Player review]. Retrieved from URL https://steamcommunity.com/profiles/76561198117660846/recommended/1003090/

Horchert, J. (2018, August 9). Hakenkreuze in Videospielen: USK kann verfassungsfeindliche Symbole in Games erlauben. [Swastikas in videogames: USK can allow unconstitutional symbols in games.] Der Spiegel. https://www.spiegel.de/netzwelt/games/hakenkreuze-in-videospielen-usk-kann-verfassungsfeindliche-symbole-in-games-erlauben-a-1222433.html

Huizinga, J. (1981). Homo Ludens: Vom Ursprung der Kultur im Spiel. [Homo ludens: A study of play-element in culture.] Rowohlt Taschenbuch Verlag.

Juul, J. (2005). Half-real: Video games between real rules and fictional worlds. MIT Press.

Kemp, L. (2020, February 11). Through the Darkest of Times review: Lead a resistance group in 1933's Berlin in this uncomfortable and important strategy game. PC Gamer. Retrieved from URL https://www.pcgamer.com/through-the-darkest-of-times-review/

Kidd, D., & Turner, A. J. (2016). The #GamerGate files: Misogyny in the media. In A. Novak & I. J. El-Burki (Eds.), Defining identity and the changing scope of culture in the digital age (pp. 117–138). Information Science Reference, An Imprint of IGI Global.

Mannelig1. (2020, January 31). [Player review]. Retrieved from URL https://www.gog.com/game/through_the_darkest_of_times

Mayring, P. (2010), Qualitative Inhaltsanalyse: Grundlagen und Techniken. [Qualitative content analysis: Basics and technique.] Beltz Verlag.

Mongoloid Mike. (2020, August 9). [Player review]. Retrieved from URL https://steamcommunity.com/profiles/76561198108148057/recommended/1003090/

Murray, S. (2018). On video games: The visual politics of race, gender and space. I.B.Tauris.

Pfister E. (2018). Der Politische Mythos als diskursive Aussage im digitalen Spiel: Ein Beitrag aus der Perspektive der Politikgeschichte. [The political myth as discursive statement in the digital game: A contribution from the perspective of political history.] In T. Junge, C. Schumacher (Eds.), Digitale Spiele im Diskurs. [Digital Games in Discourse.]. FernUniversität in Hagen. Retrieved from URL https://ub-deposit.fernuni-hagen.de/servlets/MCRFileNodeServlet/mir_derivate_00001443/DSiD_Eugen_Pfister_politi sche_Mythos_digitale_Spiele_2018.pdf

Pfister E. (2018, September 26). „Where the line of decency is drawn": Imaginationen des Holocaust in digitalen Spielen. ["Where the line of decency is drawn": Imaginations of the Holocaust in digital games.] Retrieved from URL http://spielkult.hypotheses.org/1962

Rosenbrock, H. (2012). Die Antifeministische Männerrechtsbewegung: Denkweisen, Netzwerke und Online-Mobilisierung. [The antifeminist men's rights movement: Ways of thinking, networks and online mobilisation.] Schriften des Gunda-Werner-Instituts 8. Retrieved from URL https://www.boell.de/sites/default/files/antifeministische_maennerrechtsbewegung.pdf

Rules and guidelines for GOG.COM. (n.d.). Retrieved from URL https://support.gog.com/hc/enus/articles/360010188177?product=gog

Rules and guidelines for Steam: Discussions, reviews, and user generated content. (n.d.). Retrieved from URL https://support.steampowered.com/kb_article.php?ref=4045-USHJ-3810

Seedy. (2020, February 8). [Player review]. Retrieved from URL https://steamcommunity.com/profiles/76561198099855030/recommended/1003090/

Sigl, R. (2020, February 5). „Through the Darkest of Times" im Test: Widerstandskampf gegen die Naziherrschaft. ["Through the Darkest of Times" test: Resistance vs. the Nazi regime.] derstandard.at. Retrieved from URL https://www.derstandard.at/story/2000114157098/through-the-darkest-of-times-im-test-widerstandskampf-gegen-die

Slayer_Gauge. (2020, February 3). [Player review]. Retrieved from URL https://steamcommunity.com/profiles/76561198073255560/recommended/1003090/

StevenWolfe. (2020, August 9). [Player review]. Retrieved from URL
https://steamcommunity.com/id/MrStevenWolfe/recommended/1003090/

Wayne. (2020, January 31). [Player review]. Retrieved from URL
https://steamcommunity.com/profiles/76561198000076113/recommended/1003090/

Wildcard. (2020, July 7). [Player review]. Retrieved from URL
https://steamcommunity.com/id/fantasticalhybrid818/recommended/1003090/

Zero. (2020, February 17). [Player review]. Retrieved from URL
https://steamcommunity.com/profiles/76561197970143059/recommended/1003090/

BECOMING THEIR ENEMY – ANTIFASCIST GAMING NETWORK "KEINEN PIXEL DEN FASCHISTEN!" AND ITS RIGHT-WING BACKLASH

Pascal Marc Wagner

When several dozen bloggers, podcasters, developers, streamers and other content creators from the German-speaking video gaming space launched their antifascist network "Keinen Pixel den Faschisten!" in April 2020, hatred from German-speaking right-wing outlets was sure to follow. Indeed, it did: A counter account was formed on Twitter by German GamerGate followers to ascribe an anti-free-speech stance to "Keinen Pixel". English translations of antifascist texts were seeded onto international messaging boards of the GamerGate campaign to rile up resistance from the English-speaking right.

In this article, one of the co-founders of the network outlines the steps taken against "Keinen Pixel" and puts them in context of well-known right-wing activist methodology, with fear mongering through dissemination of conspiracy myths and socially accepted antifeminism as hooks for less socially accepted utterances being the two most prominent examples. After introducing the work of "Keinen Pixel", the article will look back at what the GamerGate movement was and discuss what lingering influence of it remains specifically in German-speaking video gaming communities and the further German-speaking political field. It ties right-wing extremist organisations from German-speaking countries to these influences, as exemplified by attempts made to claim fascist spaces in mainstream video game culture.

The article references common propaganda theory and practical knowledge of antifascist organisation methodology to explain the desire of right-wing organisations to infiltrate online spaces. The article also explains, with the aid of cognitive models of information processing and neurolinguistics as well taking into consideration common extremist strategies of arguing and debating, why unopposed propagandist utterances are important for recruitment undertakings of the right in online spaces. Conversely, it also explains why opposing the right in these spaces is antifascist in nature as well as how and why counter speech can be done effectively on social media.

Keywords: Antifascism, Alt-Right, Counter Speech, Online Discourse, Cognitive Linguistics

———

INTRODUCTION

The activist network "Keinen Pixel den Faschisten" (Eng. 'No Pixel for Fascists', "Keinen Pixel" for short) is a grassroots initiative started by academics, bloggers, podcasters, streamers and other content creators from the video gaming sphere in April 2020, one of them the author himself. The German-language network found together after the right-wing terrorist act of Halle, Germany in 2019 (cp. bpb, 2020), the perpetrator of which used online streaming and gaming terms to share the crime with his ideological colleagues. The act was later controversially tied to a perceived "gaming scene" by German politicians with no regard to pop-cultural influences of popular media at large or the intricacies of the video game industry or internet gaming communities (cp. Kogel, 2019)[1]. "Keinen Pixel" established itself to educate about problematic tendencies inside gaming communities and in the games industry, but also to directly stand against and debunk fascist tendencies in online video games, commentary pages and web forums (cp. Keinen Pixel, 2020 [1]). Concentrating on direct action on social networks like Twitter as well as on the forums and group pages of the biggest digital video game seller in the world, Steam by Valve Corporation, the network has also written guides and given interviews on how to take action against misogynistic and racist harassment, how to counter right-wing recruitment efforts in gaming spaces and how to debunk fundamentalist conspiracy

---

[1] The author stresses the distinction that right-wing organisations are not interested in gaming spaces due to an interest in gaming as a topic, but in weakly moderated spaces in general, where their utterances stand unopposed the longest, and that gaming spaces such as they are right now are merely fitting this profile well enough. Nevertheless, various big publishers and video game journalism sites actively avoid political discussions around highly political games such as those covering civil war or imperialism up until this day, acting with disregard in terms of the openings they provide for right-wing operators.

myths in conversations on social media (e.g. Keinen Pixel, 2020 [3]). It also connects websites and channels that were involved in antifascist activism before the network was formed and provides thematically curated collections of articles for research and journalistic purposes (cp. Keinen Pixel, 2020 [3]).

In its runtime so far, "Keinen Pixel" has achieved goals that attracted the ire of right-wing organisations and individuals. Along with the information work provided on the website and in the network, itself a cause for aggression from the right as will be discussed further below, two substantial changes in the gaming landscape have been reached with participation of "Keinen Pixel". In August 2020, "Keinen Pixel" and other activists informed the devcom, gamescom's developer-centred panel conference, that a panel on antifascism was headed by Destructive Creations, a development studio infamous for making games such as *IS Defense* and *Ancestors Legacy*, which catered to right-wing audiences. The panel was cancelled on short notice and replaced with an interview discussion with Paintbucket Games, the decidedly antifascist development studio behind the game *Through the Darkest of Times* (cp. Suessmeier, 2020). In September, the Austrian "*Identitäre Bewegung*" and the German "*Ein Prozent für unser Land e.V.*", both organisations monitored for right-wing extremism by the German intelligence service, aimed to release a homophobe, sexist and antisemitic video game on Steam, along with releases on the indie game store page *itch.io* and their own website. Even before the release of the game, the store page made clear the intention of reproducing hate against marginalised groups of people and antifascists, and recruiting new, potentially young people for right-wing extremist groups[2]. "Keinen Pixel" directly contacted the *Freiwillige Selbstkontrolle Multimedia* (FSM) and *Bundesprüfstelle für jugendgefährdende Medien* (BpjM) of Germany, Valve, the corporation running Steam, and itch.io and published the topic on the "Keinen Pixel" Twitter account. While FSM and BpjM were not able to audit the game until it was officially released, the resulting public attention presumably moved Steam and itch.io to remove the game's shop pages before that, making the game solely available on its homepage without a single download on the digital storefronts (cp. Strobel, 2020).

---

[2] Cp. Wagner, 2020 for an in-depth analysis of conspiracy myths and stereotypes in the store page text and trailer footage used on the Steam shop site.

It is worth disclaiming that none of this was achieved by "Keinen Pixel" alone. In fact, we cannot be sure if the network's part was substantial at all in reversing the decisions of devcom and Valve. However, with the network's online appearances in both cases being the first and most vocal informing sources, and by staying in contact with corporate accounts and the FSM, it stands to reason that it helped to foster recognition for both problems along with other vocal agents.

RIGHT-WING ATTACKS AND COUNTERMOVEMENTS

When "Keinen Pixel" went public on Twitter in April 2020, it came as no surprise that it attracted the attention and rage of right extremist accounts and right-wing trolls. Tweets likened the network outspoken against fascism to the GDR and so called 'left-wing censoring dictature'. The statement of action, made public as the first piece of writing on the network's website (Keinen Pixel, 2020 [1]), was spun to allegedly demand stronger censoring of video games, shared on right-wing platforms and closed forums like "the almost-official GamerGate subreddit"[3] *r/KotakuInAction* and *8chan* and used to mobilise trolls on Twitter. It bears noting that all action taken against "Keinen Pixel" was in German, like the network's publications. The main spokesperson was the collective account of the German-speaking GamerGate community 'gamergateblogde'. These organisers and followers subsequently made a Twitter account to mimic the aesthetics and informational publications of "Keinen Pixel" with the sole intent of framing the network as "the real fascists" (cp. Twitter account 'KeinenV'[4]). When the counter-account garnered no more than 150 followers in two months, as opposed to the "Keinen Pixel" account with over a thousand followers in the first week, the account was discontinued in June, only to be reactivated on specific public occasions to agitate for pro GamerGate causes.

A Brief Excursion on GamerGate

---

[3] Tagline of the subreddit, quoted from reddit.com.
[4] No link will be provided to both Twitter accounts due to them promoting racist hate, homophobic and misogynistic behaviour and antisemitism. Interested researchers are encouraged to search for themselves, albeit with caution.

In this article, GamerGate is recognised as the driving movement for much of the discriminatory hate in online gaming spaces from 2014 onward. While a lot of specific literature has already been written on this topic and will give interested researchers all the information they need (e.g. Schwarz, 2020 p. 127ff; Keinen Pixel 2020 [4]; The Guardian, 2019 as well as further reading in the Keinen Pixel 2020 [4] literature list), a brief overview is in order for clarification. GamerGate is a movement that started in gaming communities with misogynist hatred at its core. It began consolidating in 2014, when the ex-boyfriend of game developer Zoë Quinn accused them in a blog post of paying for a positive review of their autobiographic free game *Depression Quest*. Even though this statement has been debunked by Quinn and Kotaku, the review outlet in question (since no such review even existed and thus could not have been paid for in any way), the blog post garnered the attention of antifeminist and right-leaning youtubers that picked up the topic, making it more public. In the course of GamerGate's consolidation, several other developers and journalists, most of them women such as Anita Sarkeesian or Brianna Wu, but male allies like Tim Schafer as well, were harassed online for standing with the victim. GamerGate concentrated in forums such as r/KotakuInAction, cementing an in-group bias that was effectively used to recruit right-leaning video gamers to vote for Donald Trump by the Republican campaign team under Steve Bannon in 2016 (cp. Green, 2017; Snyder, 2017).

AIMS OF RIGHT-WING ATTACKS ON ANTIFASCIST NETWORKS

When analysing right-wing attacks on antifascist online efforts, the question for possible goals needs to be addressed. Why do fascist movements try to discredit antifascist movements instead of trying to disband them and stop their information campaigns altogether?

The main reason is, most likely, one of practicability. Different to in-person rallies or demonstrations, disbanding an online network cannot be done by physical intervention or threats. Even when threatened, a well-organised online network will have created safe spaces for members suffering doxxing attacks or other threats, providing them support via affirmation, consultation, or financial aid. A functioning online network will most likely remain active and functional even if singular members are picked out by right-wing online trolls and scared or doxxed into silence. It is thus imperative for fascist countermovements not to stop an antifascist network from informing, but to discredit it in such a way that not only convinced right-wingers will not believe its information, but confused or misinformed bystanders as well. Ideally for the right-wingers, a misinformation campaign produces conspiracy myths about the corruption of a network or institution. These myths can spread wider than the reputation of the network or organisation itself, as can be seen in the case of the conspiracy myth created by right-wingers about Anetta Kahane, founder of the *Amadeu-Antonio-Stiftung* (AAS) against antisemitic violence in Berlin, Germany. Kahane worked with the *Staatssicherheit* of the GDR for some years, but then defected and started her commitment for human rights, leading to the founding of the AAS. Her work within the GDR is well-documented by historical evaluators, completely removing the AAS from accusations of GDR romanticisation (cf. AAS, 2017; Müller-Ernberg, 2017, p. 9f). Nevertheless, the right-wing myth that puts the AAS in connection with the GDR's inhumane practices of espionage and border control is well known in Germany, often more so than the efforts of the AAS itself, effectively tainting the image of the foundation's efforts of antisemitism.

A right-wing disinformation campaign has been attempted by the 'KeinenV' GamerGate Twitter account as well, rallying its followers against "Keinen Pixel" with tweets like "Who are these gaming fascists actually?" and "Have they spent all their money on tweets already? Was Kahane stingy?"[5], trying to imply that "Keinen Pixel" was paid by Anetta Kahane and therefore part of the conspiracy myth fabricated against the AAS. Articles and conference presentations from "Keinen Pixel" and associated network members have frequently been connected by GamerGaters to alleged payments from the federal government of Germany and the AAS, none of which existed, to further alleged causes like "censoring all games in Germany" made up by German GamerGate proponents. Because the Twitter account in question was not able to mobilize any support from German-speaking GamerGate proponents, they turned to international forums for help with

---

[5] Screenshots of the tweets in German language are accessible for reference here: http://keinenpixeldenfaschisten.de/wp-content/uploads/2021/01/kvhatespeechproof.png

translations of entire "Keinen Pixel" articles to garner interest in the concept of a new enemy. Their allegations reached as far as conspiracy theorist David Icke, orchestrator behind the reptiloid conspiracy myth, who tried to rally for them on his blog[6]. None of their attempts succeeded in keeping "Keinen Pixel" from their activist work.

FABRICATED CREDIBILITY: GAMERGATE'S PSEUDOSCIENTIFIC COUNTER-SOURCES

In the most active days of the GamerGate hashtag, some effort was made by spearheading content creators of the movement to give the appeal of an ethical movement for a just cause. Most were prominently debunked by progressive gaming sites or wider media outlets (cp. Valenti, 2017) and subsequently dropped from vocal GamerGate defendants' arguments except for the most secluded in-group discussions. One of the very few pieces of content that remain prominent outside of GamerGate in-group discourse is a survey conceived by Brad Glasgow, himself an outspoken GamerGater on YouTube and Twitter, in December 2015 and January 2016 that asked 725 GamerGate supporters for their own political self-description. The survey shows that GamerGaters perceive themselves as politically liberal and progressive, while still coining themselves as "the anti-progressive, anti-social justice, anti-feminist movement" (Glasgow, 2016). As an in-group self-survey, it stands to reason that the survey has no academic declarative as a secondary source at all but would need to be analysed as a primary source with due regard to the psychological and social biases that come to play in such self-observance. Such analysis was never conducted from within GamerGate. Instead, the survey is frequently cited in online arguments to persuade outsiders as 'proof' that GamerGate was a progressive movement. Up until today this survey remains one of the rare cases where a GamerGate proponent used sufficiently academic data acquisition methods and presentation to trick bystanders into ascribing merit to it. It should thus come to no surprise that a

---

[6] Screenshot of the article on Icke's blog: http://keinenpixeldenfaschisten.de/wp-content/uploads/2021/01/DIproof.png The blog will not be credited in the bibliography directly to not decorate the conspiracy theorist with a link. Interested researchers are advised to search with the information provided in the screenshot.

2020 journal article by C.J. Ferguson and the very same Glasgow[7] references the exact same 2015/2016 survey (cp. Ferguson & Glasgow 2020, p. 2) and comes to the same conclusion as Glasgow in 2016, albeit within even more formalised academic standards, appearing more credible to bystanders and colleagues without knowledge of the GamerGate discourse. The article is a paid closed-access paper and thus will mostly be shared in the form of the abstract (Ferguson & Glasgow, 2020, Abstract), which perpetuates the myth originally disseminated in 2016. Interestingly, the abstract does neither lay open that the 2020 article references the 2016 survey and that a new one was not conducted, nor does it make explicit that all claims of progressiveness are self-descriptions only. The abstract is thus effectively used by GamerGaters to obscure their right-wing attitudes in conversation with undecided bystanders, who are deceived into believing in the existence of different credible studies showing the progressiveness of GamerGate, when in reality only one such in-group survey exists[8]. It is thus no surprise that the 2020 survey article is frequently referenced by GamerGate accounts on social media (albeit never actually quoted) to contradict the articles published by "Keinen Pixel".

Combining this pseudoscientific self-affirmation with articles reflecting positively on GamerGate and negatively on a perceived 'corrupt games journalism' from GamerGate proponent websites such as *nichegamer.com* serves as the foundation for many of GamerGate's contemporary argumention strategies. Information from sites not explicitly pro-GamerGate such as *polygon.com* or wider media outlets such as the British *theguardian.com* are mostly omitted from GamerGate's canon of argumentation. Notably, ties to alt-right websites have been denied by GamerGate proponents at least since 2016 (cp. Glasgow, 2016), when investigative media such as The Guardian have started linking GamerGate to alt-right voter recruitment efforts via

---

[7] Regarding the Twitter accounts of Glasgow and Ferguson, see Footnote 1.

[8] Other surveys indeed exist, but all of them are connected by the fact that they are made by GamerGate proponents themselves. See for example The View from the Other Side: The Border Between Controversial Speech and Harassment on Kotaku in Action by Amy Bruckman et al. (2018), which notably was produced in some cooperation with Glasgow (see Twitter thread of Bruckman and others, <https://twitter.com/asbruckman/status/715658477146935298>). Bruckman herself has posted right-leaning opinions as well (cp. Bruckman 2017) but seems to have developed a different stance since.

breitbart.com as early as 2016 (Lees 2016). GamerGate was swiftly recognised by Donald Trump's campaigning strategist and former Breitbart executive chair Steve Bannon, and, consequently, gamers with alt-right tendencies were marketed to by the Trump campaigning team. The fact that GamerGate did not reach out to Breitbart for moral assistance, but the other way around, is a void argument, considering that the alt-right's recruitment efforts fell on soil so fertile.

POSSIBLE ACTIONS FOR COUNTERING RIGHT-WING ATTACKS

Organising a network or activist group is in itself a viable tactic against right-wing attacks, making it harder to doxx and dogpile[9] single people that gather behind it and publish under the label of the network instead of their own given names. However, while this makes it harder for the right to single out perceived protagonists, it may not prevent it entirely. This is where the actual networking component of an activist group comes into play, providing backing by like-minded group members against public attacks. Group members may employ counter-speech against large right-wing troll attacks or even financial support in cases where homepages need to be moved due to doxxing attempts, or when attorney fees need to be advanced.

Not all victims of extremist online abuse are part of support networks, however. Authors or political analysts often get harassed for their academic or journalistic publications, not for their activism in antifascist organisations. Prominent German-speaking examples of the last years include attacks of a right-wing columnist on Austrian politics analyst Natascha Strobl (cp. Funk, 2020) and a trolling shitstorm against Karolin Schwarz, author of *Hasskrieger: Der neue globale Rechtsextremismus* (cp. Schwarz, 2020).[10] In such cases, networks should make it their priority to provide support outside of their

---

[9] The internet term 'dogpiling' figuratively describes "[a] situation in which criticism or abuse is directed at a person or group from multiple sources." (OED 1, 1.1), much like a group of dogs lunging at a bone thrown at them.

[10] It should be noted that a significant amount of victims are women, even if this is not further discussed in this article.

usual reach by employing counter-speech and contacting victims for their needs.

Grouped Action: Counter-Speech and Deplatforming as Network Efforts

In general, no single action on social media is as impactful as organised action by several actors. Having realised this early, right-wing trolls gather in forums such as 8chan or subreddits, where a large number of readers with the will to action are readily available. When a link is posted, even if just a fraction of message board users activate and swarm the victims social media profile, it will most likely be enough to inspire fear or occupy the victim with defending their political position or written article long enough to wear them out. The platitude of "strength in numbers" does not only work in one direction, though. Counter-speech by several active network members is less stressful and more effective than counter-speech only made by the victims themselves. Studies have shown that online harassers tend to stop posting sooner if opposed by users, because one of the main drivers of online harassment is the perceived freedom of consequences (cp. Bojarska, 2018, p. 7). Counter-speech, as Benesch et al. (2016, p. 2) assess, "positively affect[s] the discourse norms of the 'audience' of a counter speech conversation: all of the other Twitter users or 'cyberbystanders' who read one or more of the relevant exchange of tweets" and thus may lead to empowering those bystanders to support counter-speech efforts themselves.

The most permanent way to remove harassers from social media, *deplatforming*, almost entirely relies on group action. Social media networks such as Twitter or Facebook only take initiative to remove harmful accounts if there is overt social interest in having them deleted – such as with the recent removal of Donald Trump's Twitter account after he incited the terrorist attack on the Capitol. Getting those networks to remove smaller harmful accounts is in some sense even more difficult. What the global political climate cannot do for those cases must be done by collected information campaigns by activist networks that are followed by as many reports as possible to the social media website.[11] Social media networks tend to react to mass reporting, sadly often even without probable cause, because their report centres tend to be understaffed and overworked and tend to be put under heavy mental strain because the content they moderate is often abusive or full of violence (cp. Solon, 2017; Newton 2019). Such deplatforming has, among others, cut the reach and funding of Steve Bannon and Alex Jones' *Infowars* as well as the Identitäre Bewegung on most social media platforms.

Blocking and Debunking

Not all right-wing attackers can be effectively deplatformed, though. An increasing percentage of harassers, guided by figureheads such as Martin Sellner or inspired by pseudoscientific methods like those explained above, have taken to harassing and trolling without explicitly violating the terms of service of a given social media platform. Because they are disguised through so called 'dog whistles', innocuous words that stand in for harmful contexts, an increasing number of attacks cannot be discovered by moderation staff without sufficient training. Right-wing extremists have taken to several prominent dog whistles, such as the 'Great Exchange' to denote the racist conspiracy myth of refugees replacing the white inhabitants of Europe or Northern America.

Also, as idealistic as it would be for every victim of online abuse to be part of or in reach of a support network, it can be assumed that the vast majority of online abuse happens to persons who cannot rely on external support. In such cases, successfully reporting an abuser is difficult. Because people interested in an honest discussion often tend to feel otherwise, it should be stressed that blocking the abusers is always an option for personal protection. One of the most toxic properties that bled out of troll culture into the more general consciousness is that trolls, and extremists with them, perceive being blocked as "winning an argument". While this may be true in their extreme in-group, the more general truth is that blocking deprives trolls not only of a

---

[11] While some argue that this way of action is a slippery slope towards the same methods right-wing extremists use, I would like to oppose with the following point: promoting harassment and hate speech is, in addition to being morally wrong, a violation of almost all social media sites' terms of service as well as most countries' laws. Being a Person of Colour, a woman, LGBTQIA+ or generally against fascism is neither.

victim, but of the victim's social media reach. Blocking is not only self-preservation, but an effective tool of silencing attacks. However, due to most trolls possessing several sock puppet accounts and most often being part of a trolling community, blocking every account manually can be enervating and often hard to achieve. Social media plug-ins such as 'Twitter Block Chain' may ease that burden when supplied with a viable reference list such as the followers of the aforementioned 'gamergateblogde' account.

For counter-speech purposes, blocking is one half of an effective strategy to positively affect the discourse norms of the social media room. The other is *debunking*; the act of putting the faulty information or propaganda posted by attackers right, not to *re*form the attacker (who is probably well aware of the falsehood of their claim) but to *in*form bystanders. Debunking best practises are an ongoing discussion topic between political scientists, linguists, neuroscientists and many others, since effective debunking has to take into consideration various neurologic, cognitive, societal and cultural mechanisms (cp. Lakoff, 2014; Schmid, 2018). Cognitive linguist George Lakoff has attempted to provide a minimum model for effective debunking on social media platforms in the wake of the 2016 U.S. presidency election. This *Truth Sandwich*, as Lakoff names it in a podcast (Lakoff & Durán, 2018) and subsequent tweet, takes into account the most recent states of psycho- and neurolinguistic speech processing findings as well as social media quirks:

Truth Sandwich:

1. Start with the truth. The first frame gets the advantage.

2. Indicate the lie. Avoid amplifying the specific language if possible.

3. Return to the truth. Always repeat truths more than lies.

(Lakoff 2018, Tweet)

The model is fit for social media use for several reasons. Schmid's (cp. 2018) entrenchment theory states that linguistic information 'entrenches' itself into the brain the more often it is perceived, like a shovel digging into the ground repeatedly. Repeating the truth more often than the lie will maximise truth entrenchment and minimise (albeit not eradicate) entrenchment of the fake information. Journalism teacher Roy Peter Clark (cp. 2020) contributes to this the observation that "emphatic word order" leads attentiveness within the three layers of a truth sandwich as much as the order of the layers does. He suggests to "[o]rder words for emphasis. Place emphatic words in a sentence at the beginning and the end." (Clark, 2020). It follows then that the factually correct information in the first and third layer should frame more insignificant parts in both layers, while the metainformation "this part is a lie" should frame the actual lie in the second layer. An effective truth sandwich within Twitter's 280 character restrain may then look like this, with emphasis marked bold:

1. **GamerGate is responsible for misogynistic harassment**. This is documented by journalists and academics.

2. **GGers claim this is not true**. They authored studies with non-viable methods. **They are lying**.

3. GG's methods **serve as blueprints for alt-right campaigns around the world**.

CONCLUSION

This article started with an overview of the antifascist online network "Keinen Pixel den Faschisten!", their strategies for pro-democratic information and discourse and the backlash from fascist online presences, especially the right-wing proponents of GamerGate. The article then elaborated on right-wing online harassment methods and possible counterstrategies, ending with an example of the application of counter-speech theory in the context of social media. The author hopes that this practical combination of case study and application frameworks serves to inspire more dedicated research on and from within antifascist networks and organisations, who are often deeply devoted to academic methods and facts, but still rarely recognised in such a manner.

REFERENCES

Amadeu-Antonio-Stiftung (AAS). (2017). Anetta Kahane. Retrieved from https://www.amadeu-antonio-stiftung.de/ueber-uns/kontakt-team/anetta-kahane/

Benesch, Susan, Ruths, Derek, Dillon, Kelly. P., Saleem, Haji. Mohammad & Wright, Lucas. (2016). Considerations for Successful Counterspeech. Dangerousspeech.org. Retrieved from https://dangerousspeech.org/wp-content/uploads/2016/10/Considerations-for-Successful-Counterspeech.pdf

Bojarska, Katarzyna. (2018). The Dynamics of Hate Speech and Counter Speech in the Social Media: Summary of Scientific Research. Centre for Internet and Human Rights.

Retrieved from https://cihr.eu/wp-content/uploads/anti2018/10/The-dynamics-of-hate-speech-and-counter-speech-in-the-social-media_English-1.pdf

Bundeszentrale für politische Bildung (bpb). (5.10.2020). Der Anschlag von Halle. Retrieved from https://www.bpb.de/politik/hintergrund-aktuell/316638/der-anschlag-von-halle

Bruckman, Amy. (22.05.2017). People with Your Politics are Not Welcome Here. Retrieved from https://asbruckman.medium.com/people-with-your-politics-are-not-welcome-here-757d2636f83d

Clark, Roy Peter. (18.08.2020). How to serve up a tasty 'truth sandwich?'. Poynter.org. Retrieved from https://www.poynter.org/reporting-editing/2020/how-to-serve-up-a-tasty-truth-sandwich/

Ferguson, Christopher J. & Glasgow, Brad. (2020). Who are GamerGate? A descriptive study of individuals involved in the GamerGate controversy. In: Psychology of Popular Media. Advance online publication. Retrieved from https://psycnet.apa.org/record/2020-15670-001

Funk, Victor. (12.08.2020). Natascha Strobl und „Don Alphonso": Kolumnistin erwägt, rechtliche Schritte einzuleiten. Retrieved from https://www.fr.de/politik/als-wuerde-ich-permanent-in-den-abgrund-schauen-90016007.html

Glasgow, Brad. (21.11.2016). No, Gamergate is Not Right Wing. Retrieved from http://www.gameobjective.com/2016/11/21/no-gamergate-is-not-right-wing/

Green, Joshua. (2017). Devil's Bargain: Steve Bannon, Donald Trump, and the Storming of the Presidency. Penguin Press: New York, London.

Guardian, The. (20.08.2019). The Guardian view on Gamergate: when hatred escaped. Retrieved from https://www.theguardian.com/commentisfree/2019/aug/20/the-guardian-view-on-gamergate-when-hatred-escaped

Keinen Pixel den Faschisten [1]. (22.04.2020). Was ist "Keinen Pixel den Faschisten"?. Retrieved from https://keinenpixeldenfaschisten.de/2020/04/22/pressemitteilung/

Keinen Pixel den Faschisten [2]. (15.07.2020). Wie man gegen Rechtsradikale in Videospiel-Communitys vorgeht. Retrieved from https://keinenpixeldenfaschisten.de/2020/07/15/wie-man-gegen-rechtsradikale-in-videospiel-communitys-vorgeht/

Keinen Pixel den Faschisten [3]. (2020-). Artikelsammlung. Retrieved from https://keinenpixeldenfaschisten.de/artikelsammlung/

Keinen Pixel den Faschisten [4]. (16.11.2020). GamerGate, eine Retrospektive. Retrieved from https://keinenpixeldenfaschisten.de/2020/11/16/gamergate-eine-retrospektive-download/

Kogel, Dennis. (14.10.2019). Seehofer und die Gaming-Szene: Ein Sieg für den Täter von Halle. DLF Kultur. Retrieved from https://www.deutschlandfunkkultur.de/seehofer-und-die-gaming-szene-ein-sieg-fuer-den-taeter-von.2165.de.html?dram:article_id=460968

Lakoff, George. (2018). The ALL NEW Don't Think of an Elephant!: Know Your Values and Frame the Debate. Revised version of Don't Think of an Elephant!: Know Your Values and Frame the Debate. (2004). Chelsea Green Publishing: Vermont.

Lakoff, George. (01.12.2018). Tweet from his profile @GeorgeLakoff. Retrieved from https://twitter.com/georgelakoff/status/1068891959882846208?lang=en

Lakoff, George & Durán, Gil. (2018). Truth Sandwich Time. FrameLab Podcast. Retrieved from https://soundcloud.com/user-253479697/14-truth-sandwich-time

Lees, Matt. (01.12.2016). What Gamergate should have taught us about the 'alt-right'. Theguardian.com. Retrieved from https://www.theguardian.com/technology/2016/dec/01/gamergate-alt-right-hate-trump

Müller-Enbergs, Helmut. (2017). Zusammenfassende gutachterliche Stellungnahme zu Anetta Kahane und die DDR-Staatssicherheit. Retrieved from https://www.amadeu-antonio-stiftung.de/wp-content/uploads/2018/11/gutachten_anetta_kahane.pdf

Newton, Casey. (19.06.2019). Bodies in Seats. Theverge.com. Retrieved from https://www.theverge.com/2019/6/19/18681845/facebook-moderator-interviews-video-trauma-ptsd-cognizant-tampa

Oxford English Dictionary (OED). DOGPILE. Lexico. Retrieved from https://www.lexico.com/definition/dogpile

Schmid, Hans-Jörg. (2017). Entrenchment and the Psychology of Language Learning: How We Reorganize and Adapt Linguistic Knowledge. De Gryuter: Berlin.

Schwarz, Karolin. (2020) Hasskrieger: Der neue globale Rechtsextremismus. Bpb: Bonn.

Snyder, Mark. (18.07.2017). Steve Bannon learned to harness troll army from 'World of Warcraft'. Retrieved from https://eu.usatoday.com/story/tech/talkingtech/2017/07/18/steve-bannon-learned-harness-troll-army-world-warcraft/489713001/

Solon, Olivia. (25.05.2017). Underpaid and overburdened: the life of a Facebook moderator. Theguardian.com. Retrieved from https://www.theguardian.com/news/2017/may/25/facebook-moderator-underpaid-overburdened-extreme-content

Strobel, Benjamin. (20.10.2020). Von Humor zu Hass. Neues-deutschland.de. Retrieved from https://www.neues-deutschland.de/artikel/1143366.computerspiele-von-humor-zu-hass.html?sstr=steam

Suessmeier, Christian. (29.08.2020). gamescom 2020 – Kontroverse rund um Show mit rechtspopulistischem Entwicklerstudio. Dvd-forum.at. Retrieved from https://www.dvd-forum.at/gamescom-2020-kontroverse-rund-um-show-mit-rechtspopulistischem-entwicklerstudio

Valenti, Jessica. (24.09.2017). Zoe Quinn: after Gamergate, don't 'cede the internet to whoever screams the loudest'. Retrieved from https://www.theguardian.com/technology/2017/sep/24/zoe-quinn-gamergate-online-abuse

Wagner, Pascal Marc. (11.09.2020). Wie Rechtsradikale den Steam-Shop zum Werben nutzen. Languageatplay.de. Retrieved from https://languageatplay.de/2020/09/11/wie-rechtsradikale-den-steam-shop-zum-werben-nutzen/

REAMIFTON NORTH: A GAME ABOUT THE UNITED STATES POSTAL SERVICE AND THE COMPLEXITY OF 2020 AMERICAN POLITICS

Michael Black, Jared Derry, Kathryn Friesen, Josey Meyer, Montserrat Patino

Over the course of a semester, our team of five undergraduate Visualization students from Texas A&M University set out to create a game that explores the current political climate of the United States through the lens of the postal service. We created a prototype called Reamifton North, a resource management simulation game. You play as Cas, the new hire at the post office, and it is your job to keep the place up and running by organizing and delivering packages.

In the face of a global pandemic and the 2020 U.S. presidential election, the national government took deliberate steps to sabotage the United States Postal Service (USPS). Our game is a critical response to this that also addresses a variety of social injustices, including economic inequality, racial inequality, inadequate pandemic response, and voter suppression.

Although it is currently only in the prototype stage, the full plans for the game include a story with multiple narrative paths and endings. The player would face critical decisions throughout the game that reflect events based on real-life United States politics. Gameplay becomes progressively harder as the story goes on and the post office the player works at is slowly stripped of necessary equipment as the government withdraws support. The increasingly difficult gameplay and narrative developments would aim to capture the frustration and despair felt by many people during the real-life events happening during the development of this game.

Our goal was to create a game that would be representative of what is currently happening with the USPS and the overall political climate of the United States, and to encourage the player to empathize with those most affected by these events.

Keywords: Game Studies, American Politics, Pandemic

————

INTRODUCTION

We developed this game over the course of a semester as part of a study abroad program held by Danube University Krems. At the beginning of the program, we were tasked to create a serious game; a game created not just for entertainment, but also to teach players something. With this task in mind, we set out to make a game that explores the current political climate of the United States through the lens of the postal service.

Amidst the 2020 presidential election and a worldwide pandemic, our team was keenly aware of the political climate of our nation. The creation of this game was in part an outlet for us to vent our frustrations with the situation and an opportunity to dive into some of the issues that currently beset our country. We wanted to explore a variety of social injustices, including economic inequality, racial inequality, voter's rights, and the lack of action from the national government in providing aid for the pandemic. The crippling of the United States Postal Service is only the tip of the iceberg when it comes to current political issues in the United States, but we chose it as our focus since the damage done to the USPS had widespread effects that touched all the other topics we wanted to explore. Our goal was to create a game that would raise awareness of what is currently happening within the postal system, how that affects other current issues, and, as a result, encourage the player into thinking critically and engaging in discussions about politics and social justice.

Reamifton North is both a critical exploration of the current political environment in the United States and a snapshot of the events and feelings of this time that asks players to step into the shoes of an essential worker. It touches on issues including economic inequality, racial inequality, inadequate pandemic response, and voter suppression. We hope the experience this game provides will educate players on the unique situation here in the United States and foster a sense of empathy for others who have faced increasing injustices due to recent events.

CONCEPT STAGE

At the beginning of the project, our team met with Natalie Denk and Alexander Pfeiffer to brainstorm topics for our serious game. We used the Miro software to work collaboratively, adding moodboards, notes about concepts, and reference images to figure out what direction we wanted to take. A common theme among the initial concepts was the idea of mental health, politics, and the complexity of the pandemic in navigating both. After some deliberation as a team, we decided the postal service game would allow us to touch on more of these topics in an engaging way. Next, we sought out references to build the foundational inspiration for our game.

A game that caught our eye in one of Alex's lectures was *Papers, Please*, a simulation game where the player performs increasingly tedious tasks while a political narrative is spun around them. *Papers, Please* became a key reference for our game as it had an excellent core game loop, including an external narrative beats as well as mechanically-driven narrative elements.

Another game we referenced heavily was *Wilmot's Warehouse*, a puzzle game focused on sorting colorful boxes in order to deliver them to customers. An element of *Wilmot's Warehouse* we particularly liked was that players had complete freedom to sort the packages in any way they wanted. The warehouse existed only as a boundary for them to organize within.

Once we had our references in order, it was time to parse out the world that it was set in. Patino and Friesen met separately from the group to nail down the major plot points and establish the characters. With a little back and forth on concepts, we soon had written the alternate reality of Samailica. In our alternate reality, the United States Postal Service (USPS) is the United Alliance Postage Assistance (UAPA). Reamifton North is the location of the post office where our main character begins their career in postal service. These elements were important for our players to implicitly understand as lampshading the current state of America.

Figure 1. Our cast, drawn by Montserrat Patino. From left to right: Zuri, Harriet, Ramiro, and Cas.

The cast of *Reamifton North* was developed by Patino and accompanied by their stellar artwork (see Figure 1). Cas is our main character, a non-binary postal worker who is friendly yet shy. Having a non-binary player character allows players to identify with them more closely as they play the game, further investing them in the narrative we build. Cas' manager is Harriet, a middle-aged woman with a firm hand and a fiery attitude. Ramiro is a postal carrier who directly works with Cas in their day to day. Finally, Zuri is a spirited young woman who works in the front of the office, handling visiting customers and other logistics. This diverse cast of characters provides us with a base with which to build the heavier narrative elements in the future.

DEVELOPMENT STAGE

With our game concept underway, we were ready to begin development. Meyer spearheaded the game design document while Patino tackled the art bible. Derry and Black immediately began scripting out initial gameplay within our game engine of choice—Unity 2020. We chose Unity for its comfortable creation of two-dimensional gameplay in its canvas elements. Alongside Unity, we employed the following software:

- Discord
- GitHub
- Miro
- Trello
- Google Drive
- Google office suite
- Adobe Creative Suite

The semester was held online due to the pandemic. While all our members remained in College Station for the duration of the project, we strictly adhered to the online format and did not meet in person. Discord was the perfect method of communication for working online. Voice channels in our server allowed us to work together in a virtual space at any time. Spontaneous meetups in our "work and chill" voice room were a delightful element of our development process. Miro and shared Google documents also helped us work collaboratively with one another while separated. Trello served as our task-tracking method for the project. And lastly, GitHub seamlessly worked with Unity to maintain our files.

GAME DESIGN

Already having found our main references of *Papers, Please* and *Wilmot's Warehouse,* we knew we wanted to create a hybrid between the two. We wanted to develop a resource management simulation game that was threaded together by story. Firstly, we set up game pillars to aim for throughout development, such as keeping the game short and sweet so players can finish the game in one sitting, but still learn from the experience. Next, we pinned down our aesthetics of play based on the MDA Framework, which was focusing on Narrative and Sensation. Narrative we already knew would be important because the characters would be our method for delivering information to the player. We wanted to make sure our characters were likeable for the player to want to keep talking to them. Then we have sensation, which was making the player feel something through the gameplay, visuals, and audio to get an emotional response from the player. This could be interpreted in a variety of ways, but for this project we specifically wanted the player to feel empathy for the postal service enough to spur them to action out-of-game.

After establishing our MDA Framework, we went to work on our core game loop. We have four main sections of the game: the newspaper display for world-building narrative elements, the breakroom as the main narrative element for the player to talk to their co-workers, the warehouse as the main gameplay element sorting mail, and the office as the results screen and a space for the player to upgrade character abilities. Within the warehouse, we only had one level design, which was an open floor plan with four plus-shaped pillars blocking out sections of the main room. Items would come in on the left and the player would stack them throughout the room before delivering a random number of them out the right.

Within our game design document, we detailed all of the above along with our basic mechanics such as camera movement and control schemes. Since we had a small team, it was not necessary to write an in-depth document detailing every possible interaction when we could simply take that same time to implement and playtest it in-game. This was a major benefit of having a small team.

2D ART DEVELOPMENT

Art development for this game drew from a variety of sources, many of them introduced during our study abroad lectures, including *Papers, Please, The Westport Unlimited, We The Revolution*, and more. We agreed early into development that we wanted a simple art style that felt a little drab and weary, without being outright depressing to reflect the feelings we hoped to put into our game. This meant a warm but desaturated color palette, and nothing visually bold. These guidelines were applied to all the game assets, UIUX, backgrounds, and character sprites.

One of our main game design pillars was that the game should be story-driven through the perspectives of multiple characters. This pillar manifested in the breakroom portion of the game loop. Inspiration was taken from visual novel games such as Monster Prom and Coffee Talk. Most visual novels have dialogue boxes across the bottom of the screen with character sprites that change expression to reflect what is being said. Thirty-five separate sprites were drawn in total for the four characters to cover the range of emotions that could be used throughout the story.

PROGRAMMING

One of the main hurdles of programming was merging all our standalone ideas into one fluid game loop. Since we had three core game stages—the narrative, warehouse, and office elements—we needed to make them fit together and communicate to the player that decisions made in each stage would affect the other stages. We were able to do this using a handful of scripts under the direction of a PersistentGameManager which was made to both manage scenes and maintain variables when transitioning between scenes. Within each game stage, there was a sub-managing programming component.

In the breakroom, a DialogueManager was used to construct how the characters' speech appeared on the screen in conjunction with their animations. A scriptable object contained the dialogue written by our story team and was then parsed through and allocated to each character along with animations, timing, and other data by class within a script of code named 'Dialogue'.

In the warehouse, the WarehouseGameManager was created to control the main gameplay stage of our game. This script certainly had the heaviest load as it was made to handle everything from player movement, timers, spawns, UI interaction, package attributes, and more.

As the final stage, the office was less taxing on our programmers but was no less important. This was the bridge between the narrative and gameplay elements. We wanted the player to feel more connected to the characters of our story and immersed in the world overall. Scripts made for the office stage included controls to activate upgrades and display important data from the warehouse gameplay. Animations were also programmed to supplement certain office design decisions in order to create an interface that could represent a post office. Programming was an integral part of our game prototype and was implemented so that future development would require small tweaks in conjunction with more narrative material.

CHALLENGES FACED

Reamifton North was an ambitious project from the beginning and we ran into several challenges along the way. A particular issue was scope creep. With a full load of classes, it was vital that we keep our expectations for the project reasonable. At times it was difficult to keep our sights set on what *must* be included in the game versus what would be nice to include. Meetings as a team and with our mentors in the program helped us to stay focused on presenting a solid core game loop. Extraneous elements we desired to include in the game we kept note of for future development, if desired.

Being solely online also proved to be challenging in its own way. Many of us had previous experience navigating online courses due to the rising pandemic in the Spring semester. But the online classes, coupled with a culture change from American undergraduate education to Austrian graduate education, gave us some new hurdles to face. Working on *Reamifton North* allowed us to coalesce around a central guiding piece of the program amidst changing courses every few weeks. The Discord server was a place where we could assist one another with homework. Fostering a sense of camaraderie over the internet took time, but it paid off. We effectively balanced the fluctuating workload of the abroad program and the consistent effort on *Reamifton North* to present something we were proud of at the conclusion of the semester.

FINAL PRODUCT

Currently, *Reamifton North* exists as a prototype presenting a vertical slice of what would be included if this game were fully developed. All three stages are presented with their proper art assets: warehouse sorting, upgrading/investing in co-workers, and the breakroom. The game was developed so that each aspect is scalable, meaning that we could later add levels to the game without having to recreate elements of it. While we do not intend to further development on *Reamifton North*, we are grateful for the experience it has given us in creating a serious game. Concepting the game was a thrilling venture that led us to provide words for real issues facing

Americans daily. Development was faced with the inevitable threat of scope creep, but we managed to stave it off with consistent team updates and meetings with our mentors. Presenting *Reamifton North* at the FROG 2020 conference was an honor we will not forget and learning from excellent minds in the games industry through the Austria Abroad Program was a rare treat.

ACKNOWLEDGMENTS

We would like to thank the following for their feedback this semester as we have developed Reamifton North. Our mentors Natalie Denk and Alexander Pfeiffer were invaluable this semester as our mentors. We would also like to thank professors Nikolaus König, Geoffrey Long, and Eugen Pfister for their feedback at our final internal presentation prior to the conference. Our presentation at the conference would not have been what it was without your help!

GAMES, PLAY
& A BETTER FUTURE
FUTURE
LOADING

GAMES, PLAY AND A BETTER FUTURE

GAMES@SCHOOL... DOES IT REALLY WORK?

Daniela Hau

Despite the great scientific interest in the learning potentials and limits of digital game-based learning, there are still – at least in Luxembourg's educational landscape – hardly any empirically proven implementations thereof in teaching practice. This discrepancy between anticipated benefits and lack of practical implementation was investigated in 13 teaching projects, which were accompanied by media pedagogical coaching. On the basis of existing theoretical DGBL models and empirical findings, an input-process-output model was developed that takes into account the characteristics of the learners, teachers, games and didactic settings with regard to learning outcomes, the impact of the games and acceptance of the DGBL approach. In the sense of a methodological triangulation, quantitative and qualitative data were collected from the students' and teachers' viewpoints. The results of the study show that the age of the students, type of games used, play time in class and the transfer from game to reality are of central importance. Overall, a positive assessment of the learning outcomes and a high acceptance of digital game-based learning by learners and teachers was observed.

Keywords: digital games, school, empirical study

INTRODUCTION AND AIM OF THE RESEARCH

The scientific interest in the effects of digital games on learning processes and performance has been very high in recent years, which is reflected in numerous studies, publications and books on the topic (van Eck, 2006; Kerres, Bormann & Vervenne, 2009; Gee, 2009). However, so far, the empirical validation of the integration of games in class has been the exception. Although there are some individual and short-term studies that examine digital games in specific didactic contexts, they lack a synopsis and broad empirical basis. In addition, the studies are often divided into two "camps": empirical research on the use of serious or educational games and studies on commercial off-the-shelf games.

Nevertheless, the existing scientific impact analyses show that digital games can offer added value for knowledge acquisition compared to traditional teaching-learning contexts (Boyle et al., 2016), especially in the context of self-directed and autonomous learning processes. This is because, in terms of contemporary pedagogy, approaches from all currently recognised learning theories can be found in digital game-based learning (Le & Weber, 2011):

- active learning - through the continuous play cycle
- constructive learning - through acting in complex systems, testing alternative actions according to the trial-and-error principle and through the individual interpretation of the collected experiences
- self-directed learning - through individual procedures and feedback
- social learning - through cooperation, coordination, competition or through the exchange of experiences between players
- emotional learning - through deep involvement in the action, personal identification, social interaction and the experience of self-efficacy
- situated learning - by placing oneself in different roles and game settings with corresponding problems and tasks.

For a positive learning experience, it seems crucial that the actors see their involvement as effective and efficient. The associated feeling of self-efficacy motivates them to face challenges, overcome setbacks and continue to dedicate themselves to an objective. The importance of self-confidence and self-efficacy is thus a characteristic that links digital games and learning at the core.

So far, however, this potential has hardly been used in formal educational contexts. Especially in secondary school, the prevailing opinion seems to be that games – and the associated notion of fun – do not fit the curriculum and "serious" pedagogical contexts. Digital games as "new" media are often denied their status as "valuable" tools because their mass media distribution means they are suspected of promoting precisely those values and behaviours in children that run counter to "serious" educational values. Despite this apparent contradiction between gaming fun and serious learning, digital games can promote a number of skills and competences when used in the classroom (Fromme et al., 2010; Ganguin, 2010). The corresponding literature specifically focusses on the development of cognitive and sensorimotor skills as well as personality-related and social skills. A positive access to technology and reflection on one's own media use can also promote the media competences of children and young people.

Figure 1. Areas of competence and skills that can potentially be promoted through digital games (source: Hau, 2019)

Consequently, in formal educational contexts, digital games can be used to learn declarative and procedural knowledge (e.g. cause-effect relationships) as well as problem-solving skills, which serve to increase creativity or present practical examples of concepts and rules that would be difficult to illustrate in the real world. In addition to these learning-relevant considerations about games, the cultural dimension also plays a central role: For many adolescents, digital game worlds are an integral part of their lives and leisure activities. Making digital games a topic in class is the necessary prerequisite to conduct an open, "eye-to-eye" dialogue and initiate a multi-layered, critical discussion and guided reflection on the subject.

Against this background, this paper pursues two main objectives. On the one hand, it aims to contribute to the practical implementation of digital game-based learning (DGBL) scenarios, and on the other, the study collected and analysed concrete teaching-learning data in formal school contexts. The focus is on the core factors influencing motivation and learning outcomes from the perspective of teachers and learners.

METHODOLOGY

The present study explores the determinants and potential of digital game-based learning in everyday lessons. Specifically, the potentially relevant characteristics on the part of the learners and teachers, the didactic frameworks and the chosen games were recorded, and the assessments of the use of digital games were determined (e.g. learning outcomes, acceptance of the DGBL approach).

In view of the limited empirical findings regarding the object of study, especially in Europe, and the problem of the generalisation of the findings in DGBL settings, an explorative research approach was pursued. It aims at recording possible input, process and output variables broadly and multimodally. In the sense of broad measurements, method triangulation was applied, including both qualitative and quantitative methods (Wrezien and Raya, 2010; Watson, Mong and Harris, 2011). Specifically, quantitative (questionnaire, Triple E framework) and qualitative survey instruments (interviews, observation) were used. Since the object of the study is an exploratory field study, i.e. scientific research under practical conditions, the evaluation does not aim to collect reliable, valid or objective values on the use of digital games in schools as a whole. The aim is rather to generate initial insights into the impact structure of digital game-based learning and analyse the interrelationships between its determinants.

Derivation of parameters

At present, there are relatively few recognised theoretical models and concepts of digital game-based learning. Thus, van Eck (2010) advocates using interdisciplinarity to bring together several relevant theoretical approaches. In this sense, the following models which have guided DGBL research were analysed:

- Input-Process-Output Game Model (Garris, Ahlers & Driskell, 2002)
- Recursive Loops of Game-Based Learning (Kearney & Pivec, 2007)
- Experimental Gaming Model (Kilii, 2005)
- Four-Dimensional Framework (de Freitas & Oliver, 2006)
- Game-Based Learning Framework (van Staalduinen & de Freitas, 2011)
- PENS Model (Rigby & Ryan, 2011)
- General Learning Model (Gentile & Groves, 2014)
- Experiential learning (Kolb, 1984)
- Theory of autodetermination (Deci & Ryan,1993)

The following central aspects for the present research project were derived from the synopsis of these theoretical models: Since most DGBL models represent learning on the basis of experience-based learning (Mitgutsch & Rosenstingl, 2008, de Freitas and Oliver, 2006), which does not concretely locate the occurrence of the learning effect, the learning outcomes are considered as an overall result of the learning process according to Garris et al. (2002). The characteristics of the learners, teachers, games chosen as well as the didactic context are considered as influencing factors.

Operationalisation of parameters

There is little doubt about the theoretical expected learning effectiveness of digital games (van Eck, 2007; Kerres, Bormann & Vervenne, 2009; Gee, 2009; Wouters et al. 2013). It is generally assumed that there are positive educational benefits of gaming technologies. However, the empirical evidence paints a mixed picture of the learning outcome of digital game-based learning (Ke, 2009): The few existing findings contradict each other and can hardly be summarised. The links between learning contents, motivation and learning effects also remain largely unclear (de Freitas, 2006).

Overall, the evidence shows that different success criteria and methods are chosen depending on the subject and underlying learning theory. Still, the synopsis of relevant empirical studies (Ke, 2009; Virvou et al., 2005; Motyka, 2018; Hawlitschek, 2013; Mitgutsch & Wagner, 2009; EUN: Games in schools, 2009; Hoblitz, 2015) suggests that the learners' gender, age, digital gaming experience as well as their interest in the topic and general academic performance may influence learning outcomes in DGBL settings. With regard to the didactic framing, the introduction phase, including the clarification of objectives, rules and tasks, the social form and the debriefing seem relevant.

Design of an IPO model

Just like all learning processes, digital game-based learning is subject to a variety of influencing factors (Motyka, 2018, p. 83). According to Helmke (2004), a rough distinction can be made between input, process and outcome variables. The figure below shows an overview model based on the derivation and operationalisation of parameters.

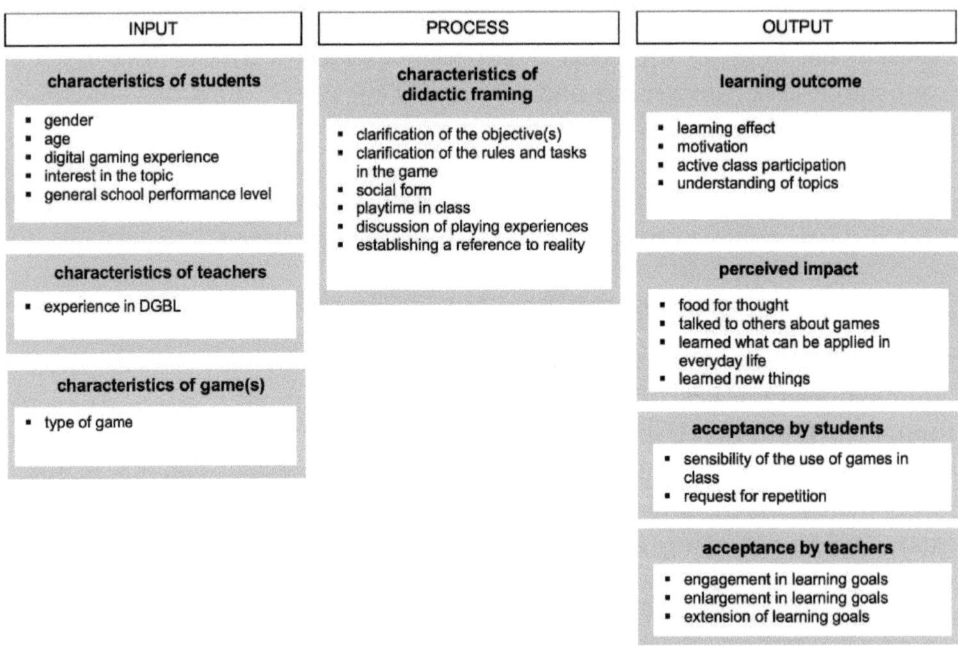

Figure 1. IPO model of digital game-based learning (Hau, 2019)

Empirical study

To test the IPO model, 13 case studies (DGBL projects) were conducted between November 2017 and October 2018. A total of 135 pupils aged between 9 and 23 years took part in the field study.

Overall, it should be noted that the teaching projects were based on the suggestions and ideas of the individual teachers, who were coached and accompanied individually and/or in group meetings. As a result, the project covered different subjects and used different types of games, including serious and commercial off-the-shelf (COTS) games, gamification and game creation applications (e.g. Bloxels). In addition, the social form applied as well as the duration of the project varied.

Table 1. Overview of the empirical study (Hau. 2019)

Nr.	Subject	Game(s)	Type of game	Social form	Duration TU*	Students
01	Communication	Fake it to make it	*Serious Game*	Partner	6 h	7
02	Social studies	Minecraft	*COTS-Game*	Group	4 h	13
03	Media studies	Minecraft	*COTS-Game*	Individual	3 h	11
04	Media studies	Bloxels	*Game Creation*	Individual	3 h	9
		SketchNation	*Game Creation*			
05	Sports	Vier gewinnt	*COTS-Game*	Group	1 h	11
06	Sports	Zombie Run	*Gamification*	Individual	5 h	1
07	Sports	Memory	*COTS-Game*	Group	1 h	10
08	Career guidance	Sims Mobile	*COTS-Game*	Individual	4 h	3
09	Mathematics	Dragon Box Elements	*Serious Game*	Individual	6 h	13
10	German	Lernabenteuer Deutsch	*Serious Game*	Individual	2 h	6
11	English	Wizadora	*Serious Game*	Individual	3 h	21
12	German	A normal lost phone	*COTS-Game*	Group	4 h	19
13	Cross-disciplinary	Kerbal Space Programm	*Serious Game*	Group	7 h	11
		Space Invaders	*COTS-Game*			
		Sam the Sumbot	*Serious Game*			
		Elegy for a dead world	*Serious Game*			
		Star Walk Kids	*Serious Game*			
TOTAL						135

* TU = teaching unit

Evaluation and data analysis

In keeping with the explorative nature of the present study, which intends to provide initial indications of correlations between relevant parameters and learning-relevant outcomes, a method triangulation was applied.

Quantitative data from the questionnaire survey

In order to obtain an overview of the total sample as well as of the individual DGBL projects, the data available were first analysed using descriptive methods. For this purpose, the respective sample size as well as the effective and percentage distributions were given. In addition, location and dispersion parameters were calculated to characterise the distribution of values.

In a second step, the input, process and output variables were examined for correlations. These were determined with the help of a two-dimensional chi-square test, a statistical significance test. In this case, the experimental design is in the form of cross-tables. In terms of the research focus, the percentages within the characteristics can provide important information on potential group differences (e.g. girls vs. boys). In the context of interpretation, the strength and direction of the observed correlations are of particular interest. Here, so-called contingency or correlation coefficients serve as statistical indicators. The usability or interpretability of these coefficients depends on the scale level of the variables: Since mainly nominal data are available, we speak of contingency. In the case of two nominal variables, the contingency coefficient Cramer's V was usually considered, and in the case of two dichotomous nominal data, the Phi coefficient (φ).

Quantitative data from the Triple E-framework

The teachers' reflection of the DGBL projects was recorded with the help of the Triple E-framework. It examines whether a meaningful use of technology (digital games in this case) in class is given. Positive evaluations result in a higher point value concerning the relationship between technology, didactic setting and learning objectives, and negative evaluations result in a lower point value.

In a first step, this analysis was performed for every single teacher and in a second step, the average values of all teachers' assessments were calculated in order to analyse in which aspects of the model the use of digital games are perceived as particularly helpful or less helpful.

Qualitative data

Qualitative data were collected during the semi-structured interviews with the participating teachers and during the lesson observations.

The teachers' assessments complement or detail the data on the use of games collected in the Triple E-framework. The lesson observation aimed to gain a general insight into what happens in class when digital games are used. Against this backdrop, an open, non-participatory observation was carried out, based on the following observation criteria:

- Students' activities
- Activities of the teacher
- Atmosphere in the class.

For organisational reasons, the observation was only carried out on a random basis and by one observer. In some cases, video recordings are available, which enable a retrospective analysis. The observations and comments of the pupils and teachers were taken into account in the analysis at an appropriate point ("didactic framing").

FINDINGS

How do the learners assess the effects of digital game-based learning?

The results of the descriptive evaluation as well as of the contingency analysis show that the use of digital games in lessons is positively evaluated and accepted by the students.

Digital game-based learning is found to be particularly beneficial in the areas of "understanding topics" and "motivation". Abstract topics become tangible through interaction in the game, and the game experiences help set "memory anchors" for topics and content. In terms of motivation, especially quieter pupils participate actively in the lessons. Overall, a high willingness – compared to other teaching methods – to do voluntary homework in the form of continuing to play or watching corresponding tutorials can be observed.

However, only 65% of the respondents reported a learning effect. This may seem surprising at first glance, but it is partly due to the different ways of interpreting the question and the survey methodology: For example, the debriefings revealed that some students interpreted the question concerning learning effects in terms of "learning for the game" (e.g. control, navigation, tasks in the game). This leads to a rather negative assessment of the learning effect, especially among frequent players. In addition, the survey methodology of the present study is based on subjective assessments collected by means of questionnaires. There was no objective recording in the sense of knowledge tests or transfer tasks. This should be investigated in more detail in further studies.

In this context, it is very interesting to note that despite the relatively low learning effects, more than 80% of the pupils consider the use of computer games in class to be useful and would like to see it repeated. In conjunction with the other observed effects, these high approval ratings may indicate that the use of digital games is also perceived and appreciated as a methodological change from "normal lessons", in addition to the reported learning outcomes.

How do the teachers assess the effects of digital game-based learning?

All teachers assessed the results of the DGBL project as positive or very positive. This assessment is in line with the results of the Triple E-framework, which showed a strong relationship between learning objectives, didactic setting and the use of digital games.

The teachers see a great potential in games to increase students' engagement: Teachers believe that digital games motivate the students to start the learning process and concentrate on the task or topic (time on task).

Furthermore, the teachers emphasise the positive effect on social competences (especially in partner or group work), as well as the willingness of some pupils to continue working or playing voluntarily at home.

However, the participants are less convinced of the ability of digital games to create a connection between school learning and the students' living environments. In the eyes of most participants, this does not happen automatically through the game; instead, it is essential to the didactic framing of the use of the game. They also tend to disagree with the statement "Digital games provide ways for students to demonstrate their understanding of learning objectives that are not possible in traditional methods". This is also reflected in the fact that the evaluation of the learning objectives in all teaching projects took place outside the game (e.g. vocabulary test with pictures from the game).

What features can influence the effects of digital games in the classroom?

The findings indicate that the age of the students, their interest in the topic, the type of games used, the play time in class and the establishment of connections with reality are significant factors influencing the learning outcomes.

With regard to age, it is interesting to note that young pupils (9-12 years) in particular report a learning effect. Among other reasons, this could be due to the overall high fascination for digital games in this age category. It is also striking here that the 16 to 19-year-olds show the highest approval ratings with regard to the desire to repeat the activity. This suggests that they appreciate the methodical change from "traditional teaching".

As expected, there is a strong correlation between the students' interest in the topic and the subjective learning effect and motivation. The moderate correlation between the interest in the topic and the assessment of the meaningfulness of games could indicate that overall, the interest of the students is a central factor in the assessment of (game-based) learning. It could be concluded that pupils interested in the topic evaluate the lessons as meaningful and that the use of games only plays a subordinate role. However, further comparative studies are necessary to examine these correlations more closely.

The type of game used is of great importance. It is particularly noticeable that the use of serious games received higher approval ratings in almost all the output variables mentioned. Only for the "thought-provoking" variable do commercial entertainment titles achieve higher values.

The "play time in class" parameter also plays an important role in the DGBL setting. However, no concrete time recommendation can be derived from the available data. Rather, decisions regarding the appropriate time approach seems to depend on the learning objectives, learner groups and the games used.

Debriefing plays a key role in connection with learning outcomes. Especially the creation of connections with reality or the topic is connected with higher values for learning effect, motivation and class participation. This implies that sufficient time should be provided for this phase of the lesson.

CONCLUSION

The aim of the present study was to collect and analyse experiences in the implementation of digital game-based learning against the backdrop of contemporary pedagogy that combines play-fun and learning-seriousness in the sense of positive teaching-learning experiences.

A significant finding is that numerous characteristics of the learners, the games and the didactic framing can influence the learning outcome in DGBL settings. Empirical evidence shows that especially the age of the learners, the type of games used, the time spent playing the game in class and the creation of reality references (debriefing) matter. Thus, it becomes clear that the interrelationships in digital game-based learning settings are much more complex than sometimes presented in the theoretical models.

In summary, the empirical findings of the present study suggest:

- Digital games can promote motivation, understanding and time on task.
- Learning through digital games is a social and active experience.

- The acceptance of digital games among students and teachers is high.
- Learning outcomes from digital games are complex and not easy to assess.
- Several parameters seem to be relevant: the age of the students, type of games used, playtime in class, interest in the topic and transfer from the digital to the real world (debriefing).

Still, it must be taken into account that generalising statements in digital game-based learning are difficult due to the interactivity of the medium. Thus, the present study only collected short-term, subjective effects. The findings offer many starting points for future research projects, especially in the area of detailed analysis of influencing factors and their interrelationships, comparative research on didactic scenarios and the measurement of learning outcomes.

FUTURE RESEARCH

Digital game-based learning in all its manifestations and didactic settings is a complex field of study. Overall, the current findings can be described as deficient; moreover, the existing studies do not paint a uniform picture of the influencing factors and learning effects (e.g. measurement of subjective versus objective learning effects). Since the potential of generalisation of empirical findings on the effect of digital games is in doubt, comparative and overview studies are required. Another deficit of the current state of research is that almost exclusively short-term effects have been proven (Sherry & Dibble, 2009) and convincing long-term studies are still lacking.

This opens up a variety of starting points for future research, ranging from the detailed analysis of possible influencing factors to concrete didactic scenarios and the measurement of learning outcomes. The topic of "stealth assessment" or the integration of test questions into the game environment were not considered in the present study. However, this topic is certainly relevant to practice, as evaluation is an integral part of teaching and learning. In that sense, further practice-oriented research projects on methods of evaluation in and with games should be carried out.

REFERENCES

Boyle et al., 2016. An update to the systematic literature review of empirical evidence of the impacts and outcomes of computer games and serious games. In: Computers & Education, Volume 94, 03/2016, 178-192.

De Freitas, S. & Oliver, M. (2006). How can exploratory learning with games and simulations within the curriculum be most effectively evaluated? Computers & Education, 46(3), 249-264.

De Freitas, S. (2006): Learning in immersive worlds: a review of game-based learning. Retrieved from URL: https://www.webarchive.org.uk/wayback/archive/20140615100504/http://www.jisc.ac.uk/media/documents/programmes/elearninginnovation/gamingreport_v3.pdf [23.03.2021].

Deci, E. & Ryan, R. (1993). Die Selbstbestimmungstheorie der Motivation und ihre Bedeutung für die Pädagogik. In: Zeitschrift für Pädagogik 39 (1993) 2, 223-238.

European Schoolnet (2009). Digitale Spiele im Klassenzimmer. Ein Handbuch für LehrerInnen. Retrieved from URL: http://games.eun.org/upload/GIS_HANDBOOK_DE.pdf [1.03.2021]

Fromme, J., Biermann, R. & Unger, A. (2010). „Serious Games" oder „taking games seriously"? In: Hugger, Kai-Uwe/Walber, Markus (Eds.): Digitale Lernwelten. Konzepte, Beispiele und Perspektiven. Wiesbaden: VS Verlag für Sozialwissenschaften, 40-57.

Ganguin, S. (2010). Computerspiele und lebenslanges Lernen. Eine Synthese von Gegensätzen. Wiesbaden: VS Verlag für Sozialwissenschaften.

Garris, R.; Ahlers, R. & Driskell, J. E. (2002): Games, Motivation and Learning. Research and Practice Model. In: Simulation & Gaming, 33. Jg., H. 4: Newbury Park (USA): Sage Publications, 441-467.

Gee, J.P. (2009). What video games have to teach us about learning and literacy. New York: Palgrave Macmillan.

Gentile, D.; Groves, C. & Gentile, J. (2014). The General Learning Model: Unveiling the Teaching Potential of Video Games. In: Learning by Playing – video games in education. Oxford University Press, New York, 2014, 121ff.

Hau, D. (2019). Didaktisch gestützter Einsatz digitaler Spiele im Schulunterricht - eine explorative Feldstudie im luxemburgischen Bildungssystem. Master thesis. Donau Universität Krems, Austria.

Helmke, A. (2004). Unterrichtsqualität: Erfassen, Bewerten, Verbessern. Seelze: Kallmeyersche Verlagsbuchhandlung.

Hawlitschek, A. (2013). Spielend lernen. Didaktisches Design digitaler Lernspiele zwischen Spielmotivation und Cognitive Load. Berlin: Logos Verlag.

Hoblitz, A. (2014). Spielend lernen im Flow – Die motivationale Wirkung von Serious Games im Schulunterricht. Wiesbaden: Springer.

Ke, F. (2008). Computer games application within alternative classroom goal structures: cognitive, metacognitive and affective evaluation. In: Educational technology & Research & Development 56(5/6), 539-556.

Kearney, P., & Pivec, M. (2007). Recursive loops of game based learning. In: Proceedings of World Conference on Educational Multimedia, Hypermedia and telecommunications 2007. Chesapeake, Vancouver BC: AACE.

Kerres, M.; Bormann, M.; Vervenne, M. (2009). Didaktische Konzeption von Serious Games: Zur Verknüpfung von Spiel- und Lernangeboten. In: Medienpädagogik - Zeitschrift für Theorie und Praxis der Medienbildung. Themenheft Nr. 15/16 (2011): Computerspiele und Videogames in formellen und informellen Bildungskontexten. Retrieved from URL: http://www.medienpaed.com/article/ viewFile/194/194 [28.03.2021]

Kiili, K. (2005). Digital game-based learning: Towards an experiential gaming model. In: Internet and Higher Education, 8(1), 13-24.

Kolb, D. A. (1984). Experiential learning: Experience as the source of learning and development (Vol. 1). Englewood Cliffs, NJ: Prentice-Hall.

Le, S. & Weber, P. (2011). Game-based learning. In: M. Ebner & S. Schön (Eds.), Lehrbuch für Lernen und Lehren mit Technologien (S. 219-228). Norderstedt: Books On Demand.

Mitgutsch, K. & Rosenstingl, H. (2008). Faszination Computerspielen: Theorie - Kultur – Erleben. New academic press.

Motyka, M. (2018). Digitales, spielbasiertes Lernen im Politikunterricht - Der Einsatz von Computerspielen in der Sekundarstufe. Wiesbaden: VS Verlag für Sozialwissenschaften.

Rigby, S., & Ryan, R. M. (2011). New directions in media. Glued to games: How video games draw us in and hold us spellbound. Santa Barbara, CA, US: Praeger/ABC-CLIO.

Sherry, J. L., & Dibble, J. L. (2009). The Impact of Serious Games on Childhood Development. In: U. Ritterfeld, M. Cody & P. Vorderer (Eds.), Serious Games: Mechanisms and Effects (145-166). New York/London: Routledge.

Van Eck, R. (2006). Digital Game-Based Learning: It's Not Just the Digital Natives Who Are Restless. In: EDUCAUSE Review, Vol. 41, Nr. 2. 16–30. Retrieved from URL: https://www.researchgate.net/profile/Richard_Van_Eck/publication/242513283_ Digital_Game_Based_LEARNING_It's_Not_Just_the_Digital_Natives_Who_Are_Restless/lin ks/0a85e53cd61cf43e29000000.pdf [9.03.2021]

Van Eck, R. (2010). Interdisciplinary models and tools for serious games. Emerging concepts and future directions. Hershey, PA: Information Science Reference.

Van Staalduinen, J.P. & de Freitas, S. (2011). A Game-Based Learning Framework: Linking Game Design and Learning Outcomes. In: Myint Swe Khine (Ed.): Learning to play, exploring the future of education with video games. New York u.a.: Peter Lang, 29-54.

Virvou, M; Katsionis, G. & Manos, K. (2005). Combining Software Games with education: Evaluation of its Educational Effectiveness. In: Educational Technology & Society 8(2), 54-65

Wagner, M. & Mitgutsch, K. (2010). Endbericht des Projektes „Didaktische Szenarien des Digital Game Based Learning". Retrieved from URL: https://www.donau-uni.ac.at/imperia/md/content/department/imb/acgs/endbericht_dsdgbl.pdf [15.02.2021]

Wrzesien, M; Raya, M. (2010). Learning in serious virtual worlds: Evaluation of learning effectiveness and appeal to students in the E-Junior project. In: Computers & Education 55 (2010). 178–187. Retrieved from URL: https://pdfs.semanticscholar.org/095b/1e4806a1355680a5c4d9c9b4c792194859c7.pdf [20.03.2021]

Watson, W.; Mong, C.; Harris, C. (2011). A case study of the in-class use of a video game for teaching high school History. In: Computers & Education 56 (2011). 466–474. Retrieved from URL: https://s3.amazonaws.com/academia.edu.documents/32623408/A_case_study_of_the_in_class_use_of_a_video_game_for_teaching_high_school_history.pdf?AWSAccessKeyId=AKIAIWOWYYGZ2Y53UL3A&Expires=1514893919&Signature=%2F7VB7tELcBruD1WGDBy8UgI5jrc%3D&response-content-disposition=inline%3B%20filename%3DA_case_study_of_the _in-class_use_of_a_vi.pdf [12.03.2021]

ADDRESSING VIDEO GAME CULTURE PHENOMENA IN EDUCATION - OPPORTUNITY, CHALLENGE AND NECESSITY

Natalie Denk, Alexander Pfeiffer, Thomas Wernbacher

As the majority of people in Austria and all over Europe play video games - regardless of gender, age, social or cultural background - video games have become an important part of today's society. Around this fascination, a "video game culture" has developed. However, in this context, it is not only the mere consumption of video games that plays a role, but rather the participation in a media landscape shaped by video games. Examples of phenomena that have emerged from video game culture are gameplay video productions (Let's Play videos and streams) and e-sports, the organised competitive playing of video games. In these fields, especially young people participate taking on different roles - for example as fans, spectators, content producers or competitors. A closer look at these activities shows that they on the one hand open up numerous spaces for (informal) learning. One the other hand, they also bring challenges to our society. However, there are hardly any school projects that address the various phenomena and challenges arising from today's video game culture. The article sheds light on the relevance of video game culture activities - beyond gaming - for the educational field and provides insights into two ongoing projects: the research project *StreamIT!* as well as the *E-Sports School League Floridsdorf+* in Vienna with the accompanying web platform *esport-schulliga.at*.

Keywords: video game culture, media education, streaming, e-sports, let's play

---

INTRODUCTION

Building on a common interest, a number of phenomena have emerged in today's video game culture that manifest themselves, for example, in gaming events, e-sports tournaments, online communities, fandoms, or the phenomenon of streaming and gameplay video production. These gaming culture activities can be characterised by active participation in a media landscape shaped by video games; they bring people together online and offline, encourage to engage with new media forms, can enrich our everyday lives, but also pose new challenges to society. While video games have already found their way into classrooms and their potential for education is widely studied and recognised (see e.g. Annetta, 2008; Olson, 2010; Tüzün et al., 2009; König & Pfeiffer, 2020), there are hardly any school projects that address the various phenomena and challenges arising from today's video game culture. Following up on an apparent need for media pedagogical research in this field (Iske & Fromme, 2021), in this article we pose the question of what relevance video game culture activities - beyond gaming - have for the field of education and how we can build on this in pedagogical work.

Exploring video game culture is an extensive, multi-faceted task that many authors have dedicated themselves to (see e.g. Shaw, 2010; Muriel & Crawford, 2018; Horban et al., 2019). These efforts turn out to be complex, shed light on the subject from different perspectives, are based on various interests and do not lead to a common definition (Elmezeny & Wimmer, 2018). The latter only underlines the heterogeneity of video game culture. In the process of addressing video game culture, we first need to be aware that video game culture is (a) not a new phenomenon; at the latest when people begin to exchange and copy game disks or meet in arcades to play games together (which was the case at the beginning of the 1980s at the latest), one can speak of the dawn of a video game culture. Nowadays video game culture is (b) not a niche phenomenon that only affects a small group of people, but has long since become an essential part of our society. And video game culture is (c) not a static phenomenon either, but is from a Cultural Studies point of view subject to permanent change, is constantly being reshaped, created and in a state of constant debate and flux (Shaw, 2010).

While a comprehensive discussion of video game culture is beyond the scope of this article, in the following section we will highlight characteristic aspects of video game culture and outline their relevance for pedagogical considerations. Next, we provide insights into two new school projects in Austria that embed video game culture activities in a formal educational context: the research project *StreamIT!* and the project around the *E-Sports School League Floridsdorf+* in Vienna along with the platform *esport-schulliga.at.*

ON THE RELEVANCE OF VIDEO GAME CULTURE ACTIVITIES FOR EDUCATION

Video game culture is built on a common interest in video games. If we look at numbers and statistics in Austria (ÖVUS, 2019) or throughout Europe (Isfe, 2020), we can conclude that this interest in video games is of utmost importance for our society, because the majority of the population plays video games - regardless of gender, age, social or cultural background. However, in this context, we need to keep in mind that there is no such thing as *the* videogame. Rather there is a broad variety of games that facilitate a broad variety of experiences. Playing *Candy Crush* (King, 2012) on the subway, putting on sophisticated VR gear to work out with *Beat Saber* (Hyperbolic Magnetism, 2018), getting engaged in lively online discussions to identify the "Imposter" in *Among Us* (InnerSloth, 2018), investing months and years in an MMORPG or immersing in an artistic indie game that enables experiences of life's bigger questions – these are just a few examples of the wide range of experiences video games can enable, and of the broad spectrum of play practices they can entail. The "diversity of them [videogames] creates the complex structure of videogame culture. Being different in so many characteristics and being partially different from games in common culture, videogames have formed special space with norms, rules and traditions, lifestyle and way of thinking, art and literature based on them. The common culture produced videogames; now culture is reflected and reproduced by videogames." (Horban et al., 2019) **If we now intend to address video game culture in an educational context, we need to consider the multi-layered nature of video game culture itself as well as the interplay between video games and society. This, in turn, brings with it a variety of opportunities to engage with video game culture in an educational setting.**

Considering the multi-faceted nature of the medium of video games, it is not surprising that so many people are enthusiastic about them, resulting in a highly diverse player base. However, the ever-growing video game culture not only holds potential and new opportunities, but also challenges for society. One of these challenges is closely linked to the question of who can find easy access to gaming culture activities and who faces entrance barriers. Looking at the phenomena of video game culture and the practices that have developed around the enthusiasm for video games, one can observe that playing video games is not a "free ticket" to be part of the community. As a significant example, we want to highlight issues concerning gender-related exclusion mechanisms. Even though statistics show that the fundamental interest in video games is similarly high across all genders (see e.g., ÖVUS, 2018; Isfe, 2020), it is evident that the "playground" of video game culture is strongly male-dominated. There are for example still less women than men engaging in the field of Streaming (StreamScheme, 2021) and in e-sports tournaments female players are still the rarest of exceptions (Scheer, 2019). The reasons for this imbalance lie primarily in gender-specific socialisation and the perpetuation of stereotypical gender role clichés, which lead to gender-specific harassment, sexual assault and discrimination being a reality in gaming culture (Eckers, 2010; Gildemeiser, 2008; Disalvo, 2016; Consalvo, 2012; Todd, 2015). **By addressing people at a young age and creating a pedagogically guided framework to meet these challenges, school projects have great potential to make a long-term difference towards a gender-inclusive and -diverse video game culture.**

Another gaming culture related challenge we need to consider within an educational context, is the fact that children and young people increasingly consider a career in the gaming scene for themselves (Banyai et al., 2020; The Lego Group, 2019). Becoming a YouTuber, Streamer or Gaming Influencer is a very common career aspiration - not surprisingly, as these professions are omnipresent in the everyday lives of children and young people. However, as with many career aspirations, these professions are often associated with unrealistic expectations. **Therefore, it is all the more important that gaming professions are addressed in the context of career orientation or in topic-specific projects at school.**

While video game culture has emerged from the popularity of video games, this does not mean that being part of the video game culture has to involve excessive video game consumption or that the act of gaming is always in the focus (Horban et al., 2019). In this context, we need to consider that a gaming community is also present when the game is over; it is formed with a common interest, can be formed in the process of play but goes way beyond playing. This applies not only in the context of video games. Johan Huizinga already noticed that the shared experience of playing can generate a feeling of community - even beyond the actual act of playing: "A play-community generally tends to become permanent even after the game is over. Of course, not every game of marbles or every bridge-party leads to the founding of a club. But the feeling of being 'apart together' in an exceptional situation, of sharing something important, of mutually withdrawing from the rest of the world and rejecting the usual norms, retains its magic beyond the duration of the individual game." (Huizinga, 1950) **In educational work, we can build on this feeling of community, which stems from a common interest, and benefit from its motivating character. And we have to accept that even if we ban video games from the classroom, video game culture is always present.**

Last but not least, video game culture opens up numerous spaces for learning, not only in the process of gaming itself, but also in taking part in game-related activities beyond gaming. In these activities people may become content creators, who make their own "Let's Play" videos, stream their game play online or produce other game related content (e.g. fan art, mods or stories). They often act as experts on games or game-related topics, exchanging and sharing their knowledge and news in everyday conversations or online, in forums, social media platforms or gameplay chats. They read game magazines or even write articles and game reviews themselves. They may participate in e-sports tournaments or other gaming events - as spectators or in more active roles such as organisers, managers, commentators, or competitors. In all these activities, learning takes place; skills and competences are acquired, improved, and tested. Typically, in this context, learning happens mostly in an informal way. As a key characteristic of informal learning, we can identify that it happens incidentally, is usually not goal-oriented (Europäische Kommission, 2001), is based on experience and often arises from dealing with a challenging situation as part of everyday life (Iske & Fromme, 2021). According to Jenkins et al. (2006) participatory media cultures such as the video game culture can also be referred to as "informal learning communities", which amongst other things refer to communal social practices, to common interests and goals of (informal) learning (Iske & Fromme, 2021). When we look at informal learning spaces of video game culture in an educational context, we need to focus on two questions in particular: **How can we facilitate these learning experiences, which usually originate from intrinsic motivation, in a formal setting? And how can we initiate a reflection on these learning experiences? This is where we can speak of a "didactics of informal learning"** (see e.g. Arnold, 2015; Zürcher 2007).

PRACTICAL PROJECTS

To illustrate how gaming culture activities can be addressed in educational projects, the basic concepts of two practical projects are described in the following sections.

The E-Sports School League Floridsdorf+ and the platform esport-schulliga.at

In 2019, Austria's first E-Sports School League was launched. Thanks to the support of the district council and school administration, the project was piloted together with 12 secondary schools and several youth centers in the district of Floridsdorf in Vienna. The number of registrations showed great interest: 49 teams with 148 girls and 146 boys, aged 10 - 14, took part in the tournament. They played the games *Rocket League* (Psyonix, 2019), *Overcooked 2* (Team17, 2018) and *FIFA19* (Electronic Arts, 2018) on PlayStation 4. Each team consisted of 6 students and had to master all three games. In the end the group ranking determined the winner. One of the goals of the *E-Sports School League Floridsdorf* was to bring the schools and youth centers of the district closer together. Therefore, the organization took place in the school, the training and qualification games were handled in the youth centers in the district. To make this possible, the youth centers were equipped with PlayStations. After an exciting qualification phase in the youth centers, the grand final was held in June 2019 with a supporting programme in the Marco Polo youth center and streamed live via Twitch.

From the organizers' point of view, the *E-Sports School League Floridsdorf 2019* was a great success. Not only did we succeed in setting up a memorable school project, but the social skills of the students were also noticeably strengthened. This was confirmed by the conversations with the students and involved educators. Thus, new friendships were formed through the participation in the League (even beyond the students' own school) and teamwork and communication skills as well as problem- and conflict-solving skills were proven. Thanks to the introduced girls' quota (at least three of the six team members had to be female), the project achieved something that is unfortunately a rarity in conventional e-sports: there were only mixed-gender teams and girls became visible as gamers. Ultimately, the students were able to bring a piece of their everyday lives into the school environment. (see Denk, 2021)

Following on from this success and further research into the subject area (see e.g., Wimmer et al. 2021; Denk, 2020; Pfeiffer et al. 2020), in September 2021, the *E-Sports School League Floridsdorf+* will take place, with a pre-season in June 2021. In the process, the organizing team will benefit from the experience and knowledge gained from the pilot phase in 2019.

Figure 1. The logo of the project centered around the E-Sports School League Floridsdorf+

Accompanying the planning of the *E-Sports School League Floridsdorf+*, a web platform was also developed (https://esport-schulliga.at), which not only contains updates on the League, but is also intended to be a central hub for all e-sports projects in German-speaking countries. The platform contains beginner-friendly articles about e-sports, tips for students and educators on the topic, as well as suggestions and teaching ideas for the implementation of own e-sports projects in education. Furthermore, educators are explicitly invited to participate as guest authors and to report on their own experiences or share teaching material about e-sports.

With this approach, we hope that the potential of e-sports projects in education will become more evident, that further e-sports projects in the educational sector will be launched and that a professional pedagogical discourse on the topic will be stimulated.

In the context of the accompanying research of the *E-Sports School League Floridsdorf+*, the acceptance, effect as well as possible improvement possibilities are to be evaluated with a mixed methods approach, both qualitatively and quantitatively. In the research process, the participating students, teachers and representatives of the youth centers will be involved.

The school project StreamIT!

In the project *StreamIT!* a participative educational concept will be developed that focuses on the active creation of gameplay videos (Let's Play videos and live streams). With reference to previous projects and studies (e.g., see Burwell & Miller, 2016; Smith & Sanchez, 2015; Vasilchenko et al., 2017; Orus et al., 2016), it can be assumed that producing gameplay videos can foster a range of competencies at the same time as building on students' intrinsic motivation. This concerns not only technical skills in handling the soft- and hardware, but also creativity, social-communicative skills, and competences in the field of media literacy. The project *StreamIT!* will investigate how the creation of gameplay videos can be integrated into a teaching context - addressing different age groups from 8 to 18 - and which learning goals can be achieved within a didactic framework.

The theoretical foundation of the project is the constructivist approach, which, in contrast to frontal teaching, focuses on self-directed and situated learning. According to the initial concept of *StreamIT!*, students will work directly in class with streaming and game recording tools. The tools and game platforms to be used for this purpose are still being worked out as part of the project. In particular, the technical possibilities in schools, which are still very heterogeneous in Austria, must be taken into account. In the context of a lifeworld orientation, the selection of the games used for this purpose is to be made by the students themselves (taking age ratings into account). The students will work in teams on the video productions with the aim of triggering social dynamics.

In addition, special consideration is given to gender aspects throughout the entire project development, and the project activities are implemented in a gender-inclusive manner. This involves developing strategies to encourage girls to actively participate in the usually strongly male-dominated scene surrounding the creation of gameplay-related content. Another focus of the project is career orientation with the aim to create a realistic image of today's extremely popular "gaming professions" (such as YouTuber or Streamer). Here we work closely with role models from the gaming scene, with a focus on filling the gap of female role models.

The *StreamIT!* teaching concept will be developed based on a literature review and needs analysis that provides insights into the interests of students and teachers as well as general conditions in the school. Following an iterative approach, the teaching concept will be tested in a hands-on-phase in autumn 2021 in five partner schools in Austria with students aged 8 to 18. The final teaching concept will be made available as an open educational resource via the project website (https://stream-it-talente.at).

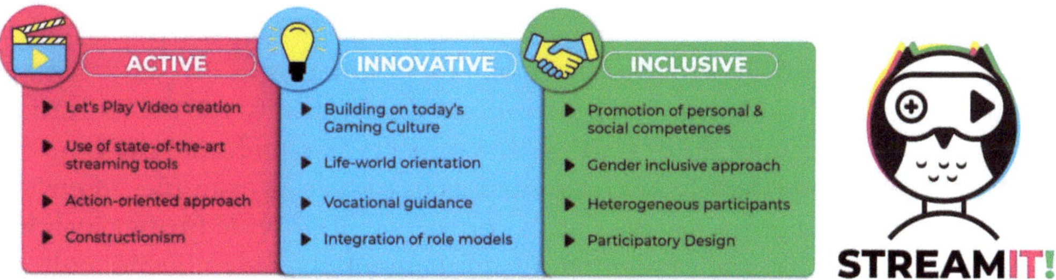

Figure 2. The framework of the project StreamIT!

CONCLUSION AND OUTLOOK

The examination of video game culture and the resulting phenomena is a necessary first step towards taking up its potentials and challenges in educational work. Considering the multi-layered character of video game culture, which is always in a state of flux, there is an ongoing need for research in media education. A first - and certainly not yet comprehensive - pedagogical analytical look at characteristic features of video game culture in this article has shown in summary the following findings:

Video game culture is of utmost importance for our society and plays an important role in the everyday life of many children and young people. Therefore, it is high time to pay attention to these developments in the field of education. The majority of people play video games, but this does not mean that everyone has easy access to video game culture phenomena. To preventively address social challenges in this context, such as gender issues, it is important to reach out to children from an early age. The video game culture has given rise to numerous new career profiles that are desirable for many young people (e.g., YouTuber, Streamer or Gaming Influencer). Therefore, this needs to be addressed in the context of career guidance at school. Video game culture opens spaces for informal learning, which children and young people usually turn to with joy and intrinsic motivation. These learning spaces can also be opened in a formal learning setting, such as school. In a pedagogical framework, reflection processes on the acquired knowledge and competences can be initiated with appropriate methods.

A central task of media education is now to develop suitable methods to address video game culture in an educational framework, to use potentials and to face the associated challenges. Considering the importance of video game culture for our society and especially for young people, this is - from the authors' point of view - not only an option or opportunity for the education system, but a necessity. In this article, the projects *StreamIT!* and the *E-Sports School League Floridsdorf+* were presented as examples as examples of efforts in this direction. Future research in the projects will provide further insights into the feasibility and potential of implementing video game culture activities in an educational setting. In this process, we hope to encourage others to set up educational projects that engage with video game culture beyond gaming activities.

ACKNOWLEDGMENTS

Innovative pedagogical projects can only emerge from a motivated and professional team and the necessary financial support. We would like to take this opportunity to thank the entire team around the *E-Sports School League Floridsdorf+* (https://www.esport-schulliga.at/team/) as well as their supporters who made the project possible (https://www.esport-schulliga.at/unterstuetzer_innen/). Furthermore, we greatly acknowledge the financial support for the project *StreamIT!* from the Austrian Federal Ministry for Climate Action, Environment, Energy, Mobility, Innovation and Technology (BMK) within the Call "Talente regional 2019" (Project number: 878696 - "StreamIT!: Streaming als pädagogisches Tool zur Interessenvermittlung & zum Kompetenzerwerb"). And we would like to thank all project team members, our partner schools and participating role models and advisors of the project *StreamIT!* (https://www.stream-it-talente.at/ueber-streamit/).

REFERENCES

Annetta, L. A. (2008). Video Games in Education: Why They Should Be Used and How They Are Being Used. Theory Into Practice, 47(3), 229–239. https://doi.org/10.1080/00405840802153940

Arnold, R. (2015). „Didaktik" informellen Lernens. In M. Rohs (Ed.), Handbuch Informelles Lernen (pp. 1–8). Springer Fachmedien Wiesbaden. https://doi.org/10.1007/978-3-658-06174-6_33-1

Arnold, R. (2015). „Didaktik" informellen Lernens. In M. Rohs (Ed.), Handbuch Informelles Lernen (pp. 1–8). Springer Fachmedien Wiesbaden. https://doi.org/10.1007/978-3-658-06174-6_33-1

Bányai, F., Zsila, Á., Griffiths, M. D., Demetrovics, Z., & Király, O. (2020). Career as a Professional Gamer: Gaming Motives as Predictors of Career Plans to Become a Professional Esport Player. Frontiers in Psychology, 11, 1866. https://doi.org/10.3389/fpsyg.2020.01866

Burwell, C., & Miller, T. (2016). Let's Play: Exploring literacy practices in an emerging videogame paratext. E-Learning and Digital Media, 13(3–4), 109–125. https://doi.org/10.1177/2042753016677858

Consalvo, M. (2012, November 11). Confronting Toxic Gamer Culture: A Challenge for Feminist Game Studies Scholars. Ada New Media. Retrieved from https://adanewmedia.org/2012/11/issue1-consalvo/

Daniel, M., & Garry, C. (2018). Video games as culture: considering the role and importance of video games in contemporary society. Routledge.

Denk, N (2021). Das war die E-Sport Schulliga Floridsdorf. Retrieved from https://www.esport-schulliga.at/das-war-die-e-sport-schulliga-floridsdorf-2019/

Denk, N. (2020). E-Sport im Schulalltag?!. In: SchulVerwaltung aktuell Fachzeitschrift für Schulentwicklung und Schulmanagement. Heft 6/2020: 170, Wolters Kluver

Disalvo, B. (2016). Gaming Masculinity: Constructing Masculinity with Video Games. In Kafai, Y. B.. (Eds.). Diversifying Barbie and Mortal Kombat: Intersectional Perspectives and Inclusive Designs in Gaming (Version 1). (pp.105–117). Carnegie Mellon University.

Eckes, T. (2010). Geschlechterstereotype. In R. Becker & B. Kortendiek (Eds.), Handbuch Frauen- und Geschlechterforschung: Theorie, Methoden, Empirie (pp. 178–189). VS Verlag für Sozialwissenschaften. https://doi.org/10.1007/978-3-531-92041-2_21

Elmezeny, A., & Wimmer, J. (2018). Games without Frontiers: A Framework for Analyzing Digital Game Cultures Comparatively. Media and Communication, 6, 80–89. https://doi.org/10.17645/mac.v6i2.1330

Europäische Kommission (2001). Einen europäischen Raum des lebenslangen Lernens schaffen. Mitteilung der Kommission vom 21. November 2001. Retrieved from https://ec.europa.eu/transparency/documents-register/detail?ref=COM(2001)678

Gildemeister, R. (2008). Doing Gender: Soziale Praktiken der Geschlechterunterscheidung. In

R. Becker & B. Kortendiek (Eds.), Handbuch Frauen- und Geschlechterforschung: Theorie, Methoden, Empirie (pp. 137–145). VS Verlag für Sozialwissenschaften. https://doi.org/10.1007/978-3-531-91972-0_17

Horban, O., Martych, R., & Maletska, M. (2019). Phenomenon of Videogame Culture in Modern Society. Studia Warmińskie, 56, 123-135. https://doi.org/10.31648/sw.4314

Huizinga, J. (1950). Homo Ludens: a Study of the Play Element in Culture. New York, NY: Roy Publishers.

ISFE - Interactive Software Federation of Europe (2020). Key Facts 2020. Retrieved from https://www.isfe.eu/wp-content/uploads/2019/08/ISFE-Key-Facts-Brochure-FINAL.pdf

Iske, S. and Fromme J. (2021). Informelles Lernen mit digitalen Medien, In Sander, U. et al. (eds) Handbuch Medienpädagogik. Neuauflage. Wiesbaden, Springer.

Koenig, N., & Pfeiffer, A. (2020). Outputs and Insights From 12 Years of Game-Based Learning Research at the Danube-University Krems' Center for Applied Game Studies. https://doi.org/10.1184/R1/11926884.v1

Olson, C. K. (2010). Children's Motivations for Video Game Play in the Context of Normal Development. Review of General Psychology, 14(2), 180–187. https://doi.org/10.1037/a0018984

Orús, C., Barlés, M. J., Belanche, D., Casaló, L. V., Fraj, E., & Gurrea, R. (2016). The effects of learner-generated videos for YouTube on learning outcomes and satisfaction. Comput. Educ. https://doi.org/10.1016/j.compedu.2016.01.007

ÖVUS (2019). 5,3 Millionen Österreicher spielen Videospiele. Retrieved from https://www.ovus.at/news/ueber-fuenf-millionen-oesterreicher-spielen-videospiele

Pfeiffer, A., Wernbacher T., Denk, N., Kriglstein, S. (2020): Do You Love It Already or Do You Still Ignore It? The Two Faces of the Phenomenon Spectatorship in Esports. in: Workshop proceedings "Be Part Of It: Spectator Experience in Gaming and eSports" in conjunction with CHI 2020. Retrieved from https://seegamesws.wordpress.com/accepted-papers

Wimmer S., Denk N., Pfeiffer, A., Fleischhacker. M. (2021). ON THE USE OF ESPORTS IN EDUCATIONAL SETTINGS. HOW CAN ESPORTS SERVE TO INCREASE INTEREST IN TRADITIONAL SCHOOL SUBJECTS AND IMPROVE THE ABILITY TO USE 21ST CENTURY SKILLS? INTED2021 Proceedings, 5782–5787. https://doi.org/10.21125/inted.2021.1168

Scheer, Y. (2019). Frauen in der Gaming und E-Sport Szene. In W. Elmenreich, R. R. Schallegger, F. Schniz, S. Gabriel, G. Pölsterl, & W. B. Ruge (Eds.), Savegame: Agency, Design, Engineering (pp. 313–325). Springer Fachmedien Wiesbaden. https://doi.org/10.1007/978-3-658-27395-8_21

Shaw, A. (2010). What Is Video Game Culture? Cultural Studies and Game Studies. Games and Culture, 5(4), 403–424. https://doi.org/10.1177/1555412009360414

Smith, P. A., & Sanchez, A. D. (2015). Let's Play, Video Streams, and the Evolution of New Digital Literacy. In P. Zaphiris & A. Ioannou (Eds.), Learning and Collaboration Technologies (pp. 520–527). Springer International Publishing.

StreamScheme (2021). Twitch Demographic & Growth Statistics 2021. Retrieved from https://www.streamscheme.com/twitch-statistics/

The Lego Group (2019). LEGO Group Kicks Off Global Program To Inspire The Next Generation Of Space Explorers As NASA Celebrates 50 Years Of Moon Landing. Retrieved from https://www.prnewswire.com/news-releases/lego-group-kicks-off-global-program-to-inspire-the-next-generation-of-space-explorers-as-nasa-celebrates-50-years-of-moon-landing-300885423.html

Todd, C. (2015). COMMENTARY: GamerGate and Resistance to the Diversification of Gaming Culture. Retrieved from http://www.wsanz.org.nz/journal/docs/WSJNZ291Todd64-67.pdf

Tüzün, H., Yılmaz-Soylu, M., Karakuş, T., İnal, Y., & Kızılkaya, G. (2009). The effects of computer games on primary school students' achievement and motivation in geography learning. Computers & Education, 52(1), 68–77. https://doi.org/10.1016/j.compedu.2008.06.008

Vasilchenko, A., Green, D. P., Qarabash, H., Preston, A., Bartindale, T., & Balaam, M. (2017). Media Literacy as a By-Product of Collaborative Video Production by CS Students. Proceedings of the 2017 ACM Conference on Innovation and Technology in Computer Science Education, 58–63. https://doi.org/10.1145/3059009.3059047

Zürcher, R. (2007). Informelles Lernen und der Erwerb von Kompetenzen: Theoretische, didaktische und politische Aspekte. Bundesministerium für Unterricht, Kunst und Kultur, Abt. Erwachsenenbildung V/8.

RIDE2PARK

Constantin Kraus, Simon Wimmer, Thomas Wernbacher

When talking about individual mobility the negative effects on the environment and health such as congestion, urban heat, bad air quality and climate relevant emissions are well known and regularly discussed within the political as well as scientific community. Costs related to these factors include certificates for carbon dioxide on the state level as well as costs for infrastructure such as parking space provided by companies. The modal and behavioural shift needed is not on the horizon, in cities such as Judenburg up to 72% of citizens still use the car to get to work or to school. With Ride2Park we offer a gamified solution for solving the climate crisis by motivating employees of medium to large companies to use carpools for getting to work. Drivers and co-drivers can collect rewards as well as can take part in weekly lotteries to win prizes provided by the employer or respective company. Next to principles of gamification our framework also includes nudging as motivational boost. Ride2Park is the first step in our plan to use insights from psychology, game studies and social anthropology to achieve a positive impact on mobility choices of employees. To achieve this goal, we plan to enhance existing business solutions by building an additional layer in terms of design, metrics and incentives. For the quantification of our reward system, we will include the results from established services such as "BlaBlaCar" or "Bike Citizens" (as application for tracking active mobility) on the one hand and user-driven inputs on the other hand.

Keywords: Carpooling, Gamification, Nudging

INTRODUCTION

When it comes to individual mobility, the negative effects on the environment and health, such as congestion on the main traffic routes, poor air quality and climate-relevant emissions, are well known and are now discussed daily in the media, politics and science. Social resonance has also increased massively since 2019 thanks to the climate protests by schoolchildren ("Fridays for Future"). The costs associated with motorised individual mobility include penalties for exceeding threshold values for emissions at the federal level as well as costs for infrastructures such as parking spaces at the municipal level. The necessary behavioural changes or a modal shift are not in sight - in cities like St. Pölten, about 56% of citizens use their car to travel to work or school (bmvit, 2015). Existing carpooling solutions such as "BlaBlaCar" or "Carsharing 24/7" show a saturated market. However, these services primarily offer the possibility of finding communities, taking out insurance and processing payment. In addition to established players on the scene, start-ups such as "Carployee" are also working on solutions, but so far only relies on a rudimentary reward system.

With "Ride2Park" we focus on a comprehensive incentive system that relies on gamification on the one hand and nudging on the other. Drivers and passengers can take part in weekly lotteries, unlock company-specific incentives and redeem discounts at partner companies after a mutual confirmation of the ride. In addition to a playful framework, we also include insights from behavioural economics, so-called "nudging", for which Richard Thaler (2008) is responsible. Thaler received the Nobel Prize in Economics in 2017 for his book "Nudge". Informative as well as reflective "nudges" are used in the form of descriptive pictograms together with short messages to promote a change in behaviour - away from individual mobility and towards sharing mobility. This combination of intrinsic (perception of positive effects) and extrinsic motives (raffles, discounts) to promote carpooling is unique in relation to existing mobility campaigns.

THEORETICAL BACKGROUND

Gamification

The massive distribution and market penetration of digital games (approx. 100 billion euros in sales in 2016 in the USA alone) is impressive. Currently, 510 million people (5 million in Austria) play games, with an average age of 35. Gamification is based on the use of game mechanics in contexts that are, by nature, unrelated to games (Deterding, 2011). Its aim is to set specific desired behavioural impulses. The aim is to apply the motivational and feedback techniques that have been tried and tested in games. Games provide clear goals (Hunicke et al. 2004), they reward (Vorderer et al., 2004), they allow to compete or cooperate with others (Yee, 2006) and they provide an interactive framework for different experiences and skills (Ivory & Kalyanaraman, 2007; Jansz, 2005).

Gamification has already been successfully used in various application areas to promote participation, such as in the context of civic courage (Coronado & Vasquez, 2014), citizen participation (Thiel & Lehner, 2015), e-learning (Barata et al., 2013) and e-government (Al-Yafi & El-Masri, 2016). The application of gamification has also yielded positive results in the mobility sector, e.g. in terms of promoting sustainable forms of mobility (Kazhamiakin, Raman et al., 2015).

Nudging

The term "nudge" or "nudging" originally derives from the field of behavioural economics and describes a soft type of influence, with the goal to elicit a certain behaviour. Thaler and Sunstein (2008) defined nudging as a positive intervention that stimulates a voluntary change in behaviour without including external (negative) consequences. The idea of nudging has been booming in the USA in recent years, whereas very few initiatives are known in Europe, where the focus is generally on the creation of politically motivated interventions, and is particularly anchored in the field of health prevention (Quigley, 2013). Nudging, in itself, is based on motivational psychological models and shows parallels to the principle of gamification. However, in contrast to gamification, stimulus-response chains in the sense of incentives and penalties fade into the background, and subtle strategies and positive interventions for decision optimization come to the fore. Nudging strategies make behavioural alternatives more visible, in the area of physical activity; for example, by making stairs more attractive than a lift (Hollands et al., 2016).

Nudging has also been successfully used in the context of mobility; for example, to promote cycling in cities (Wunsch et al., 2016). The methods used here are similar to the "persuasive technology" principle. In both cases, the aim is to promote certain behaviours or options through targeted measures in terms of the design of places, processes and graphic interfaces (Fogg, 2009). Nudges can further be seen in a similar way to gamification, or as a relevant part of gamification, in the form of the game mechanics of achievements (Pfeiffer et al., 2020; Wernbacher et al., 2020).

METHODOLOGY

Target Group

The main goal should be to target as large a user base as possible Therefore, in a first step, we aim at a B2B model: car park operators, medium to large companies with a (significant) need for parking infrastructure to make room for motorised mobility. Large companies in Austria account for up to 35% of registered companies, 27% are medium to large companies (www.statista.com). The primary target market therefore comprises around 1.5 million employees. In addition to the focus on companies, Park & Ride operators are addressed in the medium term. In the long term, B2Gov market is also relevant, as costs can be saved especially at municipal level (less parking spaces, less infrastructure costs, less emissions = better air quality & general health & lower costs for CO2 certificates...).

In terms of scalability, our approach can be adapted to a wider range of application scenarios, such as the use of public transport or cycling to work in terms of the next expansion stage.

Concept

Our concept includes mechanics derived from nudging, users are offered small nudges, which are supposed to make riding together more attractive. Ride2Park aims to motivate people to ride together more often with game mechanics such as a point-system, customizations and rewards. People who drive together more often can form a carpool and level up with this pool (see figure 1).

Figure 1. Nudging & Gamification Framework

The more people drive together, the higher the chance of winning prizes. Apart from smaller prizes such as a coffee or a public transport ticket or maybe within the company discounts in the mensa, there is also the possibility to participate in a lottery. This lottery also works reverse. In other words, the more often you lose, the higher the chance of winning (see figure 2).

Figure 2. Reward System

AIM OF THE CONCEPT

If greenhouse gas emissions in transport are not drastically reduced, the climate catastrophe can no longer be stopped - this is now common knowledge. However, the consequential costs of individual transport cannot only be limited to greenhouse emissions; the increased material consumption of valuable resources for production (10 tonnes of material per car), the high space requirement in the form of parking spaces and the associated sealing of surfaces as well as the costs for the car owner for operating a motor vehicle are also relevant factors. Assuming that "Ride2Park" motivates a person to take 3 passengers to the workplace instead of driving alone in the car used every day, this has favourable effects on the environment. The environmental effects can be mapped based on the CO2 and NOx savings. The emissions per km for petrol/diesel are 186/179 g CO_2 and 0.75/0.70 g NOx. For 100 km, the savings are thus approx. 13.5 kg CO_2 and 52 g NOx (based on www.co2online.de).

Traffic jams and long waiting times in the car lead to stress, demotivation and, in the worst case, accidents. An increase in the efficiency of car traffic brought about by this project will therefore have a positive effect on commuter traffic as a whole. Standing time, which according to the Ellen MacArthur Foundation is 92%, can thus be decisively minimised. For the target group of companies, the ecological factor can also be used economically, for example through CO2 savings values in employee mobility in the sense of applicable standards for sustainability reporting (e.g. Global Reporting Initiative GRI: www.globalreporting.org).

The chosen research approach should result in the provision of an incentive model that favours a sustainable change in behaviour (increasing the occupancy rate of cars). By rewarding the saved costs in the form of tokens, an image enhancement of carpooling should be perceived as a serious and effective alternative in the fight against complex social but also ecological and economic problems of the 21st century (e.g. influx to conurbations, climate change, increasing emissions and associated fines).

With our case study we aim for:

- the efficient use of car parks & parking spaces.
- savings of municipal costs (less infrastructure costs, less emissions = better air quality & general health & lower costs for CO_2 certificates on federal level).
- freeing up time: co-drivers can use the time in the car for social exchange - so companies can offer a (partial) shift from active (non-working time) to passive travel time (working time).
- rewarding car poolers: special incentives or vouchers can be activated for drivers and passengers. Incentives can range from vouchers for neighbouring restaurants, hotels and cultural institutions to public transport tickets.
- saving costs: Last but not least, both drivers and passengers save costs (maintenance, fuel) through carpooling.

CONCLUSION

With "Ride2Park" we want to offer a convincing solution to this problem by motivating car drivers to use carpooling for work and leisure trips. Drivers and passengers can collect rewards and participate in weekly lotteries to win prizes or redeem discounts. In addition to playful mechanisms, our approach also includes "nudges" as a motivational boost. Ride2Park is the first step in our plan to use insights from game science, psychology and social anthropology to make a positive impact on everyday and leisure mobility.

ACKNOWLEDGMENTS

The team in the form of cultural anthropologist and game designer Simon Wimmer, graphic designer and educator Constantin Kraus and media psychologist Thomas Wernbacher researches and teaches at the Centre for Applied Game Studies at Danube University Krems. The implementation of the idea will be carried out by Dundees GesbR (**www.dundees.at**), based on the economic usability and the detailed specification of a dissemination strategy.

REFERENCES

Al-Yafi, K., & El-Masri, M. (2016). Gamification of e-government services: A discussion of potential transformation.

Barata, G., Gama, S., Jorge, J., & Gonçalves, D. (2013, October). Improving participation and learning with gamification. In Proceedings of the First International Conference on gameful design, research, and applications (pp. 10-17).

bmvit. (2015). Masterplan Gehen. Wien: Bundesministerium für Land- und Forstwirtschaft,Umwelt und Wasserwirtschaft.

Deterding, S., Sicart, M., Nacke, L., O'Hara, K., & Dixon, D. (2011). Gamification. using game-design elements in non-gaming contexts. In CHI'11 extended abstracts on human factors in computing systems (pp. 2425-2428).

Fogg, B. J. (2009, April). A behavior model for persuasive design. In Proceedings of the 4th international Conference on Persuasive Technology (pp. 1-7).

Petrescu, D. C., Hollands, G. J., Couturier, D. L., Ng, Y. L., & Marteau, T. M. (2016). Public acceptability in the UK and USA of nudging to reduce obesity: the example of reducing sugar-sweetened beverages consumption. PLoS One, 11(6), e0155995.

Hunicke, R., LeBlanc, M., & Zubek, R. (2004, July). MDA: A formal approach to game design and game research. In Proceedings of the AAAI Workshop on Challenges in Game AI (Vol. 4, No. 1, p. 1722).

Ivory, J. D., & Kalyanaraman, S. (2007). The effects of technological advancement and violent content in video games on players' feelings of presence, involvement, physiological arousal, and aggression. Journal of Communication, 57(3), 532-555.

Jansz, J. (2005). The emotional appeal of violent video games for adolescent males. Communication Theory, 15(3), 219-241.

Kazhamiakin, R., Marconi, A., Perillo, M., Pistore, M., Valetto, G., Piras, L.& Perri, N. (2015, October). Using gamification to incentivize sustainable urban mobility. In 2015 IEEE first international smart cities conference (ISC2) (pp. 1-6). IEEE.

Leonard, T. C. (2008). Richard H. Thaler, Cass R. Sunstein, Nudge: Improving decisions about health, wealth, and happiness.

Pfeiffer, A., Bezzina, S., König, N., & Kriglstein, S. (2020, October). Beyond Classical Gamification: In-and Around-Game Gamification for Education. In European Conference on e-Learning (pp. 415-XIX). Academic Conferences International Limited. DOI:10.34190/EEL.20.007.

Quigley, M. (2013). Nudging for health: on public policy and designing choice architecture. Medical law review, 21(4), 588-621.

Thiel, S. K., & Lehner, U. (2015, July). Exploring the effects of game elements in m-participation. In Proceedings of the 2015 British HCI Conference (pp. 65-73).

Vorderer, P., Klimmt, C., & Ritterfeld, U. (2004). Enjoyment: At the heart of media entertainment. Communication Theory, 14(4), 388-408.

Wernbacher, T., Pfeiffer, A., Kriglstein, S., & Bezzina, S. (2020, July). The Power of Nudging for Virtual Learning Environments. In Proceedings of EDULEARN20 Conference (Vol. 6, p. 7th).

Wunsch, M., Millonig, A., Seer, S., Schechtner, K., Stibe, A., & Chin, R. C. (2016). Challenged to bike: Assessing the potential impact of gamified cycling initiatives. evaluation, 23, 24.

Yee, N. (2006). Motivations for play in online games. Cyber Psychology & Behavior, 9(6), 772-775.

CYCLE4VALUE

Thomas Wernbacher, Alexander Pfeiffer, Alexander Seewald, Mario Platzer, Constantin Kraus, Simon Wimmer, Dietmar Hofer

Despite a variety of measures to promote cycling, the overall share of bicycle traffic in Austria has changed only slightly in favour of the bicycle in recent years (BMVIT, 2017 and 2013). As part of its mobility strategy, the Austrian government has therefore set itself the goal of doubling the proportion of bicycle traffic in 7 years. In times of COVID a total of 866,263 cyclists were registered by Vienna's automatic counting stations in April 2020. Compared to the same month of the previous year, 20 percent more Viennese have thus taken to the bike. However, the sustainability of this effect is declining, so incentives must be created for a permanent modal shift. In course of the research project "Nice Rides" we aimed at promoting bike commuting in the urban area by developing a gamification framework which involves both cooperative and competitive game elements. In order to reach a meaningful playing experience, we implemented findings from game design and motivational models. Our key goal was to achieve a change in long established behavioural patterns (choosing the car for commuting) by enhancing the safety and attractivity of biking. 10 years later, the peak of gamification has been left behind, incentive tracking apps broadly adopted gamified mechanics as well as incentive systems. With "Cycle4Value" as follow-up project a transparent and low-threshold reward model for the promotion of cycling based on the blockchain technology is being tested. The economic, health and ecological advantages/effects of cycling are converted into a real value (=cycle tokens). These value units are stored in a digital wallet and can be paid for in a marketplace. The research project surpasses conventional incentive systems, since on the one hand the storage of the value units as well as the redemption process is decentralized, tamper-proof and transparent and on the other hand the real economic benefit of active cycling is monetarised.

Keywords: Cycling, Machine Learning, Blockchain, Tokenisation

INTRODUCTION

Despite various measures to promote cycling the overall cycling share in Austria has changed only slightly in favour of cycling in recent years. In contrast, applications around blockchain are booming, and mobile cycling apps including reward systems are also developing rapidly. The combination of these disruptive developments could provide the decisive impetus to promote cycling with all its associated benefits. The presented project is thus fully in line with the Austrian federal government's ambitious goal of doubling the share of cycling in 7 years. Especially in cities, cycling as a climate- and resource-friendly alternative to car traffic makes an important contribution to achieving the ambitious national and international targets. On a collective level, cycling impresses with an excellent eco-balance in terms of hardly any greenhouse gas emissions, air pollutant emissions, noise pollution, land use and minimal energy demand. In addition, by fulfilling requirements from national action plans (BMVIT, 2013, 2017) to increase the share of active mobility of the population, a valuable contribution can be made to health promotion and consequently a budgetary relief for the overburdened health system can be achieved. On an individual level, the bicycle also offers a multitude of advantages to the individual cyclist. It combines the requirements of modern means of transport in that it is fast, comfortable, healthy and inexpensive.

Blockchain applications are experiencing early hype in many sectors such as the energy sector (Aitzhan & Svetinovic, 2018) or banking (Guo & Liang, 2016), with broad market penetration still 5-10 years away. However, only few applications exist in the mobility sector and specifically in cycling. In contrast to the conventional operation of a blockchain such as Bitcoin, "Cycle4Value" does not rely on "proof of work", in the sense of the energy-intensive solving of puzzles or the race to the next block, but on the "proof of stake" variant "Ardor". This essentially means that a Raspberry Pi or even a smartphone is enough to keep the system running. As a third-generation system or application, the Ardor blockchain requires significantly less power and prevents an overload or accumulation of data that is no longer needed (blockchain bloat) through child chains that run separately from the main chain. In general, the blockchain brings many advantages, such as high transparency and security of data as well as decentralised and efficient validation of transactions. Innovative technologies such as the blockchain in combination with incentive systems such as gamification are increasingly popular within the respective communities. However, there are still many unanswered questions - for example, how real, extrinsic rewards should be given out and what their underlying measurement could look like. This is mainly related to the problem of quantifying the costs and benefits of cycling in monetary terms and ensuring the quality of tracking data. Privacy concerns also play a significant role here and provide for scepticism and criticism in the case of sprawling tracking methods and big datasets.

THEORETICAL BACKGROUND

Incentivisation systems in the context of cycling promotion

Nowadays, a wide variety of devices (smartphone apps, heart rate monitors, etc.) connected via the internet can be used to independently and automatically record one's own movement data. In this context, cycling routing apps are also successful in reaching various target groups such as commuters and athletes. From "Komoot" (www.komoot.com) to "Bikemap" (www.bikemap.net) to Google Maps, many different apps offer the possibility to explore one's own city or the green space and nature and to track various parameters such as distance and duration of the activity.

Since 2012 the project partner Bike Citizens (www.bikecitizens.net) has been using various strategies such as crowdsourcing (use of OpenStreetMap, improvement of routing profiles based on feedback from the community), gamification (activation of map material, redemption of vouchers based on one's own kilometre performance) and visualisation techniques (creation of heat maps to display the network of paths and the routes covered, including daily statistics) to increase the cycling rate as well as its user base.

Thus, although a wide range of tracking tools and incentivisation frameworks exist, they are mainly limited to rewarding users for their activity in a playful way such as the national initiatives "Radelt zur Arbeit" and the

"Bike Benefit Model". There are hardly any performance-based incentive systems in the area of active mobility that go beyond simple competitions or contests. Exceptions are pilot projects such as "Vitality" by the Generali insurance company (https://www.generali.at/vitality/), "Triffic" (https://triffic.world/) and the app "Sweat Coin" (https://sweatco.in/), which are criticised primarily due to the lack of transparency and the cost-benefit ratio (Gigerenzer et al, 2015). Based on the individual and collective cost savings through regular cycling, there is a significant scope for action that can be represented transparently and fairly as a quantifiable benefit in the form of a digital cycling currency.

Blockchain as a key technology for the mapping of cycling data

The blockchain is currently dominating the technological and social discourse due to its decentralisation, transparency and security (Buhl et al., 2017) and is therefore being treated as a disruptive innovator for a broad field of applications: - from transaction processing to land registry entries to logistics chains, the middleman is to be eliminated in the future by means of blockchain (Hopf & Picot, 2018). In principle, the blockchain is a decentralised system of linked computers that process transactions publicly, transparently and tamper-proof. An entry on the blockchain is only made after multiple confirmations, the so-called consensus. Previous crypto technologies such as "Bitcoin" and "Ethereum" rely on "proof of work" for this and reward "miners", who keep the entire network alive and validate all transactions, for solving randomly generated computing tasks. The computing power required increases linearly with the difficulty of the computing tasks and consumes more and more resources. This is where new blockchain technologies such as "Ardor" come into play, which are based on a "proof of stake" algorithm and are much more energy-efficient. The focus is not on computing power in the form of graphics card and ASIC performance, but on the investment in the respective technology and the willingness to validate transactions. The operation of a blockchain is incentivised by means of its own currencies, the so-called "crypto-currencies". In other words, one's own invested computing power ("proof of work") or the volume held including online time ("proof of stake") are rewarded. These digital currencies, or rather trading with them, is critically reflected in the media due to speculation and the associated strong price fluctuations. It is strongly recommended to separate the technology from the incentive level here, as the blockchain is used as a technological foundation in numerous pilot projects such as the transparent representation of logistics chains or car insurance (Hackius & Peterson, 2017; Namiot et al., 2017).

Regardless of the approach, from a global perspective, the peer-to-peer principle (decentralised system structure), the strong encryption and the associated security as well as the transparent storage of information represent an opportunity for various sectors, in particular also for the mobility sector, or more specifically for cycling. In this regard, automated "if-then relationships" in the form of "smart contracts" are particularly noteworthy (e.g. condition I: if 10 cycle kilometres have been ridden, then 10 cycle tokens are stored in the digital wallet; condition II: if 10 cycle tokens are invested in a "transaction voucher", then this becomes a value voucher in the amount of 10 euros, which can be redeemed at partner businesses using a QR code).

Machine learning for track validation

Machine Learning (ML) is a sub-field of Artificial Intelligence (AI), which since 1959 has aimed to develop algorithms that learn the underlying concept from exemplary data. ML, and in particular the currently popular research area of Deep Learning (DL), are used in many fields to imitate complex human behaviour. Examples where these concepts come into play are the Japanese board game "Go" (Silver et al., 2016) or self-driving cars (Levinson et al., 2015). The current challenge is the processing of Big Data. This is large, fast-moving, complex and weakly structured data. Many aspects of concern have been among the current research areas of ML for decades, such as the scaling of elaborate learning algorithms (Seewald & Kleedorfer, 2007), the direct processing of complex structured data (Seewald, 2008) and the determination of the required set of training examples to achieve the required accuracy and enable the decision quality of such systems in realistic deployment scenarios (Seewald, 2012).

The tracking and movement data used as the primary source of information in this project, which are generated in the context of cycling, meet the above Big Data criteria. (Nitsche et al., 2020) shows that a recognition of the movement pattern "cycling" is only possible by means of features derived from GPS sensor and accelerometer via HMMs (a predecessor of recurrent DL) with an accuracy of 95%. Cheaters were not included, however, as the design of the experiment did not offer any incentives for them. Since the existing users are connected via mobile phone apps, not only the classical tracking data as a time-stamped sequence of 3D positions and common derived features such as velocity and acceleration from GPS and accelerometer sensors plus their variability can be used for plausibility detection of cycling, but it is also possible to use the additional sensors built into virtually all common mobile phones (magnetic field sensors, WIFI localisation, gyroscope, gesture sensors) , which can also be used to detect characteristic movement patterns (Shoaib et al, 2015; Nitsche et al., 2014) where all sensors were recorded but ultimately only GPS and accelerometers were used. However, the application here is different, as we must explicitly assume that a small proportion of users will want to exploit the system, whereas this is not an issue with typical motion detection systems and is not taken into account.

With the already existing system of Bike Citizens, large amounts of training data for ML can be generated in a short time - one saves the time-consuming generation of ground truth training data ("bootstrapping"). Due to the increasing computing power of mobile phones, offline plausibility checks are also possible. Optimally, this system should be continuously improved without, however, excluding legitimate users. However, it must be taken into account that the movement patterns depend, among other things, on the actual position of the mobile phone on the bicycle frame or on the cyclist.

METHODOLOGY

Framework

Figure 1. "Cycle4Value" Framework

In the project, an innovative research approach is conceived, implemented on a trial basis and tested in practice, which – in simplified terms – rewards cycling by means of so-called „cycle tokens". The key technologies used, machine learning and blockchain, represent an innovative solution for the validation of route data and transaction processing. In the sense of a proof of concept, it is to be tested whether and how a secure and transparent process of value generation for regular cycling can be created on the basis of a utility token that can translate the overall economic effects of cycling into value units. In the course of a field test in Graz, Berlin and Krems, the practical implementation, target group-specific acceptance and scalability of the developed solution will be analysed. The cost savings on an individual and collective level were evaluated in close cooperation with a stakeholder board and will be ultimately transferred into a marketplace, which is to enable the exchange of cycle tokens into public transport discounts or into vouchers for cycling accessories. In addition, the possibility of donating the cycled kilometres to charity projects is also being considered. All tokens are stored in a wallet which is connected to the Ardor blockchain.

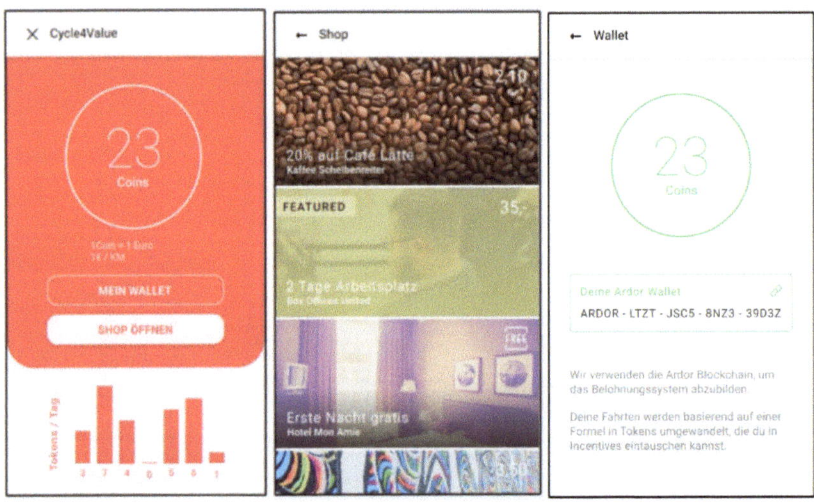

Figure 2. *"Cycle4Value" Dashboard, Market Place & Wallet*

Token Logic

The cycle tokens are realized using three different types of assets/tokens:

1. The Bike Impact Score / BIS is defined as follows: The BIS is an on-chain non-transferable, non-consumable score (like the km counter on a car). It is calculated using the following formula: number of BIS points = root (distance in km). The BIS score is not consumed. On-chain data are limited to a hash uniquely identifying the track.
2. The Bike Impact Token / BIT is defined as follows: The BIT is a fully fungible token. The BIT is a transferable global asset, which is issued proportionally to the BIS score and can be consumed.
3. A second entity/company can issue another asset and also allow minting through the BIS score. Just like some membership card could give its holder advantages at multiple places which are independent of each other. It is calculated using the following formula: number of Bike Impact Tokens = root (distance in km). The following meta data are stored: token_amount, token_receiver, track_hash, track_starttime, track_distance

4. For the city of Krems/Danube University as first entity accepting the Bike Impact Token its value is defined as 1 €. Thus, if the preconditions (see BIS score) are met, ~1 BIT is generated per 3 kilometres with the daily cap of 8 BIT.

The Unique Bike Token / UBT is defined as follows: The UBT is a non fungible token (NFT). Every track uploaded will automatically mint an NFT specific for that track. Selected NFTs can be exchanged into goodies with artistic value.

EXPECTED RESULTS

The research project aims at a mobility turnaround through new usage paradigms and behavioural changes and targets the following impact:

- Physical activity and health-promoting local and micro-mobility: By pushing/increasing cycling, a link between active and health-promoting mobility is ensured.
- New usage paradigms and sufficiency-oriented behaviour: The acceptance and functionality as well as impact of the underlying motivational research approach will be validated. The aim is to show whether habitual negative routines in mobility behaviour can be broken and new potentials for substituting car journeys can be developed.
- New publicly usable mobility offers: By incorporating innovative key technologies (machine learning & blockchain), cycling related services will be optimised.

CONCLUSION

Cycle4Value investigates the use of blockchain technologies based on a digital token to translate track data into value units. The entire process of distributing, storing, and redeeming cycle tokens can thus be mapped and automated. All accounts can also be protected against fraud and hacking thanks to blockchain technology. Because each appears once and only once as a public entry on the blockchain, it also has a one-of-a-kind, ideal value as it is the case with the unique bike token. This means that the purely digital incentive level of traditional gamification approaches is abandoned, and cycling is given real value. Another project innovation is the use of machine learning to check the plausibility of registered cycling trips or cycling routes. In contrast to ad hoc code, which typically employs highly predictive features (e.g., maximum acceleration and speed) and is thus relatively easy to defeat with appropriate countermeasures (e.g. driving slowly, controlled acceleration), the learned model in machine learning will typically use a large number of slightly predictive features, making it much more difficult to intentionally cheat the system. Another advantage of the applied machine learning system is that it can be retrained if new attack types are discovered, allowing it to be useful for a much longer period of time.

ACKNOWLEDGMENTS

The project is funded by the Federal Ministry for Climate Protection as part of the FTI programme Mobility of the Future and is managed by the Austrian Research Promotion Agency.

REFERENCES

Aitzhan, N. Z., & Svetinovic, D. (2016). Security and privacy in decentralized energy trading through multi-signatures, blockchain and anonymous messaging streams. IEEE Transactions on Dependable and Secure Computing, 15(5), 840-852.

BMVIT (2013). Der Radverkehr in Zahlen.

BMVIT (2017). Österreich unterwegs mit dem Fahrrad. Radverkehrsergebnisse der Mobilitätserhebung „Österreich unterwegs 2013/2014.

Buhl, H. U., Schweizer, A., & Urbach, N. (2017). Blockchain-Technologie als Schlüssel für die Zukunft. Zeitschrift für das gesamte Kreditwesen, (12), 30-33.

Gigerenzer, G., Rebitschek, F. G., & Wagner, G. G. (2018). Eine vermessene Gesellschaft braucht Transparenz. Wirtschaftsdienst, 98(12), 860-868.

Guo, Y., & Liang, C. (2016). Blockchain application and outlook in the banking industry. Financial Innovation, 2(1), 1-12.

Hackius, N., & Petersen, M. (2017). Blockchain in logistics and supply chain: trick or treat?. In Digitalization in Supply Chain Management and Logistics: Smart and Digital Solutions for an Industry 4.0 Environment. Proceedings of the Hamburg International Conference of Logistics (HICL), Vol. 23 (pp. 3-18). Berlin: epubli GmbH.

Hopf, S., & Picot, A. (2018). Revolutioniert Blockchain-Technologie das Management von Eigentumsrechten und Transaktionskosten?. In Interdisziplinäre Perspektiven zur Zukunft der Wertschöpfung (pp. 109-119). Springer Gabler, Wiesbaden.

Levinson, D. (2015). Climbing mount next: the effects of autonomous vehicles on society. Minn. JL Sci. & Tech., 16, 787.

Namiot, D., Pokusaev, O., Kupriyanovsky, V., & Akimov, A. (2017). Blockchain applications for transport industry. International Journal of Open Information Technologies, 5(12), 130-134.

Nitsche, M. (2020). Der Dialog mit dem Kunden in der Data Driven Economy. Digitales Dialogmarketing: Grundlagen, Strategien, Instrumente, 1-12.

Shoaib, M., Bosch, S., Incel, O. D., Scholten, H., & Havinga, P. J. (2015). A survey of online activity recognition using mobile phones. Sensors, 15(1), 2059-2085.

Seewald, A. K., & Kleedorfer, F. (2007). Lambda pruning: an approximation of the string subsequence kernel for practical SVM classification and redundancy clustering. Advances in Data Analysis and Classification, 1(3), 221-239.

Seewald, A. K. (2012). On the brittleness of handwritten digit recognition models. International Scholarly Research Notices, 2012.

Silver, D., Huang, A., Maddison, C. J., Guez, A., Sifre, L., Van Den Driessche, G., ... & Hassabis, D. (2016). Mastering the game of Go with deep neural networks and tree search. nature, 529(7587), 484-489.

IT PLAYS WHO PLAYS – THE POTENTIAL OF NON-DEFINING GAMIFICATION

Mario S. Staller, Swen Koerner

Gamification is regularly defined as the use of game elements in non-gaming contexts. This utilitarian perspective in gamification sparks controversies about the pedagogical value of gamification. While on the one hand the potential is seen in the design of joyful learning environments critics point out the pedagogical dangers. It becomes apparent that the assumptions guiding action on the subject matter of gamification in educational contexts differ. This in turn leads to different pedagogical practices.

Taking a reflexive stance towards the underlying assumptions of gamification in these contexts may allow to consolidate initially controversial positions and to open up potential for the use of gamification. With regard to the pedagogical use of gamifying elements and their empirical investigation, there are three main anchor points to consider from a reflexive stance: (a) the high context specificity of the teaching undertaken and (b) the (non-)visibility of the design elements used and (c) the potential (non-)acceptance of the gamified elements by the students. We start by providing a discussion of the definitional discourse on what is understood as gamification leading to our argument for non-defining gamification to open up its full potential. To exemplify this potential we describe a gamified concept in higher education for police recruits.

Keywords: gamification, non-definition, learning environment

INTRODUCTION

Playing is part of human nature (Callois, 1961; Huizinga, 2015). Playing is fun, motivating and engaging (McGonigal, 2011; Reeves & Read, 2009; Werbach & Hunter, 2012), promotes competition (Reeves & Read, 2009) and has a positive effect on teamwork (Vegt et al., 2014). In addition, playing games is also suitable for trying out behavioral strategies and can thus serve as a consistently reduced training and educational environment (Staller et al., 2020). Our private and professional lives seem to be permeated by gamified elements, for example when collecting bonus points when shopping, in the context of health care or when traveling (Skinner et al., 2018; Stieglitz et al., 2017). Gamified activities primarily aim to increase motivation in a wide variety of activities in order to increase the quantity and quality of the output of the corresponding activity (Morschheuser et al., 2017). The idea seems simple: the joyful and motivating aspects of the game should be used to positively influence less joyful and motivating activities, such as routine activities, in the sense of a better output (Raczkowski & Schrape, 2018). The goal is to take advantage of the positive aspects of playing and to transfer them into non-playful contexts. The instrumental use of these design elements, through which the transfer of the positive qualities of games should take place, is generally understood as gamification (Deterding et al., 2011). However, a consensual definition is still lacking (Pfeiffer et al., 2020; Schöbel et al., 2020). For example, other definitions focus on the utilitarian benefits of gamification: gamification as process improvement through playful elements in order to increase value creation (Hamari et al., 2014; Huotari & Hamari, 2012; Stieglitz, 2015) or nudge participants to perform certain actions (Pfeiffer et al., 2020). While the definition of Deterding et al. (2011) includes the use of game design elements without direct transfer to an improved output (e.g. higher motivation, more performance), this is a mandatory prerequisite for the intention of gamification in the definitions of Huatori et al. (2012) and Pfeiffer et al. (2020). The utilitarian perspective of gamification is of particular interest in educational contexts (Buckley et al., 2017; Buckley & Doyle, 2014; Córdova et al., 2017; Pill, 2013; Subhash & Cudney, 2018), but there are also critical perspectives on its use in these contexts (Buck, 2017; Woodcock & Johnson, 2018).

Empirical studies on possible positive effects of gamification show a mixed picture. An explanatory approach for the different results lies in the range of possibilities for gamification (e.g. points, levels, narrative elements) and their application in different contexts with different user groups (Hung, 2017; Schöbel et al., 2020). While much of the empirical data on gamification examine its immediate benefits (Hamari et al., 2014; Hung, 2017; Looyestyn et al., 2017), the question arises to what extent the experience of gamification per se can represent a value that does not necessarily manifest itself in behavioral change. Normatively, it is regularly stated that gamification should not impair the effectiveness of an educational setting (operationalized as learning that has taken place) (Fischer et al., 2017). However, to what extent a gamification must have positive effects on behavior, we consider worthy of discussion. Does something have to be achieved / changed or can the experience (through a gamified element) per se also be desirable? It is precisely here that the difference between different conceptualizations of gamification unfold its effects.

We argue in this article that the different concepts of gamification lead to controversies on various levels. While on one hand the potential is seen in the design of joyful learning environments (Hung, 2017) critics point out the pedagogical dangers (Buck, 2017) or the problems related to optimizing working life (Woodcock & Johnson, 2018). We argue that these different perspectives are mainly related to differing guiding assumptions about the core of gamification, which in turn lead to different derivations for pedagogical practices. Being aware of these assumptions allows for new perspectives of initially controversial positions. Being aware of these assumptions is the claim of reflexive pedagogy (Brookfield, 2017; Körner, 2009), taking a reflexive stance towards what is implemented how and why in specific contexts. Therefore – the current paper aims at reflecting the assumptions about gamification and proposes a non-definition of gamification that can guide implementation from a reflexive standpoint harnessing its context-specifity. To explain our point, we conclude by presenting a gamified learning environment that has been conducted at a German Police University and that has been evaluated elsewhere (Staller, 2020).

DEFINITIONAL DISCOURSE ON GAMIFCATION

Language allows us to differentiate between different concepts. Viewed from a systems theoretical perspective the assignment of a term towards a phenomenon A distinguishes it from what is not: not-A (Luhmann, 1981). As such gaming has been differentiated from play and play from not-play

(Walther, 2003): a game is not a play, and playing is not not-playing. Concerning the difference between game and play, there are languages that do not differentiate between these two concepts. For example, in German there is no difference between game and play. Germans use the term "Spiel". Germans "spielen" – without considering further differentiations of this term (like play and game) on a linguistic level. As such, for Germans gaming is playing and gaming is playing ("spielen ist spielen"). While one may wonder, why the aspect of non-differentiating between these concepts may be of value, we have to consider the conceptualizations of gamification, which relates to the difference between play and game. If there are different framing assumptions of what may be included within the concept of gamification, the result of what gamification in educational contexts (and others as well) looks like will likely differ. The focus on what is understood to be used excludes what is not focused upon. The definition limits its practical use.

Turning to the different conceptualizations of gamification, Deterding et al. (2011) define gamification as the use of game design elements in non-game contexts (Deterding et al., 2011), whereby the authors limit gamification to "games" and not "play". This is an important point based on Caillois' (Callois, 1961) distinction between "paidia" (playing) and "ludus" (gaming), where paidia demarcates a more anarchic mode of spontaneous interaction against the structured competition of ludus. The difference between these two concepts (play vs. game) is regularly discussed (Walther, 2003) and has implications for the concept of gamification (Woodcock & Johnson, 2018). Woodcook et al. (2018), for example, state the human desire for play (more closely paidia than ludus) as the reason for the enormous difficulties in analyzing gamification. A game is often ascribed a higher structure on several levels than a play (Callois, 1961; Juul, 2001; Walther, 2003).

These structural characteristics of a game are criticized from an educational perspective, since – according to the argumentation - it is characterized by linearity and thus offers no platform for "applying pedagogical-reflexive power of judgement" (Buck, 2017) (p. 276). Based on the assumption of a temporal and spatial framework of a game, this argumentation seems plausible. Buck (2007) continues that "Players [...] cannot transcend the rules of the game and the rules set beforehand, they are subject to a set of conditions that categorically exclude the participation, contradictions or even participation and modification of the regulator" (p. 277). This argumentation of structural containment and the resulting limited framework for action is also shared by other authors (Woodcock & Johnson, 2018). The structures given by rules and mechanics in the game undoubtedly limit, on the other hand, however, they also allow degrees of freedom that can be developed on a functional level (Körner & Staller, 2020; Torrents et al., 2020). This is, for example, where non-linear pedagogy shows its potential (Körner & Staller, 2020): It uses the setting of constraints in order for more freedom to emerge. In addition to the potential of limitation in order to open up more potential for freedom, however, the basic assumption of the pre-structuring and linearity of play also seems worthy of reflection.

In practice it becomes clear that what is played is not always a structured game. Even within the structural game frame, players find joyful activities beyond the designer's intention and beyond the intended structure (Walther, 2003). In various video game formats, the structural framework becomes blurred, which becomes clear in the example of open-world video games (e.g. GTA V or Red Dead Redemption 2). There is no rigid linearity on how to approach the game.

Irrespective of the differentiation between game and play, it is not always easy to distinguish between play and no-play (Walther, 2003). Although (Huizinga, 2015) in his seminal work "Homo ludens" points to the spatial and temporal limits as a constitutive moment of play, it is evident from the playful practices, such as Alternate Reality Games (Chess & Booth, 2013; Lynch et al., 2013; McGonigal, 2011) that this spatial and temporal structuring can be perpetuated. The question of whether a game has already begun can also be a question of perspective. In order to know that one is part of a game, the player would have to recognize that he/she is playing; which in turn raises the question of recognizing the state of play.

From a practical point of view, for example, it is possible to involve students in a game without noticing them immediately. The teacher (or puppet master in the jargon of Alternate Reality Games) is playing the game, but from the perspective of the players, this is not necessarily directly recognizable or there is also the possibility of refusing access to the game. In this case the teacher plays but not the learner. On the other hand, it is possible that a learning setting is perceived as a "game" by a learner or is redefined as such, even if the teacher would vehemently deny this. There seems to be a certain subjectivity in the concept of play: It plays, who plays.

Sutton-Smith (2001) argues in a similar direction. He argues that the definition of both play and game in positive, non-paradoxical terms is a hopeless endeavor. Instead, he suggests the use of clear examples when referring to play or game, which in turn has a subjective character. It seems like we cannot escape our paradigmatic horizon, since our observations are entangled in our understanding of the observed itself (Sutton-Smith, 2001). Sutton-Smith goes on to say that in terms of action and epistemology, we are so burdened by the game that it becomes a paradoxical task to go beyond this framework and view the game in a neutral and ontological way. It becomes apparent that it is difficult to differentiate between the two concepts (play vs. game) – and also in the differentiation from what is called non-game – and that the transitions are fluid (Juul, 2001; Sutton-Smith, 2001; Walther, 2003). Accordingly, various authors point out the transgressive character of the individual concepts (Walther, 2003; Woodcock & Johnson, 2018).

The blurring character of a conceptualization of game is evident in the Alternate Reality Games mentioned above. Their character is described as difficult to define and the design also shows that the boundaries blur. This is already evident in the immanent game principle of Alternate Reality Games: "This is not a game" –- the game as a non-game. If non-game, game, non-play and play cannot always be clearly separated, what does this mean for the use of game elements in non-game contexts? Is this subjective perspective on what is understood as play also the case with the use of game elements in non-game contexts? Woodcock and Johnson (2018) argue in that direction. They describe the playful subversion of working life, which they refer to as gamification-from-below, as the "true" form of gamification. In contrast, they refer to the application of a specific game mechanic to everyday life as gamification-from-above with a primary focus on the reinforcement of work. This "terminological foreclosing of alternative possibilities" (p. 12) restricts gamification to the linear-framed boundaries with its critiques (Bateman, 2018; Buck, 2017). By understanding gamification beyond a limiting structural framework of a playful experience beyond primarily intended playing settings, it releases its full potential. It plays, who plays. The definition as a non-definition.

THE POTENTIAL OF NON-DEFINING GAMIFICATION

With regard to teaching practice, the definitional discourses only help to the extent that they enable reflection on the own action-guiding assumptions of the own (gamified) teaching practice. From a practical perspective, the question seems to be rather whether and how gamification (in the sense of non-definition) can make a contribution in educational settings.

The literature on gamification refers – in view of a non-consensual gamification theory (Fischer et al., 2017; Matallaoui et al., 2017) - to different theories and concepts of psychology, which play a role in connection with different design elements (Huang & Hew, 2018; Richter et al., 2015; Sailer et al., 2017). For example, the concept of self-efficacy (Bandura, 1982, 1983), the needs of Maslow (Maslow, 1943), goal setting theory (Locke, 1996; Locke et al., 1981; Locke & Latham, 2002), the theory of social comparison (Festinger, 1954), flow theory (Cszikszentmihalyi, 1975; Seligman & Csikszentmihalyi, 2000) and last but not least, self-determination theory (Ryan & Deci, 2000). The latter in particular is used as a meta-theory to motivate the effects of many game elements as an explanatory approach (Rigby & Ryan, 2011; Rogers, 2017). The instrumental character of the concepts and theories in relation to the game element is striking. The central question is how a specific game element can be designed in order to ensure an optimal achievement of goals (goal theory), an optimal level of difficulty (flow theory) or an optimal degree of autonomy (self-determination theory).

However, from the perspective of a non-definition of gamification, other concepts move into focus. Concepts such as fun and enjoyment also seem to make a valuable contribution to learner-teacher settings. The role of fun in learning seems undisputed, especially in early childhood (Hromek & Roffey, 2009; Lieberman, 1977). There is also evidence for the importance of play in adulthood (Colarusso, 1993), the value of fun and humor in the workplace to increase creativity and productivity (Baldry & Hallier, 2009; Lamm & Meeks, 2009) and to reduce stress (Charman, 2013; Holdaway, 1988; Holmes & Marra, 2002).

For the specific context of higher education, recent research (Whitton & Langan, 2018) has examined what is perceived by students as a joyful learning experience. A low level of stress caused by joy and fun in the learning environment was also mentioned. In addition, the results showed that a safe learning space in which mistakes can be accepted and made is an important element. With regard to the teacher, the enthusiasm of the teacher, a high level of expertise and the desire to teach were identified as important elements. On the relationship level, the results showed that students find it joyful to share experiences with other students in playful settings. A simple contact with the teacher – by meeting at eye level – was also perceived positively. The students' reports also highlighted the stimulating effects surprises, new elements and active learning. The effect of playful settings on students is an important aspect for curriculum planning; if the effects mentioned are intended, there is the possibility to trigger effects through playful elements (whatever they may look like). However, from a practical point of view, it must be clearly stated that the strategies mentioned in the study by (Whitton & Langan, 2018) are not limited to playful settings. A recent study with a gamified learning environment – based on the non-defintion of gamification – has the potential to invoke a joyful learning experience (Staller, 2020).

On a normative level, the question of whether learning in higher education should be fun is more controversial; for example, the use of fun and playful approaches is sometimes considered inappropriate on the grounds that they undermine the academic character of higher education (Whitton & Langan, 2018). The answer to this question should – and cannot – be given at this point, since, following the (non-)definition of gamification given here, a playful element cannot always be easily differentiated and identified. When broken down to the actual practice of teachers, this means that pedagogical action is characterized by the complexity and dynamics of the teaching-learning setting. There are many possibilities to achieve specific intended effects. The art lies in transferring the processes of weighing up learners (who-dimension), the teaching content (what-dimension), the possibilities of teaching didactics (how-dimension), one's own values (self-dimension) and the specific teaching context (context dimension) into situational practice (practice dimension) against the background of the intended pedagogical goals (Staller & Körner,

2021). Gamification - as an ephemeral and unspecified tool – is thus a tool in the pedagogical toolbox.

AN EXAMPLE OF A DESIGN CASE

We want to highlight the potential of the (non-)definition of gamification by describing a case design of a learning environment in the higher education context, that was designed based on reflections around (a) the high context specificity of the teaching undertaken and (b) the (non-)visibility of the design elements – related to the non-definition of gamification and (c) the (non-)acceptance of the gamified elements by the students. The gamified concept was implemented in a psychology course of police recruits at a University of Applied Sciences in Germany. The concept centers around a potential crime that has taken place: the professor, that was indented to teach psychology to the students was kidnapped (fabricated) as media reports showed (see figure 1). The substitution to the professor is Mr. Sepur, the person standing in front of the students. He asks the students to help him developing a reflexive toolkit for police recruits, that will help them becoming more reflective practitioners. His appearance and this behavior reflect what the student refer to as "like a student". However, he will help the students to grasp the content of the psychology curriculum within the course.

1. SEPTEMBER DIESES JAHRES / 1   FINANCED BY WAYNE ENTERPRISES   1. SEPTEMBER DIESES JAHRES / 1   FINANCED BY WAYNE ENTERPRISES

Daily Planet

IN EIGENER SACHE

CHEFREDAKTEUR GESUCHT. MEHR DAZU AUF SEITE 2

"LET'S ROCK!" DIE NEUEN STUDIS DER HSPV BEGINNEN STUDIUM

NEUES RELEASE LEARNING & DEVELOPMENT MIT NEUEM SPIEL

FISK IM VISIER DIE POLIZEI IN HAGEN IST SICH SICHER: FISK STEHT FÜR TERROR

PROF ENTFÜHRT

Experte für reflektierte Praxis entführt. Hintergründe unbekannt. Die Polizei ermittelt.

Hochschule im Notbetrieb

Der renommierte Psychologie Professor Dr. Dr. Mario Staller scheint entführt zu sein. Bisher gibt es noch keine Informationen zu dem/der Täter*in. Die Polizei in Aachen geht davon aus, dass das Verschwinden etwas mit dem Expertise Gebiet des Professors zu tun hat: Er setzt sich intensiv für eine reflektierte Polizeiarbeit ein. Das Fehlen des als sehr autoritär auftretenden Dozenten führt zu massivsten Problem in der Lehre an der Hochschule für Polizei und öffentliche Verwaltung (HSPV). „Zum Glück konnten wir Herrn Sepur (L&D Games) gewinnen", so die Verwaltung der HSPV.

Die Verwaltung an der HSPV's chef ist auf dem Notbetrieb. Mit externen Dozenten wird die Lücke in der Lehre geschlossen.

Gamification in der Lehre?

Kann Gamification in der Lehre funktionieren? „Aber sicher", sagt Nr M. Sepur von Learning & Development Games. „Wenn die Story sich nahtlos in die Lerninhalte einfügt, haben Lerner*innen beim Lernen sogar Spaß. Allerdings muss auf der anderen Seite auch darauf hingewiesen gesagt werden, dass wenn nur das Spielen im Vordergrund steht - und die Inhalte zur Nebensache werden - Spiele ihren Lerncharakter verlieren."

Auch die Recherchen des Daily Planet zeigen: Von 10 Lernspielen sind 6 unbrauchbar, da die Lerninhalte sich nicht nahtlos in die Story einfügen. Auch das Game-Design hat darauf wohl Einfluss: „Wenn die Mechanik des Spiels nichts taugt", so Sepur, „dann kannst du die besten Inhalte und die beste Story haben. Das Spiel ist dann für die Tonne".

STARK INDUSTRIES   HSPV NRW

BESSER ALS BUNDESLIGA STARK INDUSTRIES GIBT GRÜNDUNG DER COP DEVELOPMENT LEAGUE BEKANNT

PROBLEME MIT HSPV E-MAIL ACCOUNT EXTERNE DOZENTEN BEKOMMEN EMAIL ACCOUNT FEHLENDER MITARBEITER*INNEN

Stellenausschreibung

Der Daily Planet sucht ab sofort einen Chefredakteur für die Abteilung „Reflexion der polizeilichen Praxis". Die Aufgabe besteht im Erstellen eines Handbuchs, mit dessen Hilfe künftige Polizist*innen das eigene Handeln reflektieren können. Bewerbungen werden ab sofort entgegengenommen.

Guru gibt Tipps für das Grundstudium

Dr. Steven Strange's Tipps für einen gelungen Start ins Grundstudium

- Gewöhne dich an das Lesen von wissenschaftlichen Texten - deine Haus-, Seminar- und Bachelorarbeiten werden es dir danken. Ist doof am Anfang - aber so ging es mir beim Erlernen der Magie auch.

- Lies zu jedem Thema 1 Sekundärquelle, 1 Primärquelle, 1 Quelle mit Polizeibezug. Das heut rein!

Figure 1. Newspaper article reporting the kidnapping of Professor Staller

Concerning the context-specificity of the learning environment, planning reflections centered around the specific organizational culture of educating police recruits in Germany. Within the police training structures of young police officers, there seems to exist mechanics of command and obedience" (p. 247) with a simultaneous willingness to submit to it unreflectively (Jasch, 2019) Tendencies, which Jasch (2019) describes as the primacy of job-specific disciplining, appear to be the cause for this. The disciplining extends on the one hand to conformist behavior and on the other hand to one's own body: tattoos, hairstyles, body jewelry and clothing. In view of this organizational-structural context within the police academy and the resulting problems for professional police practice (Jasch, 2019; Schöne, 2011), the aim of the teaching concept was to create irritation within the courses in order to provide opportunities for reflection (Brookfield, 2013), specifically concerning different roles and stereotypes towards these roles.

With regard to the (non-)visibility of the design elements the teaching concept based on reflections about the first author's (MS) own motivations and expectations towards joyful teaching and how this may influence the learning climate within the classroom (Staller & Körner, 2021). Several years of teaching experience showed MS, that being able to creatively express himself during teaching provides him with experiences of fun and pleasure, which in turn was reflected in his commitment to the corresponding teaching-learning settings. MS was always highly motivated for the teaching, which he noticed especially when he had creative and playful possibilities for shaping his teaching. The playful elements during teaching (e.g. keynote design, humor, references to pop culture, different forms of classroom settings) was perceived as an enrichment by the learners - and also for himself. Concerning the planning of the learning environment, these reflections showed that the gamification of teaching is a joyful aspect for him personally; however, this does not mean that all learners have to feel this way nor that they have to be aware that he is implanting some gamified design elements. For example, the selection of comic-shirts matching the current story line as well as hiding easter eggs on the keynote slides provided MS with feelings of fun and joy leading to a greater enthusiasm concerning the teaching (Staller, 2020). Furthermore, concerning the narrative and the mechanics of the gamified learning environment, it was important in the design of the course that there are possibilities for gamification, but that students were allowed to decide for themselves to what extent they want to engage with it (if they are aware of the elements, e.g. easter eggs).

This aspect of the potential (non-)acceptance of the gamified elements by the students was valued in the concept via the narrative and the mechanics. The procedures and rules of the (gamified) learning environment with reference to the goal that the students should achieve were introduced at the beginning of the course through the narrative of the kidnapped professor. The goal was to have the students work on a project that they would create in groups and provide – for those who want to engage with the narrative – a coherent story why this is important for them (and the kidnapped professor). In addition to the overall objectives, there were also smaller assignments (e.g. compiling a concept for de-escalating behavior in a conflict; preparation for an examination, etc.), which were introduced as "additional tasks" via narrative elements. It was important to us, that all tasks (project work, additional tasks) could also be perceived and accepted from a non-gamified perspective, since they were the regular curriculum assignments. The decision whether and how to play – and engage in the story – or not was thus left to the students on an individual level. However, even if students did not participate, they had the opportunity to experience other playful elements like keynote designs and the enthusiasm of their teacher.

A further description of the planning decisions and the evaluation of the teaching concept is presented elsewhere (Staller, 2020; Staller & Körner, 2021). In short, the results showed that an overall positive resonance towards the teaching: For the teacher, the students who wanted to play and for those who decided not to engage with the narrative. However, the results also showed that the main positive effects were to elements, that were not directly related to the gamified environment (e.g. teacher enthusiasm), but seemed to be influenced by the approach. Finally, the narrative was important for some students; but not for others. Therefore, it was up to the students to decide if there has happened a crime (the kidnapping of the professor). Some students played, some did not – and they were not always aware that their teacher played. However, they were aware of experienced positive effects concerning the learning atmosphere within the course.

CONCLUSION

Gamification in educational contexts has potential. However, its use and its perceived usefulness and pedagogical value are dependent on what is understood as gamification. Based on our analysis that the distinction between play and games and what it is not is blurred and heavily dependent on the perspective, we argued for a non-definition of gamification. From our perspective, this would allow for the integration of initial controversial positions of the pedagogical value of gamification by opening up the space for exploration, experimentation and evaluation of a context-specific use of it.

REFERENCES

Baldry, C., & Hallier, J. (2009). Welcome to the House of Fun: Work Space and Social Identity. *Economic and Industrial Democracy, 31*(1), 150–172. https://doi.org/10.1177/0143831x09351215

Bandura, A. (1982). Self-efficacy mechanism in human agency. *American Psychologist, 37*(2), 122–147. https://doi.org/10.1037/0003-066x.37.2.122

Bandura, A. (1983). Self-efficacy determinants of anticipated fears and calamities. *Journal of Personality and Social Psychology, 45*(2), 464–469. https://doi.org/10.1037/0022-3514.45.2.464

Bateman, C. (2018). Playing work, or gamification as stultification. *Information, Communication & Society, 21*(9), 1193–1203. https://doi.org/10.1080/1369118x.2018.1450435

Brookfield, S. D. (2013). Teaching for Critical Thinking. *International Journal of Adult Vocational Education and Technology (IJAVET), 4*(1), 1–15. https://doi.org/10.4018/javet.2013010101

Brookfield, S. D. (2017). *Becoming a Critically Reflective Teacher*. Jossey-Bass.

Buck, M. F. (2017). Gamification von Unterricht als Destruktion von Schule und Lehrberuf. *Vierteljahrsschrift Für Wissenschaftliche Pädagogik, 93*(2), 268–282. https://doi.org/10.1163/25890581-093-02-90000005

Buckley, P., & Doyle, E. (2014). Gamification and student motivation. *Interactive Learning Environments, 24*(6), 1162–1175. https://doi.org/10.1080/10494820.2014.964263

Buckley, P., Doyle, E., & Doyle, S. (2017). Game On! Students' Perceptions of Gamified Learning. *Journal of Educational Technology & Society*.

Callois, R. (1961). *Man, play and games*. University of Illinois Press.

Charman, S. (2013). Sharing a laugh: The role of humour in relationships between police officers and ambulance staff. *International Journal of Sociology and Social Policy, 33*(3/4), 152–166. https://doi.org/10.1108/01443331311308212

Chess, S., & Booth, P. (2013). Lessons down a rabbit hole: Alternate reality gaming in the classroom. *New Media & Society, 16*(6), 1002–1017. https://doi.org/10.1177/1461444813497554

Colarusso, C. A. (1993). Play in Adulthood. *The Psychoanalytic Study of the Child, 48*(1), 225–245. https://doi.org/10.1080/00797308.1993.11822386

Córdova, A. G., Parreño, J. M., & Pérez, R. C. (2017). *Higher education students' attitude towards the use of gamification for competencies development*.

Cszikszentmihalyi, M. (1975). *Beyond boredom and anxiety: Experiencing flow in work and play*. Jossey-Bass.

Deterding, S., Dixon, D., Khaled, R., & Nacke, L. (2011). *From game design elements to gamefulness: defining gamification*. ACM. https://doi.org/10.1145/2181037.2181040

Festinger, L. (1954). A Theory of Social Comparison Processes. *Human Relations, 7*(2), 117–140. https://doi.org/10.1177/001872675400700202

Fischer, H., Heinz, M., Schlenker, L., Münster, S., Follert, F., & Köhler, T. (2017). Die Gamifizierung der Hochschullehre – Potenziale und Herausforderungen. In *Gamification und Serious Games* (pp. 113–125). Springer Fachmedien.

Hamari, J., Koivisto, J., & Sarsa, H. (2014). Does Gamification Work? -- A Literature Review of Empirical Studies on Gamification. *2014 47th Hawaii International Conference on System Sciences (HICSS)*, 3025–3034. https://doi.org/10.1109/hicss.2014.377

Holdaway, S. (1988). Blue jokes: Humour in police work. In *Humour in society: Resistance and control* (pp. 106–122). Palgrave Macmillan. https://doi.org/10.1007/978-1-349-19193-2_6

Holmes, J., & Marra, M. (2002). Having a laugh at work. *Journal of Pragmatics, 34*(12), 1683–1710. https://doi.org/10.1016/s0378-2166(02)00032-2

Hromek, R., & Roffey, S. (2009). Promoting Social and Emotional Learning With Games. *Simulation & Gaming, 40*(5), 626–644. https://doi.org/10.1177/1046878109333793

Huang, B., & Hew, K. F. (2018). Implementing a theory-driven gamification model in higher education flipped courses: Effects on out-of-class activity completion and quality of artifacts. *Computers & Education, 125*, 254–272. https://doi.org/10.1016/j.compedu.2018.06.018

Huizinga, J. (2015). *Homo Ludens: Vom Ursprung der Kultur im Spiel*. Rowohlt.

Hung, A. (2017). A Critique and Defense of Gamification. *Journal of Interactive Online Learning*.

Huotari, K., & Hamari, J. (2012). *Defining gamification: a service marketing perspective*. ACM. https://doi.org/10.1145/2393132.2393137

Jasch, M. (2019). Kritische Lehre und Forschung in der Polizeiausbildung [Critical teaching and research in police training]. In C. Howe & L. Ostermeier (Eds.), *Polizei und Gesellschaft - Transdisziplinäre Perspektiven zu Methoden, Theorie und Empirie reflexiver Polizeiforschung* (Vol. 2014, pp. 231–250). Springer Fachmedien. https://doi.org/10.1007/978-3-658-22382-3_10

Juul, J. (2001). The repeatedly lost art of studying games. *Game Studies, 1*(1).

Körner, S. (2009). Reflexion von Kontingenz und Kontingenz der Reflexion: Ein systemtheoretischer Kommentar. In I. Lüsebrink, C. Krieger, & P. Wolters (Eds.), *Schulinterne Evaluation. Impulse zur Selbstvergewisserung aus sportpädagogischer Perspektive* (pp. 141–160). Sportverlag Strauß.

Körner, S., & Staller, M. S. (2020). Begrenzen für mehr Freiheit: Der Constraints-Led Approach als trainingspädagogische Perspektive auf das Design von Lehr-Lernsettings in- und außerhalb des Sports. In T. Vogt (Ed.), *Vermittlungskompetenz in Sport, Spiel und Bewegung: Sportartspezifische Perspektiven* (pp. 276–298). Meyer & Meyer.

Lamm, E., & Meeks, M. D. (2009). Workplace fun: the moderating effects of generational differences. *Employee Relations, 31*(6), 613–631. https://doi.org/10.1108/01425450910991767

Lieberman, N. (1977). *Playfulness: Its Relationship to Imagination and Creativity*. Academic Press.

Locke, E. A. (1996). Motivation through conscious goal setting. *Applied and Preventive Psychology, 5*(2), 117–124. https://doi.org/10.1016/s0962-1849(96)80005-9

Locke, E. A., & Latham, G. P. (2002). Building a practically useful theory of goal setting and task motivation: A 35-year odyssey. *American Psychologist, 57*(9), 705–717. https://doi.org/10.1037/0003-066x.57.9.705

Locke, E. A., Shaw, K. N., Saari, L. M., & Latham, G. P. (1981). Goal setting and task performance: 1969-1980. *Psychological Bulletin, 90*(1), 125–152. https://doi.org/10.1037/0033-2909.90.1.125

Looyestyn, J., Kernot, J., Boshoff, K., Ryan, J., & Edney, S. (2017). Does gamification increase engagement with online programs? A systematic review. *PLoS ONE, 12*(3), e0173403. https://doi.org/10.1371/journal.pone.0173403

Luhmann, N. (1981). The Improbability of Communication. *International Social Science Journal, 1*(23), 122–132.

Lynch, R., Mallon, B., & Nolan, K. (2013). Mastering the puppets: Criteria for pulling the strings in an Alternate Reality Game. *Journal of Gaming & Virtual Worlds, 5*(1), 23–40. https://doi.org/10.1386/jgvw.5.1.23_1

Maslow, A. H. (1943). A theory of human motivation. *Psychological Review, 50*(4), 370–396.

Matallaoui, A., Hanner, N., & Zarnekow, R. (2017). Introduction to Gamification: Foundation and Underlying Theories. In S. Stieflitz, C. Lattemann, S. Robra-Bissantz, R. Zarnekow, & T. Brockmann (Eds.), *Gamification, Using Game Elements in Serious Contexts* (pp. 3–18). Springer. https://doi.org/10.1007/978-3-319-45557-0_1

McGonigal, J. (2011). *Reality is Broken. Why Games Make Us Better and How They Can Change the World.* Jonathan Cape.

Morschheuser, B., Hamari, J., Werder, K., & Abe, J. (2017). *How to Gamify? A Method For Designing Gamification.*

Pfeiffer, A., Bezzina, S., König, N., & Kriglstein, S. (2020). Beyond Classical Gamification: In-and Around-Game Gamification for Education. In C. Busch, M. Steinke, & T. Wendler (Eds.), *Proceedings of the 19th European Conference on e-Learning ECEL 2020.*

Pill, S. (2013). Game play: What does it mean for pedagogy to think like a game developer? *Journal of Physical Education, Recreation & Dance, 85*(1), 9–15. https://doi.org/10.1080/07303084.2013.838119

Raczkowski, F., & Schrape, N. (2018). Game Studies. In B. Bell, T. Hensel, & A. Rauscher (Eds.), *Game Studies* (pp. 313–329). Springer. https://doi.org/10.1007/978-3-658-13498-3_17

Reeves, B., & Read, J. L. (2009). *Total Engagement: Using Games and Virtual Worlds to Change the Way People Work and Businesses Compete.* Harvard Business Press.

Richter, G., Raban, D. R., & Rafaeli, S. (2015). Studying Gamification: The Effect of Rewards and Incentives on Motivation. In T. Reiners & L. C. Wood (Eds.), *Gamification in Education and Business* (pp. 21–46). Springer. https://doi.org/10.1007/978-3-319-10208-5_2

Rigby, S., & Ryan, R. M. (2011). *Glued to Games: How Video Games Draw Us In and Hold Us Spellbound.* Praeger.

Rogers, R. (2017). The motivational pull of video game feedback, rules, and social interaction: Another self-determination theory approach. *Computers in Human Behavior, 73.* https://doi.org/10.1016/j.chb.2017.03.048

Ryan, R. M., & Deci, E. L. (2000). Self-determination theory and the facilitation of intrinsic motivation, social development, and well-being. *American Psychologist, 55*(1), 68–78. https://doi.org/10.1037//0003-066x.55.1.68

Sailer, M., Hense, J. U., Mayr, S. K., & Mandl, H. (2017). How gamification motivates: An experimental study of the effects of specific game design elements on psychological need satisfaction. *Computers in Human Behavior, 69,* 371–380. https://doi.org/10.1016/j.chb.2016.12.033

Schöbel, S., Janson, A., Jahn, K., Kordyaka, B., Turetken, O., Djafarova, N., Saqr, M., Wu, D., Söllner, M., Adam, M., Gad, P. H., Wesseloh, H., & Leimeister, J. M. (2020). *A Research Agenda for the Why, What, and How of Gamification Designs Results on an ECIS 2019 Panel.*

Schöne, M. (2011). *Pierre Bourdieu und das Feld Polizei: Ein besonderer Fall des Möglichen.* Verlag für Polizeiwissenschaft.

Seligman, M. E. P., & Csikszentmihalyi, M. (2000). Positive psychology: An introduction. *American Psychologist, 55*(1), 5–14. https://doi.org/10.1037/0003-066x.55.1.5

Skinner, H., Sarpong, D., & White, G. R. T. (2018). Meeting the needs of the Millennials and Generation Z: gamification in tourism through geocaching. *Journal of Tourism Futures, 4*(1), 93–104. https://doi.org/10.1108/jtf-12-2017-0060

Staller, M. S. (2020). *„Ich bin nur die Vertretung…": Gamifizierung in der Psychologie Lehre an einer Hochschule der Polizei.* Verlag für Polizeiwissenschaft.

Staller, M. S., Heil, V., Koch, R., & Körner, S. (2020). "Playing Doom" – A design case in self-defense training. *International Journal of Designs for Learning, 11*(2), 9–16.

Staller, M. S., & Körner, S. (2021). Game on! - Spielen in der polizeilichen Lehre. *Lehre.Lernen. Digital, 2*(1), 13-22.

Stieglitz, S. (2015). Gamification – Vorgehen und Anwendung. *HMD Praxis Der Wirtschaftsinformatik, 52*(6), 816–825. https://doi.org/10.1365/s40702-015-0185-6

Stieglitz, S., Lattemann, C., Robra-Bissantz, S., Zarnekow, R., & Brockmann, T. (Eds.). (2017). *Gamification: Using Game Elements in Serious Contexts.* Springer. https://doi.org/10.1007/978-3-319-45557-0

Subhash, S., & Cudney, E. A. (2018). Gamified learning in higher education: A systematic review of the literature. *Computers in Human Behavior, 87,* 192–206. https://doi.org/10.1016/j.chb.2018.05.028

Sutton-Smith, B. (2001). *The Ambiguity of Play.* Harvard University Press.

Torrents, C., Balagué, N., Ric, Á., & Hristovski, R. (2020). The motor creativity paradox: Constraining to release degrees of freedom. *Psychology of Aesthetics, Creativity, and the Arts.* https://doi.org/10.1037/aca0000291

Vegt, N., Visch, V., Ridder, H. de, & Vermeeren, A. (2014). Designing Gamification to Guide Competitive and Cooperative Behavior in Teamwork. In Torsten Reiners & L. C. Wood (Eds.), *Gamification in Education and Business* (pp. 513–533). Springer. https://doi.org/10.1007/978-3-319-10208-5_26

Walther, B. K. (2003). Playing and Gaming: Reflections and Classifications. *Game Studies*, *3*(1).

Werbach, K., & Hunter, D. (2012). *For the Win: How Game Thinking Can Revolutionize Your Business*. Warton Digital Press.

Whitton, N., & Langan, M. (2018). Fun and games in higher education: an analysis of UK student perspectives. *Teaching in Higher Education*, *24*(8), 1–14. https://doi.org/10.1080/13562517.2018.1541885

Woodcock, J., & Johnson, M. R. (2018). Gamification: What it is, and how to fight it. *The Sociological Review*, *66*(3), 542–558. https://doi.org/10.1177/0038026117728620

GAMES, PLAY & SOCIETY

GAMES, PLAY AND SOCIETY

CHILDREN IN AN ONLINE WORLD, VICTIMS OR PERPETRATORS? - A COLLECTION OF CASE STUDIES

Alexiei Dingli

When youngsters interact online, many guardians believe that they're interacting in a safe bubble and very few realise that their actions might also lead to serious consequences. In many cases, vulnerable children are the target of online predators. But in order cases, these young people are the ones that actually commit the wrongdoing. The most common of which is without doubt psychological harm. But some of these youngsters go to even greater lengths by associating themselves with criminal organisations whose aim is to commit real world crimes. This might include financial scams or even sexual grooming which could also lead to human trafficking. One doesn't have to be a hardened criminal to commit a serious crime; many people experienced online thefts and there have also been a few cases of murders too, all of which executed by common people. Unfortunately, these platforms have become a fertile ground for all sorts of crime. And having the dark web round the corner, one can easily get access to real weapons, thus transferring virtual issues to the real world in no time. In this paper, we will have a look at various game crimes which happened in the past years, the reasons behind them and explore different ways in which they can be prevented.

Keywords: Cyber-crime, Children, Organised Crime, Online Safety

---

INTRODUCTION

The online world has been evolving since its inception. The more time passes, the more we can see a resemblance with the physical world. But unfortunately, the good things which exist in our world are not the only things being copied online, we also find a lot of dubious or illegal content (Yar M. , 2018). Everyone who has an online identity can be a victim of these criminals but most probably, children form one of the largest cohorts

(Staksrud Elisabeth, 2009). Guardians go through great lengths to try to protect their little ones (Davidson, 2008) but there isn't an approach which is 100% effective. At most, one can try to mitigate undesirable situations as much as possible. On the other hand, living online makes it easier for someone to commit a crime (Oates B. , 2001) since digital assets can be transferred and actions can be taken at the click of a button. Because of this, the children of today have the liberty to peep outside their bedrooms onto the online world and even commit crimes some of which might have very serious repercussions. Unfortunately, online platforms have become fertile ground for all types of crime. One can easily access real weapons (Rhumorbarbe, 2018) or instantly transfer virtual issues to the real world. Because of this, the paper is divided into two sections, the first part analysis the children of today and how they can become victims of online crime. The second part looks at how youngsters are becoming criminals. Some are conscious of their acts but others do not realise the consequences of their actions until it is too late.

LITTLE VICTIMS OF THE ONLINE WORLD

Digital Natives

The children of today are normally referred as Digital Natives (DNs) (Prensky, 2008) (Thomas) (Dingli, 2015) and they can be considered as a new breed of digital citizens. Their access to technology is unprecedented; personal computers, smart phones, tablets, etc. They also have the freedom to consume digital content from anywhere using any device. The push towards ubiquitous computing (Krumm, 2018) and controllerless interfaces (Guzsvinecz, 2019) (such as the Kinect, Leap Motion, etc) allows these children to interact with a machine in a way which was unconceivable before. This interaction is made even more powerful with the availability of unlimited streams of digital content which can be downloaded from the web. The Internet provides access to educational and entertaining content which is free, thus keeping the cost of usage low. By doing so, it creates a new openness towards technology and a drive towards owning even more technological devices.

These children are not mesmerised by a handful of TV stations because now they have thousands of online channels to choose from. For them, technology is not just a tool but rather an important extension of their life. They are not afraid to use it and expose themselves online because technology is used to communicate, to learn, to express oneself, etc. From what we've seen so far, it's clear that the world is becoming more complex. Our life is shared among various realities, including the physical world, the online world, and possibly hybrid worlds. In reality, it's more of a jumble of different realities coexisting together often referred to as Mixed Reality (Speicher, 2019). To nurture our children in this digital world we need a whole new approach. We cannot expect these individuals to be treated as the other generations before them because they are very different. Traditional tuition is monotone and monochrome (limited to writing on a board). It was only in recent years that we've seen the mass introduction of technology in the class room such as the installation of the Interactive White Boards, Projectors and the more recent introduction of tablets. Now that we have technology in an educational setting, the big question is how to use it effectively. Even young educators, fresh from University, need some time to accustom themselves to these new pedagogies. When reaching adulthood, these children start interacting with the real-world and this too has to change.

Apart from their physical lives, these children have a very active digital live. First and foremost, they consider themselves as online citizens and as such, they have different rights and obligations (Oates S. , 2015). Their time on-line is as important as much as their life in the real world. Because of this, governments are very much concerned about these virtual boundaries which they find difficult to control. In most cases, they cannot be defined clearly especially when a website or web-service is provided through multiple servers geographically located in different countries around the globe. That is the reason why different countries are pushing towards digital legislations (Bokovnya, 2019). In the past, we've seen numerous attempts at granting greater control to governments (Boas, 2006) (Akdeniz, 2001) (Raustiala, 2017) (Vendil Pallin, 2017) and also various failures (Warf, 2011) (Dada, 2006).

But these people are not the only online migrants. We're also seeing other groups such as criminals (Yar M. a., 2019). The level of sophistication reached by these criminals is extremely high and governments worldwide are going to great lengths to protect their precious information (Anderson, 2019). This information can include anything from financial information to personal information. We've seen various incidents around the world where people have been kidnapped (Sorell, 2019) or even killed (Başol, 2018) as a consequence of their online activity. But even though we have these threats lurking over our head, the future does looks bright and exciting.

What's certain is that the world of today has changes. A five-year old is more versatile on the use of certain technologies than his parents and probably spends more time online than they do. And since we are fine to expose our little ones to this world, we have to teach them how to protect themselves from cyber criminals when they are interacting online.

Trust Issues

Almost half the population of the world is on social media (Dean, 2021). People on average spend over 700 billion minutes per month according to Facebook. Thus, it comes to no surprise that not everyone online is genuine and unfortunately, some of them even try to deceive other people. They do this for a number of reasons.

One of them might be to gain social notoriety (Szczurski, 2017). The primary culprit for spread fake news are people (Talwar, 2019) who seem to prefer sharing phoney stories rather than real ones. However, once an article goes viral, the algorithm kicks in and starts sharing it too, further accentuating the damage. As if this wasn't bad already, in many cases, Facebook gets paid to spread fake news (Bernal, 2018), and Mark Zuckerberg is, of course, defiant about it because as he stated in (News, 2019) "people should be able to judge for themselves". So, Facebook is fine with feeding people fake news as long as they can make money out of it.

Unfortunately, many armchair critics believe that thanks to the internet, they overnight turned into social activists. Most of these people post comments, without even bothering to read the article they're posting about. This fact came out from a recent study (Glenski, 2017) on Reddit, a social networking site, whereby 73% of posts received a rating without being read. Just because these people spend their time writing posts and replying to other people, doesn't mean that they're making any tangible contribution to society at large. At most, these keyboard worriers might be labelled as trolls[1], but nothing more than that. Let's not forget that most of these comments have no tangible benefit, only follow a particular agenda and very few of them actually get read. Eventually, the algorithm will relegate their comments to the social dump of oblivion until one fine day, the comments come back to haunt them. In fact, there have been several cases, where employees were rebuked or even fired for commenting on social media (Warren, 2011). What these people do not realise is that social media creates echo chambers (Cardenal, 2019) which accentuate polarisation (Bail, 2018). Recent studies show that the difference between Democrats and Republicans grew in the past decade only because people are becoming more polarised (Center, 2014). Thus, this leads to social unrest, similar to what the US has been experiencing in the past few months. People are so convinced about their myopic views (because the social media algorithm has nurtured them) that they are not capable of accepting contrasting views.

What these people do not realise is that just because other people agree with one's point of view, it doesn't mean that one is right. In the famous game show "Who Wants to Be a Millionaire?", it is quite common that the "Ask the Audience" helpline gives a wrong reply, especially after the 10th question (Fandom, 2021). It occurs because widespread knowledge is not necessarily the truth. Take Galileo Galilei, the famous Italian astronomer; the Roman Inquisition sentenced him because he dared go against the accepted belief of the time even though they were wrong (Kossovsky, 2020). Yet he was right! One can find a myriad of such theories on social media; the anti-vaxxers[2], the 5G conspiracy[3], the flat-earthers[4] and the list can go on forever. The social media algorithm is aware of this, and since it is a mercenary for likes, it gradually creates echo chambers (Choi, 2020) (Cinelli, 2021). The more it gets to know a person, the more it chooses posts from people who share the same

---

[1] An Internet troll is a person who intentionally upsets other people on the Internet.
[2] The group of people who are opposed to vaccinations.
[3] The 5G conspiracy centres around the belief that 5G is bad for humans.
[4] The flat-earthers believe that the earth is flat.

sentiment. Thus, these people validate each other's ideas (by pressing like) and make one believe that they're correct. Once again, it is all just an illusion!

But the problems do not stop there. Even though most social media websites are restricted to children below the age of 13, it is a known fact that 40% of children (below that age bracket) have a Facebook account (Leonard Busuttil, 2014). What's worse is that in most cases, these accounts were created by their own parents. Unfortunately, it is obvious that many guardians do not understand the risks associated with such social platforms. This is even more problematic when considering that the social media sites have very lax control over what get published online. Of course, the situation gets compounded further when we add other threads such as scams, sharing of pirated media, etc. Similar problems can be found once again in games.

Privacy Issues

Even though parents of young children are initially very concerned about their children's online activity, Eurobarometer surveys (Eurobarometer, 2008) indicate that supervision tends to decrease as children get older. During the same period, cases of cyberbullying begin to emerge (Kowalski, 2012). Furthermore, these children meet strangers online, and more than half of those children have been seen online porn (Livingstone S. a., 2005). These are also the years when children begin posting things on the internet with complete disregard for privacy (Wang, 2019). They are unaware of the magnitude of their actions and do not realise that their information is being broadcast on a global billboard. During this stage, children begin to experiment with software piracy too (Tomczyk, 2018).

What children do not realise is that once something is posed online, it cannot be undone especially if other people start sharing it. A good analogy is that of placing pictures and other personal information on a school public billboard. Nobody would want to do that. The internet is exactly the same but the digital billboard is available to all the people in the world and not just restricted to a school.

It is unfortunate that young kids do not understand enough the value of privacy at a young age (Stoilova, 2019). They do not realise that they can be spied upon. (Power, 2020) claim that many hackers target children specifically because they are easy to access, they would have clean credit histories and unused social security numbers. Then they steal their identity in order to commit other crimes. This might lead to a situation as specified in (Irshad, 2018) where the kid might be accused of a crime he did not commit.

The situation might even escalate further since a child might even get physically hurt. Most modern devices which children use can reveal the device location. This means that any perpetrator might figure out where the child is located (Winters, 2017) if he gains access to that data and subsequently decide to hurt the child. Because of this, it is important to ensure that all location enabled services are turned off, thus reducing such threats.

Children should also be taught about the consequences of their actions. By posting things on social media, they are essentially exposing their life. Some things might be innocent but others might be compromising and they will probably return to haunt them. There have been several cases where people posted or liked questionable content in their teen years. Several years later when they had forgotten about it, their employer saw it and eventually led to their dismissal from work (Warren, 2011).

These were just a few of the threats which children might face online and the list can go on. Even if guardians find a way of monitoring their social interaction, the biggest flaw in any system is always the human element (Mitnick, Kevin D., and William L. Simon) (Hadnagy, 2010) (Zambrano, 2019). Do not forget that some people specialise in social engineering which is the psychological manipulation of people into performing actions or divulging confidential information. So, the threat is still there and online portals make it much easier for these social engineers to contact their prey.

Grooming Threats

Child grooming is a threat which children face online and involves the befriending and establishing of an emotional connection with a child to lower the child's inhibitions with the final objective of sexual abuse.

First of all, let's keep in mind that this does not necessarily has to come from a human. Today, our world is governed by algorithms. These software products have one objective, that of keeping us online as long as possible because these social platforms make money through our online interactions.

According to (Ellis, 2017), they have developed various approaches to perfection that objective which might involve, showing users alternative content, recommend similar items, suggest new content and loads of other methods. The problem here is that the algorithm makes little to no distinction to the semantics of the content but bases itself on similarity metrics. Let me give you an example.

A parent gives a tablet to her child so that he can watch his favourite cartoon character on YouTube. The child is happy and quiet so the parent is relieved. As soon as the episode finishes, he will then press on the next episode or maybe on one which was recommended by the algorithm. The action simply involves a selection with his finger and it is well known that touch interfaces are easily mastered by young children (Lu, 1992). However, there have been cases on YouTube (Brandom, 2017) (What is ElsaGate?, 2017) whereby cartoons of young children have either been dubbed with foul language or else, they have been modified to show sexually explicit content. A young child has no notion of either and starts watching his favourite cartoon while been exposed to inappropriate content. The guardian is of course oblivious of what's happening because the child is quietly enjoying the YouTube content. So even though YouTube is against such content, by failing to remove them from its site, they are exposing children to algorithmic grooming. The result of which being that kids become less sensitive to such content and makes them open up to other forms of grooming which might be even more dangerous.

These digital channels provide fertile ground for this kind of content. Let's not forget that such content attracts a lot of people of any age. In fact, when a new technology was introduced in the past such as DVD formats, Virtual Reality and many others, one of the first digital contents that flourished was porn (Virtual-Reality Pornography Market Research Report, 2017). Since this digital content attracts a lot of attention, the algorithms notice such a spike in popularity and decide to share it further, thus entering into a viscous loop of sharing dubious content.

Abusive Interaction

One of the advantages of today's online platforms is the possibility of social interactions. However, these communication channels are frequently abused. There have been many cases where guardians discovered graphic messages sent to their children (Livingstone S. L., 2010) (Nielsen, 2015). In some cases, these even involved discussions on sexual activities which led to rape (Wolak, 2010). As discussed in the previous section, this is normally part of the strategy to groom children and get them to send sexually explicit material of themselves (Lenhart, 2009). Whilst some of these predators are explicit in their strategy, others are much more subtle.

There have been several cases where children are lured into installing a third-party app (Hilt, 2017) which displays similar designs to those used in their favourite game. Most people do not check who the creator of the app is and for them, a similar brand is enough to give it legitimacy and guarantee that the platform is safe. Predators are of course betting on this mistake.

Once inside, children are subjected to various abuses. There have been cases where a parent who was monitoring his child's activity saw the child's avatar being gang raped by other avatars (Roblox 'gang rape' shocks mother, 2018). On another occasion (Stonehouse, 2019), another guardian reported that one of the players told his character to lie down, then laid on top of him and began moving in a "disgusting" sexualised manner. And these criminals do not stop there but they also try to take things further. In 2008, Sky News released an investigative report (Farrell, 2011) on a 3D virtual environment called Second Life. Second Life was a virtual, online world where users create avatars which can travel to worlds and lands (called Sims), participate in role-playing games, create and sell products, and socialize with other Second Life residents. In one of the hidden islands of Second Life called Wonderland, Sky News reporters managed to discover a podophile playground. So, the level of abusive interaction is quite widespread and it is rather important to enable measures to protect our children.

BABY CRIMINALS OF THE ONLINE WORLD

Cyberbullying

Cyberbullying is bullying that occurs through the use of digital devices such as cell phones, computers, and tablets. Cyberbullying can take place via many mediums including SMS, Text, apps, as well as online in social media or gaming. On these platforms, people can view, participate and share content aimed at hurting someone else. Because of this, it involves sending, posting, or sharing negative, or derogatory content about another person.

What's worrying about cyberbullying is that it is extremely easy to do and there have been several cases where children were the perpetrators (Nassem, 2015). In fact, around 20-40% of teens declare that they suffer from some sort of cyberbullying according to (Tokunaga, 2010). However, this number might be larger since many teens feel embarrassed about the situation, they're afraid of backlash from the bully or his associates and they are worried that they did something wrong. As a consequence, guardians tend to take drastic measures and remove any digital device in order to prevent future harm. In itself, this is counterproductive because online comments stay online irrespective of whether the child has access to his machine or not. Furthermore, many children declare that one of the reasons why they report this is because they are afraid of losing access to their computer (Aboujaoude, 2015).

Cyberbullying can cause huge psychological traumas to children (Sourander, 2010). But in some cases, children move beyond the online world. This might result in fights or beatings. However, things can even get uglier. According to (Golijan, 2011), in 2011, four teens aged between 13 and 14 years were arrested accused with plotting a murder. These youth suspected a young man of being the "snitch" who got their friend, caught with a firearm at school and posted a series of threatening Facebook messages. Luckily, the law enforcements were alerted and they managed to stop the plot in time. But this proves that online bullying can quickly escalate and run out of hand, leading to more serious problems.

Violent Behaviour

The use of violence by children is often attributed to the negative influence of computer games. In the 2011 London Riots, violent games where questioned (Plunkett, 2011). The same happened after each and every school shooting in the US (Ferguson, 2008). But the truth is that even though there have been various studies on the matter, there was never a conclusive study which highlights this link once and for all. Games are different, they deal with various aspects of life and cover several domains. Violence in games too varies, ranging from cartoonish fighting to gory realistic killing. So, this makes such studies hard.

If we were to have a look at what happens in other similar domains, it is very evident that all forms of entertainment influences us in some way or another. This might range from psychological effects to physical ones. But most of the time, these effects are temporary. So, imagine you've just watched a scary movie, you feel scared for a few hours and then it wears off. The same can be said for games. The main difference is that games are more interactive and because of that, their effect might be more entrenched. Children also normally play for longer hours than the number of hours spent watching a movie. However, notwithstanding the fact that a vast number of hours are spent on gaming every day, the effect on people seems to be negligible.

This probably stems from the fact that from a young age, children are already engaged in role-playing games. Thus, they understand when a role begins and ends, thus allowing them to return back to their normal life. However, this does not hold with all people. There are some who are already disturbed, before playing a game, and playing such games may help to accentuate further their problem. Because of this, we have to be extra careful and ensure that the children playing such games understand the difference between the virtual experiences and the real world, thus allowing to have an enjoyable experience without hurting themselves.

But new technologies are emerging every day and the situation might change in the future. If we have a look at Virtual Reality (VR) system, they provide the user with total immersion which means that most of their senses are being stimulated by the game. We are also aware that exposure to VR does alter in a way our perspective on life (Camilleri, 2017). However, people are very cautious about it and the technology is only used for a few minutes at a stretch thus providing an automatic hand break which avoids overuse.

Virtual Lust

Many online citizens lust for instant wealth and quick rich schemes. The internet, doesn't disappoint here and offer numerous flavours of these schemes. Virtual objects might have no physical presence but they do have real value.

It all started more than a decade ago when games such as Farmville started selling virtual objects like potatoes, chicken feed and others. With games such as Second Life, these objects also took the form of wearables such as a new pair of shoes or a cool hairstyle to embellish one's avatar. This fuelled the rise of the virtual apparel market. Of course, people needed a way to transfer money between the real world and the virtual one, so Linden Lab, the company behind Second Life launched the Linden Dollars which had a value of $1: LD250 (Chen, 2009). With this change, people could have an almost real life on this virtual platform and some of them even managed to become millioners by creating and selling virtual objects. Of course, the rest is history and today we have various games like Roblox or even "Among Us" where they use virtual currencies to sell virtual objects and make money.

A virtual object does not exist in reality, it is a service with some properties of a real object including; the visual properties of an object, the purchasing gesture and the trophy effect. The object is something which is seen (even though it has no physical form). Normally one uses some sort of device such as a tablet or laptop to view the object. The purchasing gesture involves going through the process of buying an object i.e., viewing through a catalogue of items, selecting the object, sending the money and receiving the object. This is very similar to what we do in real life and we get out of the process the same excitement. The final property is the trophy effect which means that once the object comes in the person's possession, he has to showcase it so that others can see it, admire it and also deduce information about the person's social status. As an example, imagine if I bought a virtual sneaker, the avatar has to wear those sneakers so that others can see it. This is why big brands like Nike, Apple and Nissan decided to start selling their products on platforms like Second Life as well (Shelton, 2010). These companies play on the fact that people like to differentiate themselves and long to experience the feel-good factor. In reality, such an approach offers users a better value for their money. Traditional real-life entertainment such as going to the cinema would cost a few Euros in exchange of two hours of passive entertainment. Buying a mythical sword in an online game and for the same amount of money would unlock various possibilities thus resulting in many hours of active entertainment.

But this is not always the case and the value on an item, be it digital or real, is decided by the owner. Some people go to great lengths to buy something unique. When Crypto Kitties[5] launched, people spent millions of dollars to buy a unique kitten. In recent years, non-fungible tokens emerged in the art world. These are digital assets based on blockchain. A person recently bought a work of art called The Pixel (Pricey pixel: Digital artist Pak's NFTs auctioned off for $16.8M, 2021) (which is essentially a single pixel) that exists in digital form and he paid for it $1.76 million. Maybe not everyone sees a value in such an item but this doesn't matter as long as there are people who are ready to pay for it.

The last form of digital objects takes the shape of a person (Leinatamm, 2019). Today we have virtual influencers like Lil Miquela, Imma and many others who seem to be living a real life. These personas, post pictures on social media with real people and take part in photoshoots for big brands like Calvin Klein, Prada, Chanel and others. It is estimated that some of them earn more than $10 million per year. How this market is going to evolve is still unclear, especially with the rise of new technologies but it will heavily contribute to the creation of future online wealth.

Mixed Reality Crimes

A crime can easily span the real and the virtual worlds. In 2008, a 43-year-old Japanese woman was playing a game called Maple Story (Schulzke, 2010). The game is a free-to-play, 2D, side-scroller massive multiplayer online role-playing game, developed by South Korean company Wizet. The woman virtually met and married in the game, a 33 years old man. Unfortunately, they fought so she logged in her virtual husband's online account and killed

---

[5] https://www.cryptokitties.co

his character. Most probably, this is one of the first cases of electronic killing. The man realised what has happened, called the police and filed a real complaint. The woman was then charged with illegal computer access and manipulation of electronic data. If convicted she would face up to 5 years in prison plus a $ 5,000 fine.

At face value, this might seem like a lot but as we've seen in the previous section, people spend a lot of money on their virtual character and its possessions. In actual fact, such a fine might be much less than the value which the person actually lost. Another important thing to point out in this case is that the country where the crime happened was Japan. The Japanese police could proceed against the woman because there were electronic laws in place which protected the rights of the man. Other countries like South Korea have a Game Crime law which even covers crimes happening specifically inside games. But a lot of countries do not have these kinds of laws and because of that, when a crime is committed, the victim is not protected and cannot even refer to the local authorities for help.

Virtual Theft

Since virtual objects now have a real value, it is only natural that some people might try to steal them. This is exactly what a 17-year-old Dutch teenager did in 2007 (Keene, 2012). Essentially, he misled players of the 3D Habbo Hotel social network by creating a fake Habbo Hotel website in order to lure users into entering their account details. Once they did, he would then login with their account and steal virtual furniture from their rooms. This social network was quite popular with teenagers at the time since it boasted around 6 million monthly users.

This was probably the first time that European agencies had to deal with the theft of virtual property. The type of fraud used is rather simple and has been used in other virtual worlds. People find it problematic to protect personal information, such as usernames and password. In fact, after this case, the police all over the world started facing similar reports. In (Smith, 2010) China's police force has been dealing with virtual theft cases for years, including ones involving organized gangs who plan online robbery. Some years ago, officials in the southern city of Shenzhen arrested more than 40 suspects for stealing up to 700,000 yuan (£ 45,500) of virtual items from users of a popular website.

The virtual objects market is a booming industry, but despite the popularity of virtual environments like World of Warcraft, Fortnite, Apex Legends and many others, virtual property laws remain unavailable in most parts of the world.

Virtual Slavery

The virtual economy of several Massive Multiplayer Online (MMO) games and the exchange of virtual items into real money has triggered the rise of game sweatshops. A sweatshop is a workshop, where manual workers are employed at very low wages for long hours and under poor conditions. These workshops are normally found in developing countries such as those located in South America, Asia and Eastern Europe. The job of these people would be to spend around 12 hours a day across an entire month, performing in-game tasks. Most of the cases are usually related to the harvesting of resources or collection of virtual coins. This is what the Chinese Adena Farmers did in the game Lineage II (Dibbell, 2007). In the Star Wars Galaxies game (Denny, 2010), the creators were faced with a large-scale incident whereby these criminals managed to duplicate virtual money. Both of these issues may put tremendous pressure on the artificial economy thus requiring a robust design and frequent updates to maintain a fair pay for work / game balance.

There were also reports of collusion between the farmers and the online currency exchanges. According to (Grimes, 2006) in 2002, Blacksnow Interactive, a game currency exchange, admitted that the company was using workers in a "virtual sweatshop" in Tijuana, Mexico to harvest funds and items for Ultima Online and Camelot's Dark Age. When the game clamped down this practice, Blacksnow tried to sue it.

Virtual Mugging

The term "virtual mugging" refers to the act of attacking or robbing someone in a virtual space. This is normally achieved using pre-programmed bots. By completing in-game tasks very quickly using a bot, the criminal quickly outplays other human-controlled characters gaining an unfair advantage over other players.

One of the original reported cases of "virtual mugging" occurred in the Lineage II game (Joshi, 2008) where players used robots to defeat the characters of other players and take over their items. In Japan, the Kagawa prefectural police arrested a Chinese student on August 16, 2005, after reports that someone carried out a virtual attack and sold stolen items online.

Many game companies are using countermeasures to try to detect the activity of the bots. They do so by asking questions to the character or placing it in a particular situation. If the character is unable to give a satisfying reply, then most problem it is being controlled by a bot. Cyber criminals are becoming increasingly sophisticated since making money through virtual worlds has become very important.

Virtual Pimp

A pimp is a person who controls prostitutes and arranges clients for them while taking a percentage of their earnings in return. This task too migrated online with games like The Sims Online. The game, also referred to EA Land, was a massively multiplayer online variant of the popular computer game The Sims. It gives players total control over the lives of virtual humans called Sims. The look and personality of the Sims can be adjusted to suite the aspirations of their creators. Sims have needs, dreams and aspirations. Through the game, the player can build a home for them, give them a job, make friends with other Sims, enhance their skills and even give them hobbies. Practically they can do almost anything which one could do in real life.

In 2005, a 17-year-old boy name "Evangeline" took advantage of all the flexibility offered by the game (Wilson, 2009). After building a home, he decided to turn it into a cyber-brothel where customers would pay sim-money for minutes of cybersex. Eventually he was discovered and his account was cancelled but no legal action was taken against him mainly because he was above the age of consent. Even though the Sims is rather unique as a game, today there are many open-world games which might give criminals the opportunity to abuse of the flexibility offered by the game and take it into shady areas, beyond its intended use.

Kidnap

What happens in the online world does not always stay over there and in some cases, the issues get transferred to the real world. This is what happened in 2008 (Michels, 2009) when a person with the online name of Kimberly Jernigan started a romance with a 52-year-old person through the Second Life platform. This was an inter-species affair since one of the avatars represented a human and the other a lion. Unfortunately for them, the relationship ended and the Jernigan took it badly.

Since she couldn't accept the breakup, she identified the physical location of the person and decided to go over there. She then devised a plan to kidnap her ex-lover at his job. This did not work out so she decided to kidnap him at home. To do so, she went to great lengths by impersonating a postal worker for four days so she could find his apartment. When she did, she entered his window and hid waiting for her ex-lover. As soon as he came in his apartment, he noticed a silhouette of a person standing who was pointing a gun with a laser beam at his torso. So, he decided to run out and call the police. The police arrived, started searching the neighbourhood and arrested her an hour later in a close by location. Eventually she was arraigned for attempted kidnapping, burglary and other charges.

In this case, one can appreciate that play can have some very real consequences. Some people do not make a distinction between the online and the virtual world. Problems span the worlds thus leading to some very serious consequences.

Murder

Maybe the most heinous crime possible is the one which leads to a murder. (Madary, 2014) is another case where online issue overflow to the real world and with dire consequences. Legend of Mir 3 is another massive multiplayer online role-playing game which features heroes, villains, sorcerers and warriors, many of whom wield enormous swords.

A 41 Chinese player Qiu Chengwei and a friend had jointly won a mythical weapon called the Dragon Sabre. These digital weapons are very rare and are sought after by gamers. Subsequently, he lent it to another friend of him, Zhu Caoyuan who then sold it for around Euro 720. When Qui got to know, he was furious and went to the police. Unfortunately, the police could not help him because in China, there were no laws to regulate the selling of virtual objects (even if they were stolen). So, he decided to take the law in his own hand, and killed Zhu by stabbing him in the chest. Eventually Qiu was sentenced to life imprisonment.

This case raises a few important issues. First of all, players attribute true value to digital assets. Many of them put in numerous hours (if not days) of game play and spend loads of money to acquire these rare objects. The fact that these objects have no physical form and only exist in the virtual world doesn't matter for them. The effort to obtain them is real. So, when they feel cheated (like in this case), they seek redress. The authorities did not offer some sort of solution and this leads to a sense of unfairness. Because of this, local governments should start taking game issues seriously and enable appropriate laws to protect their citizens and avoid such situations.

Organised Crime

Since the online world is becoming so lucrative, organised crime organisations too want to own a slice of that wealth (Lavorgna, 2015). Numerous cases were reported of stolen credit card details. They do so by creating websites which give the impression that they dish out free virtual coins (Moiseienko, 2019). These coins like V-bucks (used in Fortnite) and Robux (used in Roblox) are used within the game to buy digital apparel or other addons. They help the player customise his own avatar or even to enhance his avatar's performance within the game. The websites created by these criminals claim to generate coins. They use a similar branding to the one in the game and are used to lure innocent players into obtaining these coins. Of course, it is too good to be true and once the players type in their personal and credit card details, these criminals use them to purchase virtual coins which of course they keep for themselves.

However, these virtual coins are traceable so the criminals need to launder them. They do so by creating other websites on the dark web where they sell the virtual coins at a discounted price. By doing so, they effectively clean their stolen money. When considering that games like Fortnite have millions of active monthly players, this is big business for these crime syndicates. In fact, according to XeroFOX[6], an IT security company, Fortnite scams can reach up to 53,000 cases per month. When taking into consideration all the other games available online, the global scam market is incredibly huge.

CONCLUSION

From what we've seen in this paper, it is extremely evident that the internet is turning into a world wild web where children are both victims but also perpetrators. Unfortunately, not all parents monitor their children's activities online and most of the monitoring techniques that most parents rely on are not always good enough to help keep their children safe. There is simply too much content being created both directly (uploading of photos) or indirectly (through game interaction) by children and their peers for them to control. Predators who lurk on the web, in places visited by children, smell this content in the same way that sharks are attracted to blood. With the passage of them; the evolution of the internet and the rise of tech-savvy children, this situation will only worsen.

Thus, the best approach to counter these dangers is to inform children about these threats and teach them how to handle them. Very simple instructions understandable by children can go a long way. The following are some of them;

1. Personal details, especially login information, should be kept private and never shared, not even with close friends. They should be stored in a safe place accessible only by who owns them.
2. Secure their accounts with a strong password and when possible, enable two-factor authentication which guarantees a more secure login.

---

[6] https://www.zerofox.com/blog/zerofox-finds-fortnite-scams/

3. Even though remembering passwords can be annoying, the use of similar passwords should be avoided since they can be easily guessed by hackers and their bots.
4. At the end of the online session, users should log out of their account. This will ensure that if a person uses the same machine after them (especially when using public machines), he will not have access to that account and use it for malicious purposes.
5. Information about possible threats and how they work should be shared with children so that they learn to identify them and escape before they get trapped.
6. When someone exhibits bad behaviour, he should be reported immediately.
7. Do not trust websites which propose incredible offers even if they look exactly like the website of the game they're playing. If the offer is too good to be true, then most probably there's something hidden behind it. Only buy from trusted online stores.
8. Avoid public Wi-Fi since one cannot know who is snooping the data passing through that Wi-Fi node.
9. Most programs alert the user when they need to update the software. It is important to keep all the programs up-to-date since hackers tend to exploit vulnerabilities in the software.

These guidelines are rather simple to follow but very important. If we adhere to them, a lot of issues can be avoided. As we've seen from the different cases, the severity of the issue varies. Cybercrime will never be eliminated but if we are cautious, we can reduce it.

REFERENCES

Yar, M. (2018). A failure to regulate? The demands and dilemmas of tackling illegal content and behaviour on social media. International Journal of Cybersecurity Intelligence & Cybercrime.

Staksrud Elisabeth, a. S. (2009). Children and online risk: Powerless victims or resourceful participants? Information, Communication & Society, 12(3), 364-387.

Davidson, J. a. (2008). Protecting children online: Towards a safer internet. In Sex as crime (pp. 338-355).

Oates, B. (2001). Cyber crime: How technology makes it easy and what to do about it. Information systems management, 18(3), 92.

Rhumorbarbe, D. D. (2018). Characterising the online weapons trafficking on cryptomarkets. Forensic science international.

Prensky, M. (2008). The emerging online life of the digital native.

Thomas, M. (n.d.). Deconstructing Digital Natives: Young People, Technology, and the New Literacies (2011 ed.). Taylor & Francis.

Dingli, A. a. (2015). The New Digital Natives. Springer.

Krumm, J. (2018). Ubiquitous computing fundamentals. CRC Press.

Guzsvinecz, T. a.-L. (2019). Suitability of the Kinect sensor and Leap Motion controller—A literature review. Sensors, 19, 1072.

Speicher, M. B. (2019). What is mixed reality? Proceedings of the 2019 CHI Conference on Human Factors in Computing Systems, 1-15.

Oates, S. (2015). Towards an online bill of rights. The onlife manifesto, pp. 229-243.

Bokovnya, A. Y. (2019). Study of Russian and the UK Legislations in Combating Digital Crimes. Helix 9, pp. 5458-5461.

Boas, T. C. (2006). Weaving the authoritarian web: The control of Internet use in nondemocratic regimes. ow revolutionary was the digital revolution.

Akdeniz, Y. (2001). Internet content regulation: UK government and the control of Internet content. Computer Law & Security Review.

Raustiala, K. (2017). An internet whole and free: Why Washington was right to give up control. 96(2), 140-147.

Vendil Pallin, C. (2017). Internet control through ownership: the case of Russia. Post-Soviet Affairs , 33(1), 16-33.

Warf, B. (2011). Geographies of global Internet censorship. GeoJournal , 76(1), 1-23.

Dada, D. (2006). The failure of E-government in developing countries: A literature review. The electronic journal of information systems in developing countries, 26(1), 1-10.

Yar, M. a. (2019). Cybercrime and society. Sage.

Anderson, R. C. (2019). Measuring the changing cost of cybercrime.

Sorell, T. a. (2019). Online romance scams and victimhood. Security Journal, 32(3), 342-361.

Başol, G. a. (2018). Motives and consequences of online game addiction: A scale development study. Archives of Neuropsychiatry, 55(3), 225.

Dean, B. (2021, April 1). Social Network Usage & Growth Statistics: How Many People Use Social Media in 2021? (Backlinko) Retrieved from https://backlinko.com/social-media-users

Talwar, S. A. (2019). Why do people share fake news? Associations between the dark side of social media use and fake news sharing behavior. Journal of Retailing and Consumer Services, 51, 72-82.

Szczurski, M. (2017). Social media influencer-A Lifestyle or a profession of the XXIst century?

Bernal, P. (2018). Facebook: Why Facebook Makes the Fake News Problem Inevitable. Northern Ireland Legal Quarterly, 69(Winter), 513.

News, C. (2019, December 3). Facebook CEO on political ads: People should "judge for themselves the character of politicians". (CBS) Retrieved from https://www.cbsnews.com/news/facebook-ceo-mark-zuckerberg-political-ads-people-should-judge-for-themselves-the-character-of-politicians/

Glenski, M. C. (2017). Consumers and curators: Browsing and voting patterns on reddit. IEEE Transactions on Computational Social Systems, 4(4), 196-206.

Warren, C. (2011, June 16). 10 People Who Lost Jobs Over Social Media Mistakes. (Mashable) Retrieved 2021, from https://mashable.com/2011/06/16/weinergate-social-media-job-loss/?europe=true

Bail, C. A. (2018). Exposure to opposing views on social media can increase political polarization. Proceedings of the National Academy of Sciences, 115(37), 9216-9221.

Cardenal, A. S.-P.-V. (2019). Echo-chambers in online news consumption: Evidence from survey and navigation data in Spain. European Journal of Communication, 34(4), 360-376.

Center, P. R. (2014, June 12). Political Polarization in the American Public. (Pew Research Center) Retrieved 2021, from https://www.pewresearch.org/politics/2014/06/12/political-polarization-in-the-american-public/

Fandom. (2021, April 2). Ask the Audience. (Fandom) Retrieved 2021, from https://millionaire.fandom.com/wiki/Ask_the_Audience

Kossovsky, A. E. (2020). Galileo's Trial and Imprisonment. In The Birth of Science (pp. 133-136). Springer.

Choi, D. S. (2020). Rumor propagation is amplified by echo chambers in social media. In Scientific reports (pp. 1-10). 10.

Cinelli, M. G. (2021). The echo chamber effect on social media. In Proceedings of the National Academy of Sciences (p. 118).

Leonard Busuttil, L. C. (2014). Digital and Video Game Usage in Malta. gamED.

Eurobarometer, F. (2008). Towards a safer use of the Internet for children in the EU–a parents' perspective. European Commission.

Kowalski, R. M. (2012). Cyberbullying: Bullying in the digital age. John Wiley & Sons.

Livingstone, S. a. (2005). UK children go online: Final report of key project findings.

Wang, G. J. (2019). Are Children Fully Aware of Online Privacy Risks and How Can We Improve Their Coping Ability? In KOALA Project Report 3.

Tomczyk, Ł. (2018). Digital piracy among adolescents — scale and conditions. In Proceedings New Trends And Research Challenges In Pedagogy and Andragogy (p. 63).

Power, R. (2020). Child Identity Theft. In Carnegie Mellon CyLab.

Stoilova, M. S. (2019). Children's data and privacy online: growing up in a digital age.

Irshad, S. a. (2018). Identity theft and social media. In International Journal of Computer Science and Network Security (pp. 43-55).

Winters, G. M. (2017). Sexual offenders contacting children online: an examination of transcripts of sexual grooming. In Journal of sexual aggression (pp. 62-76).

Mitnick, Kevin D., and William L. Simon. (n.d.). In The art of deception: Controlling the human element of security (p. 2003). John Wiley & Sons.

Hadnagy, C. (2010). Social engineering: The art of human hacking. John Wiley & Sons.

Zambrano, P. J. (2019). How Does Grooming Fit into Social Engineering? In Advances in Computer Communication and Computational Sciences (pp. 629-639). Springer.

Ellis, S. a. (2017). Hacking growth: how today's fastest-growing companies drive breakout success. In Currency.

Lu, C. a. (1992). Mastering the machine: A comparison of the mouse and touch screen for children's use of computers. In International Conference on Computer Assisted Learning (pp. 417-427). Springer.

Brandom, R. (2017, Dec 8). Inside Elsagate, The Conspiracy-Fueled War on Creepy YouTube Kids Videos. (The Verge) Retrieved 2021, from https://www.theverge.com/2017/12/8/16751206/elsagate-youtube-kids-creepy-conspiracy-theory

What is ElsaGate? (2017). (Reddit) Retrieved 2021, from https://www.reddit.com/r/ElsaGate/comments/6o6baf/

Virtual-Reality Pornography Market Research Report. (2017). (Market Research Engine) Retrieved 2021, from https://www.marketresearchengine.com/virtual-reality-pornography-market

Livingstone, S. L. (2010). Risks and safety on the internet: the perspective of European children: key findings from the EU Kids Online survey of 9-16 year olds and their parents in 25 countries. London School of Economics.

Nielsen, S. S. (2015). 'Pervy role-play and such': girls' experiences of sexual messaging online. Sex Education, 5, 472-485.

Wolak, J. D. (2010). Online "predators" and their victims: Myths, realities, and implications for prevention and treatment. (13).

Lenhart, A. (2009). Teens and sexting. Pew Internet & American Life Project.

Hilt, S. a. (2017). How Cybercriminals Can Abuse Chat Platform APIs as C&C Infrastructures. Trend Micro.

Roblox 'gang rape' shocks mother. (2018, July 3). (BBC) Retrieved 2021, from https://www.bbc.com/news/technology-44697788

Stonehouse, R. (2019, May 30). Roblox: 'I thought he was playing an innocent game'. (BBC) Retrieved 2021, from https://www.bbc.com/news/technology-48450604

Farrell, J. (2011, November 4). Investigation: Paedophilia And Second Life. (Sky News) Retrieved 2021, from https://news.sky.com/story/investigation-paedophilia-and-second-life-10484195

Nassem, E. a. (2015). Why do children bully? School Leadership Today, 6(5).

Tokunaga, R. S. (2010). Following you home from school: A critical review and synthesis of research on cyberbullying victimization. Computers in human behavior, 26(3), 277-287.

Aboujaoude, E. M. (2015). Cyberbullying: Review of an old problem gone viral. Journal of adolescent health, 57(1), 10-18.

Sourander, A. A. (2010). Psychosocial risk factors associated with cyberbullying among adolescents: A population-based study. Archives of general psychiatry, 67(7), 720-728.

Golijan, R. (2011, February 4). Teens arrested over bizarre Facebook death plot. (NBC News) Retrieved 2021, from http://www.nbcnews.com/technology/technolog/teens-arrested-over-bizarre-facebook-death-plot-125357

Plunkett, L. (2011, October 8). Don't Blame Grand Theft Auto for the London Riots . (Kotaku) Retrieved 2021, from https://kotaku.com/dont-blame-grand-theft-auto-for-the-london-riots-5829358

Ferguson, C. J. (2008). The school shooting/violent video game link: Causal relationship or moral panic? Journal of Investigative Psychology and Offender Profiling, 5.

Camilleri, V. M. (2017). Walking in small shoes: Investigating the power of VR on empathising with children's difficulties. 23rd International Conference on Virtual System & Multimedia .

Chen, L. a. (2009). The influence of Virtual money to real currency: A case-based study. IEEE International Symposium on Information Engineering and Electronic Commerce, 686-690.

Shelton, A. K. (2010). Defining the lines between virtual and real world purchases: Second Life sells, but who's buying? Computers in Human Behavior, 26(6), 1223-1227.

Pricey pixel: Digital artist Pak's NFTs auctioned off for $16.8M. (2021, April 14). (Aljazeera) Retrieved 2021, from https://www.aljazeera.com/economy/2021/4/14/pricey-pixel-digital-artist-paks-nfts-auctioned-off-for-16-8m

Leinatamm, K. a. (2019). Virtual avatars rising: the social impact based on a content analysis and a questionnaire in the context of fashion industry.

Schulzke, M. (2010). Defending the morality of violent video games. Ethics and Information Technology, 12(2), 127-138.

Keene, S. D. (2012). Emerging threats: financial crime in the virtual world. Journal of Money Laundering Control .

Smith, R. G. (2010). Identity theft and fraud. Handbook of internet crime, 273-301.

Dibbell, J. (2007). The life of the Chinese gold farmer. New York Times, 17.

Denny, M. L. (2010). Star wars galaxies: control and resistance in online gaming.

Grimes, S. M. (2006). Online multiplayer games: a virtual space for intellectual property debates? New Media & Society, 969-990.

Joshi, R. (2008). Cheating and virtual crimes in massively multiplayer online games. Royal University of London .

Wilson, R. F. (2009). Sex play in virtual worlds. Wash. & Lee L.

Michels, S. (2009, January 8). Virtual Relationship Leads to Criminal Charges. (ABC news) Retrieved 2021, from https://abcnews.go.com/TheLaw/story?id=5650591&page=1

Madary, M. (2014). Intentionality and virtual objects: the case of Qiu Chengwei's dragon sabre. In Ethics and information technology (pp. 219-225).

Lavorgna, A. (2015). Organised crime goes online: realities and challenges. In Journal of Money Laundering Control.

Moiseienko, A. a. (2019). Gaming the System: Money Laundering Through Online Games. RUSI Newsbrief, 39(9).

HATE SPEECH IN DIGITAL GAMES. ARE ONLINE GAMES A PLACE OF DISCRIMINATION AND EXCLUSION?

Sonja Gabriel

Hate speech has become a severe problem in every corner of the internet where communication can take place which means that online games as well as game communities are heavily affected. Although hate speech is not a new phenomenon, it has only been discussed in relation to games for about ten years. The starting point can be seen in the #gamergate controversy. Up to now there has not been a common definition of hate speech but there are some criteria that all the definitions have in common: hate speech are verbal attacks (sometimes also by using pictures) with the aim of humiliating a certain group of people because of their race, their culture, their sexual orientation, their gender, their religion and so on. The affected group of people themselves define if a statement is to be regarded as hate speech which makes it different from a phenomenon like cybermobbing. In some cases, it is not easy to find out if a statement can be regarded as hate speech as it is not only the language itself but the context that needs to be taken into account.

The number of people of all age groups playing digital games has been growing enormously in recent years. As most of the games provide the possibility to communicate with in the game, the number of hate speech incidents has risen as well. But it is not only within games where hate speech takes place, but also in affinity spaces – meaning online communities where people meet to share their common interest - like Twitch, Discord and forums. These toxic environments have called for measures to be taken against this negative development. Game companies and community providers make their players aware of the rules applying to the games, have introduced in-game possibilities to report hate speech and harassment and even make use of artificial intelligence to automatically detect hate speech. However, all these attempts have not proven to be very successful up to now. Incidents like the Black-lives-matter-movement in 2020 have shown that there is a need for making the public aware of all the hate-speech in the online-world, however, leaving the companies promising to take action without having a real solution for the problem.

Keywords: hate speech, affinity spaces, toxic environments, platforms, online games

———

INTRODUCTION

Video games have become a huge market, being worth nearly 160 billion dollars in 2020 (Video Game Industry Statistics, Trends and Data - 2020 & 2021 2020). Many of the games nowadays are online games and allow for communication within their games as well. Additionally, you can find communities around games in forums or social networking platforms like YouTube, Discord or Twitch. Thus, oral or written communication in and around games has become an important part of playing itself which brings along a phenomenon which has been discussed intensively within the last years: hate speech. Before having a look at hate speech in games, the term itself needs to be closer looked at.

Up to now there has been no agreement on a common definition of hate speech. In 1997 the Council of Europe dealt with hate speech and declared it as "covering all forms of expression which spread, incite, promote or justify racial hatred, xenophobia, anti-Semitism or other forms of hatred based on intolerance, including: intolerance expressed by aggressive nationalism and ethnocentrism, discrimination and hostility against minorities, migrants and people of immigrant origin" (Committee of Ministers, 1997). Although this definition mostly refers to hostility towards migrants, there are also other forms of hatred meant, including discrimination against people because of their sexual orientation, mental or physical abilities or being members of certain social classes. These groups have been explicitly integrated in the definition later (Mihajlova et al., 2013). The United Nations define hate speech as follows: "any kind of communication in speech, writing or behaviour, that attacks or uses pejorative or discriminatory language with reference to a person or a group on the basis of who they are, in other words, based on their religion, ethnicity, nationality, race, colour, descent, gender or other identity factor" (United Nations, 2020).

There are, however, some characteristics that are the same in all definitions or at least appear to some extent. These include:

- Hate speech includes verbal attacks (oral or written) or the use of images (memes).
- Statements of contempt for humanity that are made deliberately.
- The aim is to devalue groups on the basis of their skin colour, origin, sexuality, gender, age, religion, disability, membership of a certain social class or occupational group.
- Affected individuals or groups define whether a statement should be regarded as hate speech or not.

The patterns of hate speech are very diverse and can range from the deliberate spreading of false statements, which can also be disguised as humour or irony, to derogatory terms and insults, to the spreading of stereotypes and prejudices and conspiracy theories. Generalisations or the linguistic division into "us" and "them" are also frequent. Poster-like imagery or memes, which also reproduce stereotypes and/or reinforce prejudices, are quite often used in social networks or forum discussions. In some cases hate speech can be identified quite easily (see for example Liukkonen, 2018a), but in some other incidents hate speech is harder to detect, so can irony or sarcasm be used (which is especially harder to decipher in written language), sometimes emojis are inserted which might even contradict the statement written before or new language expressions are created (for example rapefugees). In forums or social networks also hashtags might be used to express hate speech (ElSherief et al. 2018).

According to (Schmitt, 2017), the reasons for hate speech can be attributed to numerous factors, but can be roughly divided into four motivations. These include

- Exclusion: People (groups) are excluded from social networks or other online groups. Within these groups, hate speech leads to verbal disparagement and thus to the targeted marginalisation and exclusion of these groups as well as to dehumanisation, which lowers the inhibition threshold for physical violence.

- Intimidation: People often feel threatened by other groups and try to intimidate this foreign group of people by verbalising their hatred. The focus is on the interests of one's own group, while fear, doubt and conflict are stirred up against "the others".
- Dominance and interpretative sovereignty: One's own group is seen as socially superior. Hate speech is used to devalue other groups of people and also to get other people to adopt the attitude of the hate speakers.
- Fun and thrill: Due to the (supposed) anonymity of the internet, there may well be people who insult and humiliate other groups not so much because of their convictions and values, but simply for the fun of it.

Especially through the spread of social networks and increased communication possibilities via the internet - thus the producer/consumer relationship has shifted, as everyone can consume content, but also produce it - the dissemination of news has changed greatly (Latour, 2017).

HATE SPEECH AND DIGITAL GAMES

Due to improved internet connections and the possibility of going online on the move by using smartphones as game-consoles, the number of online games has been increasing during the recent years. Of course, hate speech in and around video games mainly affects multiplayer games, i.e. those games in which playing together or against each other via the internet is an essential part of the gameplay. Especially in massively multiplayer online role-playing games (MMORPG), i.e. internet-based computer games in which up to several thousand players can be present at the same time, communication comes to the fore due to the highly complex processes in the game (Stertkamp, 2017). Avatars are used to interact with other players or non-playable characters (NPCs) in the persistent virtual game world. The increased need for communication in such games (such as World of Warcraft) results from the fact that certain quests (tasks) can only be completed as a team. This does not mean, however, that it is only online role-playing games that are affected by hate speech, as there are now numerous online-based games with integrated communication options. Free-to-play games, which are mainly playable via mobile devices, are particularly interesting for children and young people. According to the JIM Study 2020, Fortnite and Minecraft – both games playable via Smartphone or tablet-PC as well - are the most popular games among 12- to 17-year-olds (Medienpädagogischer Forschungsverbund Südwest, 2020). Especially Fortnite has been attractive for teenager and children alike since it was published. Although there is a legal age restriction, many younger children also take to the game. As per Kulman (2019), the youngest competitor at the 2019 Fortnite World Cup was 13 years old. Apart from the contents of the game which has been a topic of discussion among parents and teachers (cf. for example Berthelson, 2018; Ehmke, 2018; Ucciferri, 2018), the game has also been in the spotlight because of hate speech. As per a report by Anti-Defamation League (2019), 70 % of the Fortnite players have admitted that they had experienced some form of harassment within the game.

It is precisely these communication options, which are integrated into many online games, that also offer themselves as ideal breeding grounds for hate speech. This problem has become so widespread that it is increasingly being reported on in the international media. In 2010, for example, the NBC News website published an article about hate speech in the online version of Call of Duty: "One gamer told an opponent he presumed to be Jewish that he wished Hitler had succeeded in his mission. Many exchanges involve talk of rape or exult over the atomic bombing of Japan. There are frequent slurs on homosexuals, Asians, Hispanics and women" (Geraniols, 2010). But also actions within online games are used to humiliate other players or to impose an opinion on them: "Take the survival shooter game H1Z1: King of the Kill, currently the third-most-popular on the world's biggest online game platform. Matches in Asia are sometimes interrupted by the Red Army, a band of Chinese players who have won praise from local media for championing in-game nationalism. One tactic involves cornering rivals and forcing them to pay tribute to the motherland by saying "China No. 1." Those who fail to comply are swiftly dispatched" (Bloomberg, 2017). In some games, like for example Minecraft, users construct items that refer to Hitler and National Socialism in order to get a certain message across. Even if this is not a verbal statement, it can be regarded as part of hate speech as well. Another widespread method of hate speech within games and game communities is to choose an avatar name or clan name which humiliates a certain group of people (for example BadNewsForJews or FuckTheNiigPig as Clan names in Clash Royale).

The issue of hate speech in and around games became particularly well known through the so-called Gamergate affair. The feminist video blogger Anita Sarkeesian launched an extremely successful Kickstarter campaign in spring 2012 to raise money to create a series of videos on the topic of female stereotypes entitled Tropes vs Women in Video Games. Already in the funding phase, the blogger was called sexist names, belittled and threatened with violence. She collected all these attacks on her own website (Sarkeesian, 2012). A game was even developed based on this very bipolar discussion, in which game players could beat up a character in Sarkeesian's image (Fernandez-Blance, 2012). Under #gamergate, a discussion flared up that went far beyond gaming forums and was also heavily discussed in the media (Freidel, 2014).

In the English-speaking world, the terms "toxic environment" and "toxic game communities" are often used (Liukkonen, 2018b). Especially when playing games, many (negative) emotions are released, which can then be discharged in comments and statements towards other players. In some gaming environments, insulting and disparaging comments are the norm, even though these gamers do not have extreme political views and would not describe themselves as discriminatory. Often, such gamers state that they did not mean the comments to be taken seriously anyway (Hardaker, 2010; Craker und March, 2016). In addition, unlike in social networks, the others often simply refer to fellow gamers, without them belonging to a specific group. "Most importantly, the "other" is not me or us, the "other" is the opposite of me. And in the mind of every gamer who has ever engaged in toxic behaviour, I am better than the "other"" (Liukkonen, 2018). As (Breuer, 2017) also notes, defeat in online games can also evoke negative emotions, making aggressive behaviour more likely. The author also states that gamers who strongly identify with the hobby of online games or certain genres or games may also perceive the presence of people who are different from themselves (for example in terms of gender, ethnicity or sexual orientation) as a threat to their social identity if they believe that certain groups of people are not "true/real/right" gamers (ibid. p. 110).

Groups that want to spread their hatred of certain other groups also use computer games, among other things, by integrating this hatred into the game design. For example, computer games such as Ethnic Cleansing or Shoot the Blacks were already created at the beginning of the 20th century in which specific ethnic groups were the target (Council of Europe, 2012). Another factor in the spread of hate speech is the growing interest in e-sports and its large worldwide distribution. Spectators in particular play a major role in this. In February 2019, a tweet from a Danish Counter-Strike team caused a stir. In public chats or direct messages, the professional players received messages such as "I hope you all die" (Krishan, 2019), which were made public by the team captain. In 2018, US-based female e-sports player Spawntaneous began posting videos that provide insight into what female gamers are constantly confronted with. These range from exclamations such as "Oh my God, a girl!" to pushy flirting attempts, requests to send nude photos to sexist stereotypes or threats of rape. As a result, the topic of hate speech received media attention and was discussed, especially in the field of e-sports.

AFFINITY SPACES

Affinity spaces, as defined by Gee und Hayes (2012), are places where people voluntarily come together in groups because of a common interest or activity. These spaces can also be online, e.g., within game communities. The feeling of belonging is not determined by origin, social background, ethnicity or gender, but mainly by common interests, goals and activities. The exchange and development of knowledge and skills also takes place there (Gee 2013, 2017). For some people, online role-playing games in particular function as a so-called third place, which, in addition to home and work, offers a space that one chooses voluntarily, and with which one identifies. These places convey a feeling of home and serve to relieve stress or escape from everyday life. Communication plays a major role in these third spaces (Ducheneaut et al. 2007).

Content in affinity spaces is created primarily through social interactions and activities in this social space (Hudson et al., 2016). "This whole set of physical and virtual spaces (and the physical and digital routes among them) that characterize the comings and goings of gamers with a shared affinity is an affinity space. These sorts of affinity spaces today are often really squishy. They are fluid and ever changing and hard to strictly demarcate" (Gee 2017, p. 29). Participation can take place in different forms and ways, there are no fixed leaders of these groups and members achieve status through different activities (Gee, 2004).

This means, then, that online games as well as the forums and community offerings around computer games (such as Let's Play channels) fall under the definition of affinity spaces. This also includes services such as Discord or the community around the online game distributor Steam. Here, like-minded people exchange information about game strategies, game experiences and the like, or when it comes to communication in the game, especially about the current game situation and the roles in the team, but also about everyday topics. These affinity spaces, which basically promote learning and exchange, can also be places where hate speech arises and is carried on in various forms (Moor et al., 2010). In online games, the targets of exclusion are not only the groups generally affected, such as women, members of other cultures and religions, but also inexperienced players. A study on the well-known massively multiplayer online role-playing game World of Warcraft found that the community here is particularly hostile towards homosexuals, bisexuals and transgender people (Pulos, 2013). Another study of the Xbox Live community found that women of colour who communicate with their team members via in-game voice chat are particularly affected by hate speech (Gray, 2016). This behaviour means that women often do not see themselves as gamers, as the identity and group membership is strongly male-defined (Groen, 2017). This feeling is also reinforced above all by the content and portrayal of women in digital games - for the most part, they are portrayed in a highly sexualised and helpless manner (Chenelle, 2019). The design of games may well make hate speech more acceptable in computer game environments than in other areas (Rogers, 2016).

MEASURES TO BE TAKEN AGAINST HATE SPEECH

Hate Speech is not only a problem in and around games. In fact, it has become a topic widely discussed by various organisations as the easy access to online communities, social networks and other means of seemingly anonymous ways of communicating globally has resulted in an enormous increase of this phenomenon. The UN, for example, has dealt with the topic intensively, talking about what kind of hate speech there is and how it can / should be restricted by law and proposing a strategy consisting of 13 commitments (United Nations, 2020). Game companies and communities have been trying to tackle the problem (mostly not very successful) for years. In many communities and online networks, there is a Community Code of Conduct - a collection of rules and behavioural guidelines on how to behave in the network or in the game, which now also explicitly mention hate speech. In these Codes of Conduct, behaviours that fall under the definition of hate speech are thus explicitly described as a violation. However, there is no clear statement of what happens if the Community Code of Conduct is breached - it quite often only refers in very general terms to the fact that the company reserves the right to delete statements and messages without prior notice or to suspend accounts temporarily or permanently. For example, consequences of violating the guidelines of the Online RPG Dungeons and Dragons will also - at the discretion of the platform operator - be admonished or (especially in the case of serious or repeated violations) temporarily or permanently banned from the game. Game characters or game accounts may even be deleted without prior warning. In addition, all communications within the game (including private chats) are logged so that they can be accessed in case of doubt.

In other online games, such as the popular Fortnite Battle Royale (Epic Games), there is an option within the game to report hate speech and players harassing others by using a system within the game (Ditch the Label, 2019). The operators of Rainbow Six Siege have gone one step further: since July 2018, players have been automatically banned for a short period of time if they use racist or homophobic remarks in text chat (Park, 2018). These and similar measures were announced on the developer's blog a little earlier, as Ubisoft feels compelled to do something about the rising toxicity: "Our end goal is to track negative player behaviour, manage those that behave poorly, and eventually implement features that will encourage players to improve their behaviour" (Ubisoft, 2018).

The occurrence of hate comments may even cause players to refrain from using an online community any further, as the example from the League of

Legends community shows. For example, a user posts the following: "I'm tired of playing games and getting harassed by people. If the internet was properly policed most of your community would be rotting in jail for hate crimes. I'm tired of it. I shouldn't have to mute players every game I play. I shouldn't have to deal with the homophobic and racial slurs. Reporting players does nothing because they'll simply make a new account and keep harassing other players. I can't just play a game, and have fun? I have to be harassed and hazed for simple mistakes. I just want to play a game - but if I make even the simplest mistake I get bombarded with BULLSHIT. Delete this post Riog, idgaf. I'm done with your toxic community, and I'm done with your piss-poor policies that commend this sort of behaviour." (source of the posting: https://boards.na.leagueoflegends.com/en/c/player-behavior-moderation/xnEKAVJo-homophobic-slurs-every-game-im-done). As can be seen from this posting, gamers are not really satisfied with how the operators of online games or the associated forums deal with hate speech. Numerous postings call for a tougher crackdown and monitoring of communication in the game so that hurtful and discriminatory messages do not go undetected and, above all, unpunished. In addition, as can be seen from the above quote, the powerlessness and helplessness vis-à-vis the perpetrators of hate postings is made clear, since even the blocking or deletion of an account on the part of the operator websites does not mean that hate speakers will stop with their actions. Therefore, the function "Report Hate Speech", which now exists in numerous games or communities, is not given much importance. "And then he ended the conversation by telling me that he has 22 accounts and if I try to report him for threatening to kill me then he will create 6 more accounts to harass me. He said something like 'cut one head off and 6 more will grow back'" (source of the posting: https://boards.na.leagueoflegends.com/en/c/player-behavior-moderation/xnEKAVJo-homophobic-slurs-every-game-im-done). The problem is often that written statements (e.g. in chats or forums) can be reported, but with voice chats the evidence is much more difficult. Player or guild names cannot always be easily reported via an in-game function. In Fortnite, for example, this function is well hidden in the feedback menu (Knoop, 2018). Discord, which is rather popular with all kind of gamers for discussing games and game-play, has been accused of providing home for hate speech for quite a while. Although there are community rules stating that hate speech will not be tolerated, there do not seem to be taken enough measures against it. In Discord's transparency report from the first half of 2020 (Discord Transparency Report: Jan — June 2020 - Discord Blog 2020), nearly 37 % of all reports received were categorised as harassment, but action taken by the provider was only taken in 13 %.

In February 2021 the game Destruction AllStars (Lucid Games) was not only reviewed as having new and fun gameplay but also because of having a lot of troubles with hate speech. The voice chat which originally has been switched on by default made players listen to racial discrimination, homophobia, sexism and so on (Notis, 2021a). The game itself did not provide a lot of help for players taking action against hostile players at the beginning. Because of the complaints being overwhelming, Lucid Games released a game fix few days after the introduction of the game to switch the default chat to off (Notis, 2021b). This example shows that there is not enough awareness among game companies and community providers when it comes to tackling hate speech.

One of the problems of tackling hate speech can be seen in the fact that it is rather hard to find out if it is really harassment or if it just serves as a joke or irony. Many companies and communities use artificial intelligence to watch over chats or forum entries. The algorithms which are used on the basis of deep learning face problems when it comes to natural language processing and generating. First of all, there is the question of context – while some words are considered toxic when used (f. ex. queer), this is not the case when it is used by a member of this community. Sap et al. (2019) found out in their study that – depending on the material the algorithms are trained upon – the artificial intelligence might be biased as well: "We find strong evidence that extra attention should be paid to the confounding effects of dialect so as to avoid unintended racial biases in hate speech detection" (ibd. p. 1672). A further problem is that these models used for detection of hate speech are rather sensitive. This means that for example removing the spaces between hateful words or using leet-speech (substituting symbols or numbers for letters) might result in not detecting statements as hate speech by the software but, however, players can read the statements perfectly well. League of Legends (Riot Games) has long been seen as a very toxic environment. To tackle this the company introduced the Tribunal in 2011: Gamers who volunteered for taking part reviewed the reports by other players and decided upon the punishment. As the system was quite slow and inefficient as well as often inaccurate, Riot closed down the Tribunal in 2014 and replaced it by a machine learning solution (Crabb, 2019). Valve takes a slightly different direction by having their users decide if they want offensive language blurred on Steam. Users can even add or remove words from their personal filter. As this might seem a good solution, it means that the Chat filtering option does not get rid of offensive language and hate speech but just have people decide if they want to see it or not (Grubb, 2020).

Of course, the political level is also called upon - the topic of hate speech is currently still handled completely differently in Europe. In Austria, for example, the first law to combat hate speech in online environments was only passed in autumn 2020. Digital games in particular occupy a special position here. When the Netzdurchwirkungsgesetz (Network Enforcement Act) came into force in Germany in 2018, which is intended to curb hate content on the internet, online games were not included, although they were included in an initial draft of the law (Brause, 2018). The protection of minors from harmful media also does not classify games according to the potential of hate speech, but according to the content of the games, the graphics, the language used by the game characters, etc. This naturally raises the question of whether the games are suitable for minors. This naturally raises the question of how children and young people in particular can be protected from being exposed to such hate comments.

CONCLUSION

Hate speech is a growing problem in online games and in communities around digital games. Groups that are also affected by exclusion, discrimination and racism in real life due to gender, origin, religious or sexual orientation are also confronted with stereotypes and negative comments in game environments. Pointing out these toxic behaviours leads to their being known and discussed by a broader public, but does not yet lead to the containment of hate speech. The operators of online games and communities are increasingly called upon to take measures that raise awareness that hate comments are not a trivial offence and that such behaviour has no place in an online gaming world that is actually accessible to everyone and should focus on the fun of playing together without barriers or prejudice. The game designers of computer games are also called upon to change the strongly stereotypical portrayal of women, but also of other groups - especially in blockbuster games. The pressure on companies to do something against hate speech is getting harder as there are more and more news articles (not only in gamer-focussed magazines) deal with the topic of hate speech in games. In 2020, the case of the Black-lives-matter-movement was also mirrored in many games: There were in-game splash messages in Fortnite reminding players to be awesome to each other. Ubisoft, Nintendo, Playstation and others also promised to fight racism in their game communities, however, they did not give any exact details of how to do that (Smith, 2020). In some cases even the communities themselves try to force platforms to do more against hate speech (Powell, 2020; Statt, 2020). But it is not only the platforms and politics that should do something against an environment getting more and more toxic – it is everybody who plays an online game, takes part in a forum discussion or registers for a social media account. A community which does not accept hate speech as a form of communication is much more powerful than any rules, laws and artificial intelligence detention can ever be.

REFERENCES

Anti-Defamation League (Ed.) (2019): *Free to Play? Hate, Harassment and Positive Social Experiences in Online Games.* Retrieved from https://www.adl.org/free-to-play, last access on 12.02.2021.

Berthelson, Brandy (2018): *Teachers are Using Fortnite as a Teaching Tool.* In: *SuperParent,* 08.06.2018. Retrieved from https://superparent.com/article/111/teachers-are-using-fortnite-as-a-teaching-tool, last access on 12.02.2021.

Bloomberg (Ed.) (2017): *Racial and ethnic hate speech thrives in online games.* Retrieved from https://www.japantimes.co.jp/news/2017/05/14/business/tech/racial-ethnic-hate-speech-thrives-online-games/#.XLlzQaTgqUk, last access on 19.04.2019.

Brause, Christina (2018): *Die gefährliche Lücke im Maas-Gesetz.* Retrieved from https://www.welt.de/politik/deutschland/article173229873/NetzDG-Justizminister-Heiko-Maas-verschont-Onlinespiele-Branche.html, last access on 17.06.2019.

Breuer, Johannes (2017): *Hate Speech in Online Games.* In: Kai Kaspar, Lars Gräßer und Aycha Riffi (Ed.): Online Hate Speech. Perspektiven auf eine neue Form des Hasses. Düsseldorf, München: www.kopaed.de (Schriftenreihe zur digitalen Gesellschaft NRW, Band 4), S. 107–112.

Chenelle (2019): *Wie die Gamingindustrie den Sexismus prägt und fördert*. Retrieved from https://blog.hslu.ch/majorobm/2019/03/17/teil-2-studien-und-fakten-wie-die-gamingindustrie-den-sexismus-pr%C3%A4gt-und-f%C3%B6rdert/, last access on 09.06.2019.

Committee of Ministers (1997): *Recommendation No. R (97) 20 of the Committee of Ministers to member states on "hate speech"*. Retrieved from https://www.coe.int/en/web/freedom-expression/committee-of-ministers-adopted-texts/-/asset_publisher/aDXmrol0vvsU/content/recommendation-no-r-97-20-of-the-committee-of-ministers-to-member-states-on-hate-speech-?_101_INSTANCE_aDXmrol0vvsU_viewMode=view, last access on 10.02.2021.

Council of Europe (Ed.) (2012): *Mapping study on projects against hate speech online.* Retrieved from http://rm.coe.int/09000016807023b4, last access on 20.04.2019.

Crabb, Jacob (2019): *Classifying Hate Speech: an overview - Towards Data Science*. In: *Towards Data Science*, 28.05.2019. Retrieved from https://towardsdatascience.com/classifying-hate-speech-an-overview-d307356b9eba, last access on 14.02.2021.

Craker, Naomi; March, Evita (2016): *The dark side of Facebook®: The Dark Tetrad, negative social potency, and trolling behaviours*. In: *Personality and Individual Differences* 102, S. 79–84. DOI: 10.1016/j.paid.2016.06.043.

Discord Transparency Report: Jan — June 2020 - Discord Blog (2020). In: *Discord Blog*, 27.08.2020. Retrieved from https://blog.discord.com/discord-transparency-report-jan-june-2020-2ef4a3ee346d, last access on 14.02.2021.

Ducheneaut, Nicolas; Moore, Robert J.; Nickell, Eric (2007): *Virtual "Third Places": A Case Study of Sociability in Massively Multiplayer Games*. In: *Comput Supported Coop Work* 16 (1-2), S. 129–166. DOI: 10.1007/s10606-007-9041-8.

Ehmke, Rachel (2018): *A Parent's Guide to Dealing With Fortnite*. In: *Child Mind Institute*, 29.06.2018. Retrieved from https://childmind.org/article/parents-guide-dealing-fortnite/, last access on 12.02.2021.

ElSherief, Mai; Kulkarni, Vivek; Nguyen, Dana; Wang, William Yang; Belding, Elizabeth (2018): *View of Hate Lingo: A Target-Based Linguistic Analysis of Hate Speech in Social Media*. In: *Proceedings of the Twelfth International AAAI Conference on Web and Social Media*, S. 42–51. Retrieved from https://ojs.aaai.org/index.php/ICWSM/article/view/15041/14891, last access on 15.02.2021.

Fernandez-Blance, Katherine (2012): *Gamer campaign against Anita Sarkeesian catches Toronto feminist in crossfire*. Ed. v. The Star. Retrieved from https://www.thestar.com/news/gta/2012/07/10/gamer_campaign_against_anita_sarkeesian_catches_toronto_feminist_in_crossfire.html, zuletzt aktualisiert am 20.04.2019.

Freidel, Morten (2014): *Wenn Kritik kommt, hört das Spiel auf*. Ed. v. FAZ. Retrieved from https://www.faz.net/aktuell/feuilleton/medien/gamergate-wenn-kritik-kommt-hoert-das-spiel-auf-13232818.html, last access on 20.04.2019.

Gee, James Paul (2004): *Situated language and learning. A critique of traditional schooling*. New York, London: Routledge (Literacies). Retrieved from http://site.ebrary.com/lib/alltitles/docDetail.action?docID=10162948.

Gee, James Paul (2013): *Good video games and good learning.* Collected essays on video games, learning and literacy. 2. ed. New York: Peter Lang (New literacies and digital epistemologies, 67).

Gee, James Paul (2017): *Affinity Spaces and 21st Century Learning.* In: *Educational Technology* 57 (2), S. 27–31. Retrieved from www.jstor.org/stable/44430520, last access on 19.04.2019.

Gee, James Paul; Hayes, Elisabeth (2012): *Nurturing Affinity Spaces and Game-Based Learning.* In: Constance Steinkuehler, Kurt Squire und Sasha A. Barab (Ed.): Games, Learning, and Society. Learning and meaning in the digital age. Cambridge: Cambridge University Press (Learning in Doing: Social, Cognitive and Computational Perspectives), S. 129–153.

Geraniols, Nicholas K. (2010): *Hate speech corrodes online games. Players report frequent slurs on homosexuals, Asians, Hispanics and women.* Retrieved from http://www.nbcnews.com/id/36572021/ns/technology_and_science-games/t/hate-speech-corrodes-online-games/#.XLlxT6TgqUk, last access on 19.04.2019.

Gray, Kishonna L. (2016): *Race, gender, and deviance in Xbox Live. Theoretical perspectives from the virtual margins.* London: Routledge.

Groen, Maike (2017): *"gogo let's rape them" - Sexistischer SpracEdebrauch in Online Gaming Communities.* In: Kai Kaspar, Lars Gräßer und Aycha Riffi (Ed.): Online Hate Speech. Perspektiven auf eine neue Form des Hasses. Düsseldorf, München: www.kopaed.de (Schriftenreihe zur digitalen Gesellschaft NRW, Band 4), S. 113–119.

Grubb, Jeff (2020): *Steam chat tool removes slurs, but it should remove the slur users.* In: *VentureBeat,* 27.08.2020. Retrieved from https://venturebeat.com/2020/08/27/steam-chat-tool-removes-slurs-but-it-should-remove-the-slur-users/, last access on 14.02.2021.

Hardaker, Claire (2010): *Trolling in asynchronous computer-mediated communication: From user discussions to academic definitions.* In: *Journal of Politeness Research. Language, Behaviour, Culture* 6 (2), S. 277. DOI: 10.1515/jplr.2010.011.

Hudson, Richard C.; Duncan, Sean; Reeve, Carlton (2016): *Affinity Spaces for Informal Science Learning: Developing a Research Agenda.* Report of Workshop held July 6 & 7 at Games+Learning+Society 2015. Ed. v. TPT Twin Cities Public Television. Retrieved from https://www.informalscience.org/affinity-spaces-informal-science-learning-developing-research-agenda-report-workshop-held-july-6-7, last access on 19.04.2019.

Knoop, Joseph (2018): *What happens if you say something racist in the most popular multiplayer games.* Retrieved from https://www.pcgamer.com/what-happens-if-you-say-something-racist-in-the-most-popular-multiplayer-games/, last access on 17.06.2019.

Krishan, Marvin Rishi (2019): *"Ich hoffe, ihr sterbt alle" E-Sport und Hate Speech.* Ed. v. Spiegel Online. Retrieved from https://www.spiegel.de/netzwelt/games/e-sport-und-stoptoxicity-ich-hoffe-ihr-sterbt-alle-a-1253227.html.

Kulman, Randy (2019): *How Young Is Too Young for Fortnite?* Ed. v. Psychology Today. Retrieved from https://www.psychologytoday.com/us/blog/screen-play/201908/how-young-is-too-young-fortnite, zuletzt aktualisiert am 12.02.2021, last access on 12.02.2021.

Latour, Agata de (2017): *We can! Taking action against hate speech through counter and alternative narratives*. Revised edition. Ed. v. Celina Del Felice und Menno Ettema. Strasbourg: Council of Europe.

Liukkonen, Ina (2018a): *Hate Speech in the Gaming Community and Social Media*. BrandBastion. Retrieved from https://info.brandbastion.com/hate-speech-in-the-gaming-community-and-social-media, zuletzt aktualisiert am 13.02.2021, last access on 15.02.2021.

Liukkonen, Ina (2018b): *Managing Hate Speech in Gaming Media*. Ed. v. BrandBastion. Retrieved from https://info.brandbastion.com/hate-speech-in-the-gaming-community-and-social-media, last access on 20.04.2019.

Medienpädagogischer Forschungsverbund Südwest (Ed.) (2020): *JIM-Studie 2020. Jugend, Information, Medien*. Retrieved from https://www.mpfs.de/studien/jim-studie/2020/, zuletzt aktualisiert am 12.02.2021, last access on 12.02.2021.

Mihajlova, Elena; Bacovska, Jasna; Shekerdjiev, Tome (2013): *Freedom of expression and hate speech*. Skopje: OSCE, Mission to Skopje.

Moor, Peter J.; Heuvelman, Ard; Verleur, Ria (2010): *Flaming on YouTube*. In: *Computers in Human Behavior* 26 (6), S. 1536–1546. DOI: 10.1016/j.chb.2010.05.023.

Notis, Ari (2021a): *How To Mute Jerks In Destruction AllStars*. In: *Kotaku*, 03.02.2021. Retrieved from https://kotaku.com/how-to-mute-jerks-in-destruction-allstars-1846188716, last access on 12.02.2021.

Notis, Ari (2021b): *You No Longer Have To Hear Hate Speech While Playing Destruction AllStars*. In: *Kotaku*, 02.05.2021. Retrieved from https://kotaku.com/you-no-longer-have-to-hear-hate-speech-while-playing-de-1846205888, last access on 14.02.2021.

Powell, Steffan (2020): *Twitch: Streamers call for a blackout to recognise victims of sexual and racial abuses*. In: *BBC News*, 24.06.2020. Retrieved from https://www.bbc.com/news/newsbeat-53164395, last access on 15.02.2021.

Pulos, Alexis (2013): *Confronting Heteronormativity in Online Games. A Critical Discourse Analysis of LGBTQ Sexuality in World of Warcraft*. In: *Games and Culture* (8 (2)), S. 77–97.

Rogers, Ryan (2016): *How video games impact players. The pitfalls and benefits of a gaming society*. Lanham, Boulder, New York, London: Lexington Books.

Sap, Maarten; Card, Dallas; Gabriel, Saadia; Choi, Yejin; Smith, Noah A. (2019): *The Risk of Racial Bias in Hate Speech Detection*. In: *Proceedings of the 57th Annual Meeting of the Association for Computational Linguistics*, S. 1668–1678.

Sarkeesian, Anita (2012): *Image Based Harassment and Visual Misogyny*. Retrieved from https://feministfrequency.com/2012/07/01/image-based-harassment-and-visual-misogyny/, last access on 16.06.2019.

Schmitt, Josephine B. (2017): *Online Hate Speech: Definition und Verbreitungsmotivationen aus psychologischer Perspektive*. In: Kai Kaspar, Lars Gräßer und Aycha Riffi (Ed.): Online Hate Speech. Perspektiven auf eine neue Form des Hasses. Düsseldorf, München: www.kopaed.de (Schriftenreihe zur digitalen Gesellschaft NRW, Band 4), S. 51–56.

Smith, Noah (2020): *Video game companies vow to fight racism in their communities, but offer few details*. In: *The Washington Post*, 16.06.2020. Retrieved from https://www.washingtonpost.com/video-games/2020/06/16/video-game-companies-vow-fight-racism-their-communities-offer-few-details/, last access on 15.02.2021.

Statt, Nick (2020): *Some popular Reddit communities go private to protest the platform's hate speech policies*. In: *The Verge*, 06.03.2020. Retrieved from https://www.theverge.com/2020/6/3/21279601/reddit-dark-subreddits-protest-police-violence-racism-hate-speech-policies, last access on 15.02.2021.

Stertkamp, Wolf (2017): *Sprache und Kommunikation in Online-Computerspielen. Untersuchungen zu multimodaler Kommunikation am Beispiel von World of Warcraft*. Gießen: Gießener Elektronische Bibliothek (Linguistische Untersuchungen, 11).

Ubisoft (Ed.) (2018): *Dev Blog: Toxicity*. Retrieved from https://rainbow6.ubisoft.com/siege/en-us/news/152-321200-16/dev-blog-toxicity, last access on 16.06.2019.

Ucciferri, Frannie (2018): *Parents' Ultimate Guide to Fortnite*. In: *Common Sense Media*, 06.03.2018. Retrieved from https://www.commonsensemedia.org/blog/parents-ultimate-guide-to-fortnite, last access on 12.02.2021.

United Nations (Ed.) (2020): *United Nations Strategy and Plan of Action on Hate Speech*. Detailed Guidance on Implementation for United Nations Field Presences. Retrieved from https://www.un.org/en/genocideprevention/hate-speech-strategy.shtml, last access on 14.02.2021.

Video Game Industry Statistics, Trends and Data - 2020 & 2021 (2020). In: *WePC.com*, 13.10.2020. Retrieved from https://www.wepc.com/news/video-game-statistics/, last access on 21.01.2021.

'WHOSE FREEDOM'? THE LOSS AND RESTRICTION OF LUDIC AND POLITICAL AGENCY IN GAMES.

Tobias Unterhuber

Agency is not only a central concept in game studies but also in political discourse. What we are allowed and what we ought to do in a society is the range of our agency. It can be limited by laws, implicit and explicit rules, ethics and authority Furthermore, tactics of marginalization, often based on race, gender and class, restrict people's agency even further. The specific mediality of video games always affords players agency. The range may vary, but it is essential for games that players can choose their actions, to make decisions. This has two consequences: 1. Society and games are both rule-governed systems, which give players and people agency. Therefore, ludic agency can act as a structural analogy of political and social agency. Players' agency can represent the player character's agency in a fictional world and society. 2. Since players are accustomed to having agency, the loss and restriction of their agency can be a powerful tool to show the aforementioned analogy and to let players experience, in a safe media environment, how people's agency is not identical based on their class, their race and their gender. The paper shows how games implement situations of agency loss and restricted agency to represent lost and restricted social and political agency.

Keywords: Agency, Subjectivity, Classism, Racism, Sexual Harassment

---

INTRODUCTION[1]

So, who doesn't love agency? It is one of the defining features of video games, a part of their specific mediality. It gives us, as players, the feeling of being powerful and able to change the (game) by our actions. It is also one of

---

[1] All translations of German quotes are my own.

the key research interests in game studies since day one. Janet Murray already sees it as a central point in digital media or the computer in general.

> Agency is the satisfying power to take meaningful action and see the results of our decisions and choices. We expect to feel agency on the computer when we doubleclick on a file and see it open before us or when we enter numbers in a spreadsheet and see the totals readjust. However, we do not usually expect to experience agency within a narrative environment. (Murray, 2016, p. 159)

Murray sees the possibility for agency in narrative forms in digital media as well and this leads us directly to games:

> Agency, then, goes beyond both participation and activity. As an aesthetic pleasure, as an experience to be savored for its own sake, it is offered to a limited degree in traditional art forms but is more commonly available in the structured activities we call games. Therefore, when we move narrative to the computer, we move it to a realm already shaped by the structures of games. (Murray, 2016, p. 162)

Often mingled, confused, combined with the terms interaction and interactivity, (Mertens, 2004) agency distinguishes video games, at least to some extent, from other media which might allow us agency on an interpretative level but not on a direct action level (analogue forms like game adventure books or tabletop role-playing games might be some of the exceptions). Agency even structurally informs the many hero narratives or 'god games' most video games try to be. A very widespread, partly escapist understanding of video games sees this as the main attraction of video games: Being able to take meaningful action when you are not able to in the real world. But what happens in the cases video games deny us our formerly given agency, limiting our possible vectors of power, of choice, of change? I think these moments of agency loss should be a focus of research because they tell us a lot not just about video games on their own but how we experience agency in general and how established and conventionalized forms of media reception can be subverted and used as reflections on greater societal and political topics.

LUDIC AND POLITICAL AGENCY

First, we have to widen our understanding of agency. Murray's short definition of agency is helpful but it binds itself to the technical affordances of the medium. However, agency is a concept and phenomenon with a wider range of use, as Martin Hewson shows:

> Agency is the condition of activity rather than of passivity. It refers to the experience of acting, doing things, making things happen, exerting power, being a subject of events, or controlling things. This is one aspect of human experience. The other aspect of human experience is to be acted upon, to be the object of events, to have things happen to oneself or in oneself, to be constrained and controlled: to lack agency. As people are both actors and acted upon, the interplay between agency and context is a central issue. (Hewson, 2010, p. 12)

Therefore, agency is, as Sony Andermahr, Terry Lovell and Carol Wolkowitz for example put it "[a] crucial term in the theory and practice of feminism, as indeed any politics" (1997, p. 12). Even though I would not claim these different forms of agency are identical, I propose that there is a structural analogy between ludic and what can be called political agency. Political here means political in the broader sense of the saying "Everything is political", not just in the sense of participation in political processes.

This is based on the fact that society and games are both systems governed by rules in which players and people are afforded agency. We might be able to vote, to participate in protests or utter our opinion freely but our role in society is also defined in other ways. Which societal role we hold, which parts of society we belong to, give people different agency. This parallel allows ludic agency to be used as an analogy for political agency. Games can give players and their characters agency in a fictional world and society. We can alter the virtual world by our actions but our character might also have boundaries to their agency. It might be helpful to differentiate between player agency and character agency as they might not always be aligned (Unterhuber, 2015). For example, in *Transistor* (2014) we might be able to control Red's body and movement, but we can't make decisions for her. This is emphasized by her having no voice. We only hear the Transistor, her sword, talk and thus, are not can privy to her thoughts. While we move her through the different levels of the game, we don't know what her goal or her intentions are and might be surprised when we find out.

Nevertheless, agency in the field of games is not just connected to the way can play (in) video games. It feeds into games' function as a medium and technique of subjectivity in general. The literary scholars Christian Moser and Regine Strätling state:

> "Indeed, games seem to offer themselves as a technique and medium for the exploration, testing and shaping of our self-relationships, because in them basic parameters that structure subjective experience can be set arbitrarily and can thus be arranged and changed individually: space and time structures, but also the configuration of rules and choices. The specific setting of these parameters seems to be committed solely to the creation of play. As long as this purpose is fulfilled, they can be modified." (Moser & Strätling, 2016, p.16)

These characteristics give games power to be not only a form of enjoyment or a leisure activity. Moser and Strätling continue:

> "Thus, they offer the possibility of breaking through the habitual routines and automatisms of everyday life even by means of minimal, voluntary and intentional shifts in the parameters of action, thereby creating new ways of relating to oneself and new insights into oneself. Games allow for alternative self-relationships, different from those of everyday routines, because their structures can (apparently) be modified at will – unlike the structures of everyday life. Which mostly elude the access of the individual. In a game, the rules that guide our behaviour can be intentionally manipulated within the framework of a limited playing time and a limited scope." (Moser & Strätling, 2016, p.16)

The understanding of game and play in modernity is always already bound to this function, making ludic agency inherently political in and on itself. Because subjectivity is one of the core categories of modernity since the saddle period, a time of socio-political and conceptual transformations in Europe between 1750 and 1850 (Koselleck, 1972 & Fulda, 2016), game and play as forms of self-actualization occupy a central position in the episteme of modernity. As Josef Früchtl puts it: "The subject is only fully modern when at play" (2016, p. 50). That is why Friedrich Schiller centres his aesthetics on the ludic drive as what makes us fully human (2010). That is why 18th century pedagogue Johann Christoph Friedrich GutsMuth claims you can know a human's character not from their deeds but from their games (1796, p. 16). That is why Freud sees a basic similarity between play and literature and literature as the continuation of or substitute for play (2010). That is also why Huizinga sees play and games as the basis of culture in general, as a force, which connects all living beings (2013). However, Strätling and Moser's framing of games as a tool of subjectivity can be seen as a rebuttal to Huizinga's often-misunderstood concept of the 'magic circle' (2013, p.10 & Günzel, 2010). Games have their own space and time but they are always connected to society, to culture and ideology, as Huizinga himself formulates his idea of the 'Homo Ludens' in contrast to the rise of fascism in the 1930ies (Ascher, Unterhuber, 2018 & Otterspeer, 1997).[2]

However, what does that mean for the phenomenon of agency loss and agency restriction in games? They can be powerful tools. Since games as a medium of agency have accustomed players to have agency, restricting or denying them agency in games can be impactful experiences, sometimes even unnerving or infuriating. (Jørgensen, 2016) It shows players that agency in games is always just borrowed. (Unterhuber, 2015) Thus, it offers the possibility for people to feel, in a relatively safe media environment how agency in the real world is not distributed equally among people, especially because of class, race and gender. Even though society claims that we should all be treated as equal, we in fact are not. Be it because of systemic racism, sexism, classism or hierarchical power structures in general. Games can help us to understand this by simulating these experiences. (Schellong, 2015)

AGENCY LOSS AND RESTRICTION IN ACTION

The following three example show how games can use the structural analogy of ludic and political agency in vastly different ways, on a formal

---

[2] It seems odd that Huizinga and his 'Homo Ludens' are often seen as completely unpolitical while he argues extensively against Carl Schmitt's political theory (2013, pp. 208-211) and even (almost prophetically) describes German fascism as a force that might shatter "civilization itself". (2013, p. 211) He declared this perspective already in 1933: "Today's Europe is more exposed to a force that threats it of a return to barbarism [...]. A worrying weakening of ethical principles in the lives of nations as in that of individuals, has continued to emerge. When comparing the ideal professed by the nations of the present time to the conceptions that animated the great epochs of the past, the contrast is obvious. It is the economic well-being, the political power, the purity of race, who took the place of the generous aspirations of freedom or truth of the past. Realism, it will be said, instead of illusions and fictions. The fact remains that these old concepts had in general an obvious ethical value. It is the practice of morality, after all, by communities as by individuals, which alone can heal our poor world so rich and so infirm." (Quaglioni 2019, p. 193)

level, to engage players with different forms of discrimination, assault and injustice on a narrative level.

Sexual Harrasment & Bleed

The short-game *Freshman Year* (2014) by Nina Freeman focuses on the protagonist's experience of sexual harassment in a college bar. Even though the mechanics of the game are very simple point-and-click or visual-novel-style mechanics, we have agency. By suddenly taking away players' agency in the moment of harassment, the game exemplifies how sexual harassment and violence constrains the power and agency of the victims. We as players sit there and cannot act as the protagonist is frozen by fear herself. Even though Freeman chose a minimalistic audio and visual style, the moment is powerful because we, as players, are made to feel powerless, as we cannot stop the incident. Even the mouse cursor as a symbol of our agency vanishes. After the intervention of a friend, the game ends with the protagonists looking into a mirror with which it invites the players to self-reflect.

By taking away the players' agency in step with the agency of the protagonist, *Freshman Year* achieves the 'bleed' effect in players. Bleed describes the effect if players against their trained media competence cannot and do not separate their feelings, thoughts and experience from those of their inhabited characters. The fictional situation affect the real situation of the player, thus fiction 'bleeds' into reality. (Unterhuber, 2014)

Freeman shows us that it is not the fidelity of audio-visual representation or the inclusion of gory details that creates bleed: It is the abruptness of losing all our tools to achieve agency. With a length of not even ten minutes for a single playthrough *Freshman Year* takes us into the setting by using established forms of interaction, takes them away and leaves us with the aforementioned mirror image to reflect. Therefore, the shortness of the game enhances the experience even more as it does not give us space to resolve the experience. The assault happens and like the protagonist Nina, we are left alone with it.

It is pivotal to point out that *Freshman Year* can only work because of a tradition in video games to give players near to unlimited agency even in simple visual novel style games. Normally we can click us through such games, make decisions, and see the outcome of those decisions. Conventionalized and internalized game play traditions are the basis on which agency loss can become a powerful tool. That Freeman chose the sadly everyday experience of sexual harassment and assault makes for an even starker disparity to common hero narratives in games which tend to focus on singular and exceptional situations. The experience of everyday life situations is – to some extent – in fact a commonality of all my examples in this paper.

Class & Repetition

Richard Hofmeier's *Cart Life* (2010) is a management simulation game where you take on the role of one of three different street vendors trying to survive under capitalism. One of the three story lines focuses on Andrus, a migrant worker in the US who has to work from paycheck to paycheck, away from his family and with limited language skills. The game tries to evoke the feeling of powerlessness of a person on the fringes of society. The game uses a black and white pixel style with sometimes surrealistic images and a midi soundtrack.

Again, we have a simplistic art style. Also the game play mechanics are quite simple. Besides navigating a city our main task in the game is to work a cart stand and to make enough money to put foot on the table, to feed our pet and to not get evicted. How does *Cart Life* make labor playable? The game translates every menial task of Andrus' work into a challenge on the players' side with redundant typing or button-pressing tasks. Opening your newspaper cart for example needs you to cut the binding of the newly delivered newspapers, fold them and stock the shelves. But these tasks do not translate into pressing just three buttons. You have to repeatedly type sentences like "Cut the Bands", "Folding newspapers", "Leave a Nice Crease", "Stocking Shelves" or "Almost Done". Every task requires multiple inputs and the repetitive nature of the work is amplified by the aforementioned phrases. The game play might seem easy at first but every typo and every mistake costs you time. So every failure to fulfill the tasks turns into a considerable loss of income and therefore Andrus' ability to pay his rent, which can result in losing his job, his home and for us to lose the game. However, even if players meet all the challenges there is still no real happy end because of his migrant status and working class background.

There is no magic escape, the options we and the protagonist have are severely limited and they stay that way. This is no from rags to riches story. The American dream and the neoliberal promise that we can become anything we want if we try hard enough is revealed just as powerful illusions. We will not get rich even though we finish all our tasks in record time, save our money and deprave us of any comfort. The cards, so to say, were already dealt before we started the game.

So here, the restriction of agency is not based on abruptness but on subversion of the neoliberal ideology, most games subscribe to. (Baerg, 2009) We are accustomed in games to get what we want if we work hard enough but *Cart Life* shows us that it does not own us a happy end. It even shows us that there is no real reward for fulfilling its tasks but only punishment if we don't. This does not mean our actions and decisions do not matter (Unterhuber, 2014). It means they only matter if we fail. This might feel unfair to players and it should as the game play structure tries to simulate unjust issues of labor and systemic racism. The latter is also fundamental to the next example.

Racism & Injustice

Dontnod's *Life Is Strange: Season Two* (2018/2019) puts us in the role of two latinx boys, Sean and Daniel Diaz, in a Trump-America where discrimination and violence against minorities by neighbours, strangers and the police are normality. Already in the opening, we are confronted with slurs and hate speech from the hostile neighbour boy Brett who with a white fence and the American flag overhead represents the everyday racism of the American middle class. The thread of racism hangs over the Diaz household from the beginning. This is underlined by a neighbour's request to build a fence between their houses; a thinly veiled nod ad Trump's promise of a border wall.

Throughout the game, Sean and Daniel's agency is limited and even crimes they didn't commit are attributed to them, only based on their Hispanic heritage. Even though normally players' choices matter a great deal in the *Life-Is-Strange*-series, the game purposefully ignores some choices to show how a racist society is not fair but discriminatory and that the agency of minorities is always restricted and precarious. This works especially well because the series as part of a wave of games that centered players' decisions (Unterhuber, Schellong, 2016) has emboldened players' agency over the story. So the restriction of agency in this context has become even more powerful.

As a concrete example, in the first episode we have two situations, which show this poignantly. First, after a confrontation with their racist neighbour the cops are called. In addition, even though their father Esteban tries everything to deescalate the situation nothing works. Their minority status and the inherent racism of the police is enough to leave Esteban on the ground shot multiple times by a scared cop. There is no possibility for negotiating as there was in similar situations in the original *Life Is Strange* (2015). We are made to watch the murder of an innocent character, our protagonist's father with no way to stop it. That Daniel's supernatural power manifests in this moment also does not help. It escalates the situation further. The second season of *Life Is Strange* uses the superpower trope as a metaphor and enhancement for Sean and Daniel's otherness, making them even more separated from the rest of society. (Unterhuber, 2018)

The second situation makes the difference between season 1 and 2 even clearer. Sean and Daniel are on the run and stop at a gas station. With little to no money in their pockets, players have the choice to beg for food, steal it, or buy just the bare necessities. However, this choice does not matter. Sean always ends up beaten and zip-tied to a pipe by the racist shop-owner, something Alice Bell from *Rock, Paper, Shotgun* in her review called unfair (2018). However, this is not an oversight or a plot hole. It is intentional and it shows, especially to a white audience, its blind spot, as Gita Jackson puts it:

> "For a split second, I thought to myself, Well, that would never happen. Then I remembered the murder of Tamir Rice, a twelve year old boy playing with an airsoft rifle on a playground. The police shot him dead. I remembered Trayvon Martin, a seventeen year old walking home from the store with candy and a drink, who was killed by George Zimmerman. It does happen. People do look at brown and black bodies, especially brown and black men, and just assume that violence is necessary." (2018)

Putting us into their shoes, even in the safe space of a game, lets us glimpse the experience, where every interaction can become dangerous, in the games' "constant rhythm of discomfort", as Allegra Frank from *Polygon* puts it (2018). Thus, *Life Is Strange: Season Two* achieves to portray an atmosphere of constant fear, by denying us ways to escape it. It dissolves the relation between action and consequence, as Sean and Daniel have to face consequences, no matter how they act or if they act at all.

CONCLUSION

With these very different examples I hope I could show you how agency loss and restriction can be used as an impactful rhetoric device in games and that our analyses of games should not only focus on the agency players have but on the agency they are denied. Moreover, the parallels between ludic and political agency as well as the inherent politicality of games as a technique of subjectivity makes the idea games could be non-political even more moot than before. What we can and cannot do, not just in games but in reality is important. Or as the Austrian music collective "Die Schmetterlinge" put it:

> It is the freedom of those who have as many choices as possible. If the laborer has as many choices as possible, it is the freedom of the laborer. If capital has as many choices as possible, it is the freedom of capital. Look around you, who has as many choices as possible, and you know whose freedom it is. (Unger, 2015, p. 52)

Accepting agency as part of the specific mediality of games means to accept that games are inherently political, not only in their content but in their form as well. As game studies researchers, we should not just declare agency a central research category. We should focus on who is afforded agency and thus freedom in games, how player and character agency are not the same and how games use agency to show their underlying ideologies. Especially if we see games as tools for subjectivity the distribution of agency becomes relevant not only for video games but for our understanding of society and culture in general.

REFERENCES

Games

Freeman N., Knetzger L. (2014), Clark S. L. Freshman Year (PC). USA.

Hofmeier, R. (2010) Cart Life (PC). USA.

Dontnod (2018/2019). Life Is Strange. Season Two (PC). France: Square Enix.

Dontnod (2015). Life Is Strange (PC). France: Square Enix.

Supergiant Games (2014). Transistor (PC). USA: Supergiant Games.

Texts

Andermahr, S., Lovell, T., Wolkowitz, C. (1997). A Glossary of Feminist Theory. London: Arnold.

Ascher, F., Unterhuber, T. (2018). Editorial 2018. PAIDIA – Zeitschrift für Computerspielforschung. Retrieved from https://www.paidia.de/editorial-2018/

Baerg, A. (2009). Governmentality, Neoliberalism, and the Digital Game. symplokē, Vol. 17, pp. 115-127. Retrieved from https://www.jstor.org/stable/10.5250/symploke.17.1-2.0115

Bell, C., Caldwell, B. (2018). Life is Strange 2: Episode 1 verdict-o-chat. If you go down to the woods today. Rock, Paper, Shotgun. Retrieved from https://www.rockpapershotgun.com/life-is-strange-2-episode-1-verdict-o-chat

Frank, A. (2018). Life is Strange 2 is an intimate portrait of the Trump era. Polygon. Retrieved from https://www.polygon.com/2018/10/2/17911080/life-is-strange-2-episode-one-review

Freud, S. (2010). Der Dichter und das Phantasieren. In O. Jahraus (Ed.). "Der Dichter und das Phantasieren". Schriften zur Kunst und Kultur (pp. 101-112). Stuttgart: Reclam.

Früchtl, J. (2016). Interrationale Beziehungen. Das Subjekt ist nur da ganz modern, wo es spielt. In C. Moser, R. Strätling. Sich selbst aufs Spiel setzen. Spiel als Technik und Medium von Subjektivierung (pp. 38-50). Paderborn: Wilhelm Fink.

Fulda, D. (2016). Sattelzeit. Karriere und Problematik eines kulturwissenschaftlichen Zentralbegriffs. In E. Décultot, D. Fuldac (ed.). Sattelzeit. Historiographiegeschichtliche Revisionen (pp. 1-16). Berlin, Boston: De Gruyter.

Günzel, S. (2010). Der reine Raum des Spiels. Zur Kritik des Magic Circle. In M. Fuchs, E. Strouhal (ed.). Das Spiel und seine Grenzen. Passagen des Spiels II (pp. 189-202). Wien, New York: Springer.

GutsMuth, J. C. F. (1796). Spiele zur Uebung und Erholung des Körpers und Geistes für die Jugend, ihre Erzieher und alle Freunde unschuldiger Jugendfreuden (2nd edition). Schnepfenthal: Buchhandlung der Erziehungsanstalt.

Hewson, M. (2010). Agency. In A. Mills, J. Albert, G. Durepos, E. Wiebe (Ed.). Case Study Research. Vol 1. (pp. 12-16). Los Angeles: SAGE.

Huizinga, J. (2013). Homo Ludens. A Study of the Play Element in Culture. Boston: The Beacon Press.

Jackson, G. (2018). The First Episode Of Life Is Strange 2 Doesn't Hide Its Politics. Kotaku. Retrieved from https://kotaku.com/the-first-episode-of-life-is-strange-2-doesnt-hide-its-1829334309

Jørgensen, K. (2016). The Positive Discomfort of Spec Ops: The Line. Game Studies, 16, 2. Retrieved from http://gamestudies.org/1602/articles/jorgensenkristine

Koselleck, R. (1972). Einletung. In O. Brunner, W. Conze, R. Koselleck (ed.). Geschichtliche Grundbegriffe. Band 1 (pp. XIII–XXVII). Stuttgart: Ernst Klett Cotta Verlag.

Mertens, M. (2004). Computerspiele sind nicht interaktiv. In C. Bieber, C. Leggewie (Ed.). Interaktivität. Ein transdisziplinärer Schlüsselbegriff (pp. 272-288). Frankfurt am Main: Campus.

Moser, C., Strätling, R. (2016). Sich selbst aufs Spiel setzen. Überlegungen zur Einführung. In C. Moser, R. Strätling. Sich selbst aufs Spiel setzen. Spiel als Technik und Medium von Subjektivierung (pp. 9-27). Paderborn: Wilhelm Fink.

Murray, J. H. (2016). The Future of Narrative in Cyberspace (2nd edition). Cambridge, MA/London: MIT Press.

Otterspeer, W. (1997). Huizinga before the Abyss. The von Leers Incident at the University of Leiden, April 1933. Journal of Medieval Studies 27, 3.

Quaglioni, D. (2019). The Weakening of Judgement: Johan Huizinga (1872–1945) and the Crisis

of the Western Legal Tradition. In K. Tuori, H. Björklund (ed.). Roman Law and the Idea of Europe (pp. 181-200). London: Bloomsbury Academic.

Schellong, M. (2015). Simulation als Grenzerfahrung – Grenzerfahrung als Simulation. PAIDIA – Zeitschrift für Computerspielforschung. Retrieved from https://www.paidia.de/simulation-als-grenzerfahrung-grenzerfahrung-als-simulation/

Schiller, F. (2010) Über die ästhetische Erziehung des Menschen. Stuttgart: Reclam.

Unger, H. F. (2015). Proletenpassion ff. Wien: mandelbaum.

Unterhuber, T. (2014). What would you do? – Entscheidungsmöglichkeit als Spezifikum des Mediums Computerspiel. PAIDIA – Zeitschrift für Computerspielforschung. Retrieved from https://www.paidia.de/what-would-you-do-entscheidungsmoglichkeit-als-spezifikums-des-mediums-computerspiel/

Unterhuber, T. (2015). Walk a mile in her shoes – Die Erfahrung des Bleed im Computerspiel als Ausgangspunkt gender-theoretischer Überlegungen. PAIDIA – Zeitschrift für Computerspielforschung. Retrieved from https://www.paidia.de/walk-a-mile-in-her-shoes-die-erfahrung-des-bleed-im-computerspiel-als-ausgangspunkt-gendertheoretischer-uberlegungen/

Unterhuber, T. (2018) „Everything is political, Sean" – Verschiebungen im öffentlichen Diskurs am Beispiel ‚Life Is Strange 2'. PAIDIA – Zeitschrift für Computerspielforschung. Retrieved from https://www.paidia.de/everything-is-political-sean-verschiebungen-im-oeffentlichen-diskurs-am-beispiel-life-is-strange-2/

Unterhuber, T., Schellong, M. (2016). Wovon wir sprechen, wenn wir vom Decision Turn sprechen. In F. Ascher, R. Baumgartner, G. Grelczak, M. Kutscherow, M. Schellong, M. Schöffmann, T. Unterhuber (Ed.). "I'll remember this. Funktion, Inszenierung und Wandel von Entscheidung im Computerspiel (pp. 15-31). Glückstadt: Werner Hülsbusch.

PROMOTION OF EMPATHY WITH DIGITAL GAMES. ON THE POSITIVE IMPACT OF DIGITAL GAMES ON THE EMOTIONAL INTELLIGENCE

Bastian Krupp

Digital games have been an integral part of individual media biographies for half a century now. More than one third of Germans play regularly. For society as a whole and for media education in particular, this raises the question of the "right" way to deal with digital games in educational activities. Up to now, social evaluation has been primarily based on the consideration of ascribed risks (see discourse on "killer games" and the loss of empathy through glorification of violence and social isolation). If the potential risks are made the central origin of a social debate, they also control pedagogical action. A benevolent societal attitude toward digital games helps media education to gain greater recognition for its pedagogical use, which has already produced impressive successes for learning and educational processes. Scientifically, there are few relevant research results on the effect of digital games on the development of empathy. Emotional intelligence (EQ) is still a young field of research but is not being considered in the context of digital games. This article deals with the question of the effect of digital games on emotional intelligence by means of a quantitative survey of gamers. It is brought from the results of the final assignment for the course of study of Action-oriented Media Pedagogy (Master of Arts). The results show that gamers, in principle, have no differences in the development of emotional intelligence compared to non-gamers. Thus, they represent an important finding for a social as well as a scientific debate about their recognition as cultural assets. The identification of gender-specific differences also draws attention to the lack of diversity in gaming, which is still characterized by ideals of masculinity and in which environment women are repeatedly confronted with sexism. In society, as in media education, there is a need for a reformation that allows for a sharpening of the view for the positive effects of digital games and addresses problematic conditions in the gaming scene.

Keywords: digital games, emotional intelligence, empathy, media education

---

INTRODUCTION

Anyone who has ever played a digital game can understand that they cause emotions (Salen and Zimmerman, 2010, p. 314). They have been inspiring people for over 50 years and have captivated several generations. The generation of gamers has become caring parents and they have not lost their passion for games. Games have grown enormously in the past decade as a popular leisure activity for people of different age groups and social milieus. With the increasing relevance of the medium in public and private space, the question of the consequences and benefits and the modes of action of digital games arises. However, the social reception is not uniform. In public discourse, gamers are repeatedly discussed as the perpetrators of violent acts. This representation still shapes the opinion of those who believe in the cause and effect of violence in digital games and associate them with real violence perpetrated by assassinations and rampages. The suggested transfer from digitally perpetrated violence to the perpetration of violence in "real life" however falls short and neglects human and environmental factors (Ferguson, 2018; Anderson, Gentile, & Buckley, 2007; Klimmt & Trepte, 2003). The presentation of digital games influences public perception and the perception of the opportunities and dangers of the medium. In practical media education work, educators and parents proclaim the negative consequences of digital games on development, but these views do not usually stem from a scientific examination of digital games, but rather from the formation of public opinion.

So far, a broad discourse on digital games has evolved. What remains are heated and emotionally charged discussions about the opportunities and risks of digital games. This paper focuses on the effects of digital games on the development of emotional intelligence in order to work out their positive emotional effects and tries to contribute to an objectification of the debate about their potentials and dangers. A more positive view of the medium would help media education gain more recognition for the use of digital games in educational work. With the help of a quantitative online survey, the focus is placed on the emotions in digital games, on which emotions they trigger and how they influence the development of emotional intelligence. The aim is to clarify the tendency of gaming in general to influence emotion and empathy, a subarea of emotional intelligence, and specifically single and multiplayer preferences and certain game genres. In their diversity, digital games offer numerous application possibilities for the promotion of social, cognitive and creative competences (Bonn und Karsch Johannes, 2019, p. 29). In their diversity, digital games are an analogy to the established media landscape (e.g. books and films), whose educational effects do not differ significantly (Geisler, 2019, p. 126).

The results of the survey will provide an empirical view of the influence of digital games on the emotional intelligence of gamers. The findings will serve as a starting point for further research and focus on the positive effects of digital games on the upbringing of children and young people as well as on society.

RELATED WORK

The current scientific discourse on games is ambivalent. On the one hand, there is research that concentrates primarily on research into the violence-promoting and "aggressive-emotional" potential of digital games (cf. Anderson et al., 2004). On the other hand, there is research that is limited to educational potentials (cf. Staudacher, 2019). Overarching studies that focus on emotional benefits in the sense of socio-emotional competencies are rare. One reason for the one-sided data situation in research is the polarisation of "violent games" in public discourse. However, the results of these studies should often be treated with at least a certain degree of scepticism, for example, because they treat complex contexts one-dimensionally. Most of the publications to date approach games from a negative angle, as if a promotion of antisocial behaviour or an inhibition of empathy development is the subject matter. For example, Anderson et. al. (cf. Anderson et al., 2004) examined the effect of violence in digital games on aggressive behaviour or studies on pathological computer game consumption in the context of internet and computer game addiction (cf. Mößle et al., 2014). Anonymity and the lack of face-to-face communication that occurs in online multiplayer games favour the omission of respect towards other human players. It is also easier to insult and hurt others on an emotional level (Coeckelbergh, 2007, p. 226).

Anonymity also inhibits the development of empathy (ibid., p. 226). Other studies deal with the effects of violent media content in digital games on the development of children in the first decade of life, which are associated with increased aggressiveness and less prosocial behaviour (Greitemeyer et al., 2010, p. 796). The increased occurrence of aggressive behaviour and acts of violence in public spaces is seen here as a direct consequence of video game consumption (Coeckelbergh, 2007, p. 219). Furthermore, in contrast to more non-violent games, digital games containing violence are suspected of increasing hostility and fear, which lower the level of empathy (Anderson and Dill, 2000). So far, there is no research on the long-term effects of playing digital games on personality development.

This form of media criticism not only affects the image of gaming in general, but also that of gamers. Simply put, gamers are defamed as addicted perpetrators of violence. Assuming that these attributions are true, shouldn't the number of violent crimes increase with the increasing success of the medium, given the market power of digital games and the number of gamers in Germany? This is contradicted by current statistics on violent crime and bodily harm, which fell by 17.5% (-36,869 cases) to 181,054 cases compared to 2007 (Bundesministerium des Innern, für Bau und Heimat 2020, p. 29). The explanatory model often used in this area assumes the interdependence of virtual and analogue actions, which causes a mutual influence of actions (Coeckelbergh, 2007, p. 220). However, the individual is exposed to diverse environmental influences on a daily basis. It seems exaggerated to attribute such an important influence to the hobby of playing digital games in the development of young people (ibid.) that they can be held up as an explanation of rampages or the like. The fixation on the idea that games containing violence consistently lead to an increase in behaviour containing violence falls short from the point of view of media research, since the statement on aspects of content must otherwise also be applied to other media content, such as films and books (ibid.). This simplified model ignores complex stimulus-response chains because it assumes that an action in the digital world inevitably has an impact on analogue action and also disregards sociological and psychological aspects.

At the macro level of the societal perspective, the question is what influence digital games have on social coexistence. In order to be able to look at the societal macro level in a differentiated way, the effect on the individual must be considered at the micro level. In this respect, media are considered to play an important role as an educational and socialisation instance (Badawia, 2008, p. 37) with regard to their social relevance. Digital games as such represent a factor in media use that young people can hardly imagine without them (cf. Feierabend et al., 2019, p. 44), which is why they inevitably fulfil the established criteria of high social relevance. This explicitly does not mean that they replace the classic socialisation instances consisting of parents, school and the group of peers. Instead, digital games change traditional play and learning patterns of adolescents. Growing up with digital media in a media society also creates an important status for social participation, they enable spaces for action and are indispensable for well-being and quality of life (Süss, 2016, p. 43). In order to be able to talk about the latest game in the schoolyard or simply to play together with friends, digital games offer adolescents spaces for social and societal participation and create emotional memories. Friendships, as they form between guild or clan members, are highly relevant in times of Covid-19 and an accompanying social isolation by force. In a mediatised society, digital games represent an institution, i.e. a social system of society, which promote social integrity and offer spaces for social action (Jarren and Klinger, 2017, p. 36). They stimulate democratisation processes, such as promoting social integrity through founding or participating in a clan or community. In addition, they offer the possibility of assuming perspective in a role-playing game or as an avatar of a different gender. In the analysis of the social relevance of digital games from a macro and micro perspective, it becomes clear that these two sectors cannot be viewed in isolation from each other; rather, one must speak of a mutual influence. This insight makes perfect sense, as digital games represent images of their present (Geisler, 2019, p. 167).

Play(ing) influences human thought and emotional processes on various levels, which is confirmed by brain research, among other things. Long-term studies have shown that playing in the brain promotes the development of mirror neurons, which form the basis for empathy (Liska, 2017, chapter 2.7.1.). The development of empathy is a complicated process that is mutually influenced by several factors. In addition to the neuropsychological explanation of the development and promotion of empathy, other branches of research must therefore also be addressed. The development of empathy is not a process that can be explained on a purely neuropsychological basis. Whether a child learns to empathise with the emotional worlds of its fellow human beings is largely determined by socialisation and upbringing by parents (Bosley and Kasten, 2018, p. 50). Meanwhile, socialisation and youth media research agree that socialisation is not limited to growing up in public spaces. Thus, media worlds represent another important instance of socialisation that is recognised as media socialisation in its own right (Deinet and Derecik, 2013, p. 82). For young people, digital games are a natural part of their lifeworld, which makes them an instance of socialisation in digital spaces.

In 2020, 34.3 million people in Germany play digital games, 36% of them regularly (cf. infographic German Games Market 2020 from the German Games Industry Association), with an annual upward trend. The trend can be explained by various mutually supporting theories on the power of fascination of digital games. One of them is based on the motivation of gamers to be attracted to one or the other game and thus relates to the emotional experience of the individual. The resulting structural coupling thus has a decisive influence on the power of fascination of digital games on gamers. An increasing correspondence between analogue and digital worlds of experience means, conversely, a desired emotional value that is attributed to the games by the players (Fritz ,2011, 25 ff.). The theory is based on the assumption that digital games are a polyvalent offer structure. Thus, the spectrum of effects of a game changes depending on the players. Depending on the aspect of a digital game, whether it concerns the visual presentation, the story or the gameplay, they unfold different incentives in the players. Thus, the power of fascination is largely dependent on the extent to which players can relate to a game (Fritz and Rohde 2011, p. 10). The study thus shows that the choice of game (genre) is closely linked to personal interests. It can also be assumed that this choice has an impact on the emotional state of the players. In addition to the structural coupling of overlapping convergences to digital game worlds, digital games exert fascination through their nature as an interactive medium. In this context, the so-called flow effect was coined by Mihaly Csikezentmihhalyi (1993). The flow effect describes a state of complete absorption and growth in or with the activity of playing. In flow, a development takes place in the course of which players experience themselves as increasingly competent in their game. Flow in the digital game thus arises from the feeling of being completely absorbed in the activity of the game process. As a positive effect, players feel competent in their activity and emotions such as joy and happiness emerge (Salen and Zimmerman, 2010, p. 336). The triggering of positive emotions influences players to develop an intrinsic motivation towards playing digital games, from which the multidimensionality of the medium's power of fascination emerges. Digital games offer immediate feedback of learning progress through game-inherent rewards, which results in the release of dopamine and activates the reward centre in the brain. A lack of a sense of achievement in analogue life can thereby quickly become a catalyst for uncontrolled playing of digital games (Süss, 2016, p. 50). However, making digital games responsible as an instrument for the development of computer game addiction falls short and diverts the focus away from the actual cause. This often lies in a lack of satisfaction of needs in the analogue world, which are then sought in the digital world (Hüther and Quarch, 2016, p. 103 f.). According to Lewinsohn's model of the absence of positive reinforcement, games help to generate positive experiences, thus preventing depression, but if the positive experiences cannot be transferred to the analogue world, the impression of a gaming addiction arises (cf. Lewinsohn et al., 1976).

Games form a highly emotional experience as they symbolise a construct of values and emotions. As such, they actively exert a positive influence on self-esteem and increase positive feelings such as satisfaction and happiness (Andric 2017, chapter 3.2.). French sociologist Roger Caillois also describes the link between play activity and emotions: "He therefore distinguished play activity involving [...] (c) intelligence, emotion, and will power (games of recognition, memory, imagination, attention, reason, surprise, fear, etc.)." (Caillois and Barash, 2001, p. 164), supporting his assumptions with the work of Piaget. In his research, he came to the conclusion that children discover their sense of morality through play (cf. ibid., p. 165) and that play as such contributes to forming empathy. Caillois even goes so far as to say that a playful process can never exist detached from emotions (cf. ibid., p. 245). Qualitative research on the effects of digital games is often conducted under highly regimented and reduced conditions, which in their respective design, due to short game units with specifically selected games, are conducted under laboratory-like conditions and often do not allow for a generalisability of the results. For example, studies on increasing empathy and prosocial behaviour by playing digital games were conducted by Greitemeyer and Osswald. They found that prosocial behaviour can be promoted by playing prosocial games through playing Lemmings (DMA Design/Psygnosis, 1991) (Greitemeyer et al., 2010, p. 797). Existing studies thus only deal with empathy, a subarea of emotional intelligence. There are no long-term effects that can be concluded from regular and harmless game consumption on emotional development.

The research situation around emotional intelligence, as well as empathy research, is very clear. The concept of empathy is a part of emotional intelligence (EQ for short). EQ represents the ability to perceive one's own feelings and those of others, to assess their diversity and to use the information gained to guide one's own thinking and behaviour. The concept of emotional intelligence according to Salovey and Mayer includes three affective processes: Recognition and expression of emotions, their regulation and adaptive use (cf. Salovey and Mayer 1990). A scientific consideration of emotional intelligence makes it necessary to distinguish between emotional and social intelligence. The concept of social intelligence implies any basis for shaping interpersonal relationships. In contrast, emotional intelligence refers to the perception and understanding of feelings (Bosley and Kasten, 2018, p. 41). The term emotional intelligence was coined in 1996 by the American science journalist Daniel Goleman, so this is still a relatively young field of research, although it seems to be closely interwoven with early childhood development and play: "According to Tomasello, children are capable of shared intentionality at a very early age. His evidence: their natural ability to play a game with others, and their ability to recognize when someone isn't playing the game in a way that favors the group." (McGonigal, 2011, chapter 13). The assumption of a temporal overlap in the development of shared intention in young children, e.g. to achieve a play goal, and the development of empathy is obvious, since children's play is significantly involved in the child's learning of social skills. A recently published study by the National Literacy Trust (Picton et al., 2020) draws attention to the promotion of empathy, creativity and reading skills in the context of digital games. In this study, two out of three young people stated that they were better able to put themselves in other people's shoes by playing digital games. Many of them also stated that they were better able to deal with negative emotions through playing games or that they were able to exclude them (ibid.). Emotions also shape the choice of games we (want to) play. The choice of digital game, but also of genre, is strongly linked to the interests of the gamer. An emotional attachment to the topic of football, whether physically playing on the pitch or on the virtual pitch, thus has a clear influence on the choice of game. Through their regularity, games set a controllable framework that makes it easier for players to achieve an emotionally positive outcome than is possible in analogue reality. The rules of the game function as a simulation for life, which can often be mastered more successfully through play. In psychology, playful simulations are already used to treat phobias, for example with the help of VR applications (cf. Eichenberg and Wolters ,2012, p. 374).

The four most important areas of emotional intelligence are empathy, knowledge of human nature, persuasion and emotional self-control. Each area in itself comprises a behavioural competence that can be extracted from each other using test procedures and together result in a determinable value, the EQ (cf. Satow, 2012a). As part of the development of emotional intelligence, children learn to better understand themselves and their own feelings. They are challenged to assess situations appropriately and to develop their own guidance system. This development is often based on moral decisions, which makes moral development a decisive factor in the development of emotional intelligence (Coeckelbergh, 2007, p. 221). Digital games repeatedly show references to the real world and provide unsparing insights into social and human abysses. The form of moral exploration that is made possible in games, for example by performing socially inadequate actions, is not to be evaluated here. In addition the portrayal of gender, in which female characters are often over-sexualised, is a point of criticism to be submitted in the moral evaluation of digital games (ibid., p. 221). Empathy can help make the world more human if each individual uses his or her empathy to understand people of a different gender, religion or nationality (ibid., p. 221).

In the formation of empathy, gender-specific differences can be identified in current research. Thus, women indicate a higher developed empathy than men (Greitemeyer et al., 2010, p. 798). These findings are rooted in gender differences, as women are seen as more caring than men from a sociological perspective. If it can be assumed that emotions are anchored in our biology, in what way do they influence our socialisation? This question was also addressed by Paul Ekman, who became one of the most influential emotion researchers of the 20th century. As such, he coined the term basic emotions (cf. Ekman 2010). In his research he noticed that basic emotions such as joy, fear or anger are found in all cultures of this world and are expressed in the same way everywhere. In an intercultural comparison, it turned out that other emotions such as pride, arrogance, humility, shyness, despair, etc. show differences in cultural expression. It can thus be assumed that emotions are universal processes (Bosley and Kasten, 2018, p. 19). During puberty, adolescents should be taught the basic principles of emotional competence, such as dealing with anger, impulse control or empathy (Bosley and Kasten, 2018, p. 54). Here, a lifeworld-oriented approach to teaching emotional competencies seems promising.

METHODOLOGY

Research Questions

The guiding research question is: How do digital games affect the development of emotional intelligence? Subordinate questions are: Does gender influence the development of emotional intelligence in the context of gaming? What influence do the duration and frequency of gaming have on emotional intelligence? How does the preference for single and multiplayer experiences affect emotional intelligence? To what extent do genre (representatives) of digital games influence the formation of emotional intelligence? The evaluation question is: Do gamers have a higher EQ than non-gamers? These questions lead to a detailed description of the research methodology used, the selection of the research object and the specific game selection, as well as the approach taken to select the groups of people interviewed for the data collection and the explanation of how the survey was conducted.

Sample

A total of 234 people took part in the survey. 179 data sets could be evaluated. Due to the disproportion in gender distribution (ratio of 2:1 male to female), there are inaccuracies in the comparative gender analysis. In order to achieve better comparability, the results between the genders were given as a percentage of the respective number of women and men.

Research Instruments

To better verify the thesis of whether there is a verifiable correlation between playing digital games and the development of emotional intelligence, the partially standardised online interview is used. Through open questions, individual emotional experiences can be presented in a more suitable form. In order to reach the target group of 25 to 45-year-old gamers, the online survey is suitable, as it makes it possible to bring the survey to where this target group predominantly moves, on the internet. More specifically, certain areas in social networks and forums can be targeted, where gamers are found in a higher concentration. The empirical data collected will be selected in relation to the predefined research criteria. The data is analysed according to exploratory factor analysis, which looks for correlations in the collected data (Häder, 2015, p. 443) that relate to the predefined research questions.

The participants in the study received a standardised questionnaire, which can be filled out directly online in the browser. The questionnaire is divided into three thematic blocks. In the first part, the socio-demographic data of the participants is recorded, including gender, age and formal education. Within the framework of the survey, the emotional intelligence of the survey participants is recorded using the Emotional Intelligence Inventory (EI4) by Satow (Satow L., 2012b). The EI4 assigns an integer degree of expression from 1 to 9 to each concrete statement. In the survey, metadata is collected that allows conclusions to be drawn about the participants' actual enthusiasm for games and thus provides an important basis for answering the research questions. The EQ score is composed of four sub-scores: empathy, knowledge of human nature, persuasiveness and emotional self-control. In the last block, the metadata on individual use of digital games in recreational activities is considered. The metadata includes questions on their gaming habits, for example frequency of playing, preferred genre and the preference for single or multiplayer games. At the end of the questionnaire, up to three emotional moments in digital games can be described.

AIM OF THE RESEARCH

The study conducted here is to be classified within media education in the context of media effects research. The effect of digital games and their impact on emotional intelligence is examined. The aim is to obtain more precise insights into the potential benefits (or harms) of digital games from a media education perspective. For this purpose, a statistical analysis of data collected by means of a survey on the influence of digital games on emotional intelligence is carried out. The study records the emotional intelligence of the study participants and places this in a frame of reference to the respective individual media biographies in the context of digital games. Here, both the current engagement with digital games and the emotional intelligence quotient (EQ) are considered in the context of a changed gaming biography. The aim of the study is to highlight the connections between the development of emotional intelligence and the playing of digital games.

FINDINGS

Within the framework of the study, only the data sets of "game-affine" persons were deliberately evaluated. The EQ in Satow's EI4 is normally distributed and defined by the general average value in the reference population. This takes the value 5 here. For the German EQ, since according to a survey by the Association of the German Games Industry, almost half of all German citizens have digital gaming experience, this is sufficiently representative.

There is hardly any difference between genders in level of EQ. The frequency of playing has no visible effect on the EQ, neither on those who play less than once a week nor on those who play daily. In relation to game duration, it can be seen that there is a negative trend in the excessive consumption of computer games, which explain a reduced EQ. To a certain extent, a slightly higher EQ can be expected for those who prefer multiplayer experiences. People who prefer to play serious games have the highest EQ. The shooter games quoted in the beginning are in the upper midfield. With an average EQ value of 5, shooter games even reflect the exact standard value. The comparison of gender-specific preferences reveals clear differences. Men play shooters and sports games much more often, while women prefer party and puzzle games (see figure 1). The statistical significance of these differences was calculated using chi-square tests showing p values between .001 and .015.

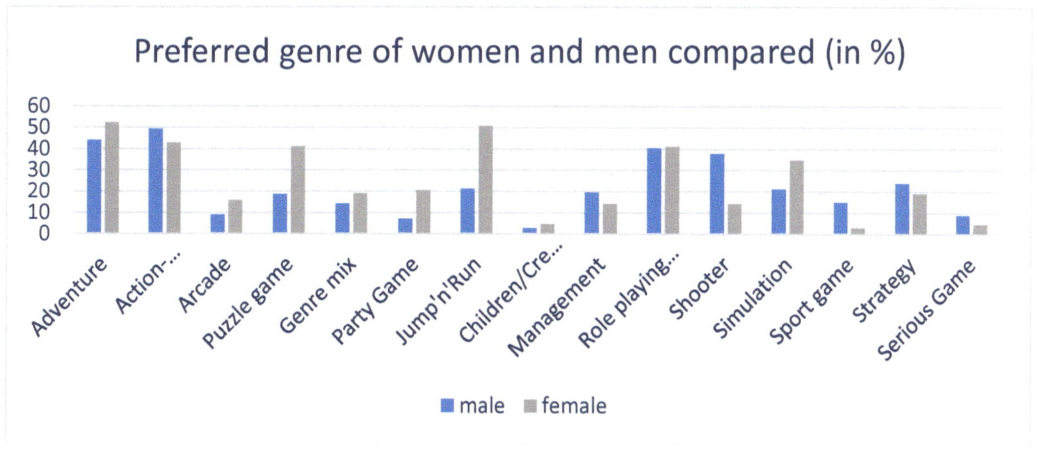

Figure 1. Preferred genre compared to gender

CONCLUSION

Women and men do not show any significant differences in the results of the study with regard to EQ in the context of gaming. This is also shown in the comparison of the individual values, consisting of the four subcategories empathy, knowledge of human nature, persuasiveness and emotional self-control. Only in the case of empathy differences between the genders can be identified. On average, women have a higher value than men, a tendency that corresponds to previous research that attributes a generally higher empathy to women. However, the results must be interpreted with caution due to the small number of cases, especially since the differences are not significant and should therefore not be overinterpreted.

The division of the participants in the study into "rare gamers", "frequent gamers" and "excessive gamers" allowed for a more valid comparison of the duration and frequency of gaming. While the differences in the results between the rare and frequent gamers are not worth mentioning, the excessive gamers showed a significant drop in EQ in contrast to the other groups. Since this group only accounted for 2% of the entire study group, these results must also be taken with reservation. However, the results suggest that excessive involvement with digital games has an inhibiting effect on the development of emotional intelligence. How strong this effect is on emotional intelligence can only be speculated at this point, since even the EQ in the group of "excessive gamers" is within the norm, albeit in the lower range. In the group of those who show excessive gaming behaviour, only men appear in the analysis of the results. Consequently, men should have an average EQ below that of women, which is not the case. This connection allows the assumption that "playing a lot" does have a positive influence on the development of emotional intelligence. However, the results do not provide any visible evidence for this thesis. The comparative examination of gaming time from the past to the present leads to the assumption that gamers devoted significantly more time to gaming in their childhood and adolescence. The theoretical discussion shows that emotional development in early childhood extends into young adulthood. This sensitive phase of emotional development has a decisive influence on the formation of emotional intelligence in adulthood. It is precisely in this sensitive phase, in which young people learn to assess and control their own and other people's emotions, that digital games take up a lot of space in emotional and cognitive experience. From the results of the study, neither a fundamentally negative nor a positive effect on emotional intelligence can be determined.

Joint play(s) offers a suitable space for learning social togetherness, as rules for social togetherness are negotiated here (McGonigal, 2011, Chapter 5). McGonigal's thesis is confirmed by the results of the study, even if they are not unambiguous. It seems logical that those who prefer to play in multiplayer modes have developed a higher EQ on average. The differences between players who prefer single-player modes and those who prefer multiplayer modes are so small that we cannot speak of a clear trend. Digital games, even in single player, are not exclusively consumed alone. Digital games also enable people to experience emotions passively, for example when watching friends or siblings play.

In the analytical processing of the study results, no specific factor can be identified that allows for a difference in the development of emotional intelligence between the preference for single or multiplayer modes. The survey also revealed that women prefer single-player experiences significantly more often than men. Thus, the differences between the sexes lie less in the game shares, but much more in the type of use. Women have a different usage behaviour, as they use other games or platforms, for example (Groen and Witting 2016, p. 179), or prefer other game modes. The question inevitably arises as to how these different preferences can be justified? One major reason can be identified in the gaming scene, which is shaped by ideals of masculinity (Groen and Witting, 2016, p. 184), which is also reflected in the content presentation of a broad mass of digital games. Many women who are active in online gaming communities have experienced sexist comments and threats, up to and including rape threats. These experiences increase the likelihood that women will withdraw from gaming communities (Groen and Witting, 2016, 184 f.), especially online-based multiplayer games. This may be a clue to the difference in performance between women and men in terms of their EQ. If women withdraw from online multiplayer games because of sexualised hostility, they miss out on important experiences with other players that would support their emotional development.

Studies devoted to the effects of digital games on emotions often look at this effect in the context of a game genre or a specific game (cf. Greitemeyer and Osswald, 2009, pp. 896-900). The influence of game frequency and game time are increasingly considered in relation to their effect of the game process on the development of aggression. These considerations are determined in qualitative studies under laboratory-like conditions. Thus, play sessions are predefined in terms of time and content in relation to the choice of game. However, this approach ignores factors that make up the power of fascination of digital games, such as the flow effect. The structural coupling, which can be seen, among other things, in the choice of a genre or a specific digital game, is also not applied here. In the evaluation of the expression of EQ in relation to the respective genre, the players of serious games have the highest EQ value. However, this is not very meaningful due to the small number of cases. Apart from the genre of serious games and strategy games, adventure games, action-adventure games and role-playing games represent the players with the highest EQs. They are among the most popular genres among the participants in the study. Not only in the choice of genre, but also in the description of emotional moments, games from these genres are particularly memorable to them. Digital games with a high narrative content, as they appear in games from the adventure, action-adventure and role-playing genres, with their profound character developments offer a space in which emotions can be learned and applied. The genre of first-person shooters, often declared as "shooter games" or "killer games", is in the upper midfield in the comparison of genres in terms of their impact on emotional intelligence. With an average EQ value of 5.0, first-person shooter games mirror the exact norm. First-person shooters have been blamed in the past for triggering analogue violence. Such evidence could not be confirmed by the study. When comparing female and male preferences in the choice of game genre, clear differences emerge in some cases. In particular, they are noticeable in the selection of game genres that are considered either predominantly male or female due to social stigmas. For example, only a few female preferences are found in the genre of sports games. This finding resonates with the world of sport, which is shaped by ideals of masculinity, especially those depicted in the field of digital games. A comparable picture emerges with the shooter genre, which is also strongly dominated by men. The reference to male stereotypes in the military is clear here. Meanwhile, women are not only said to be more top-heavy, but also more sociable. This stereotypical view is also confirmed among women, who play thinking and social games significantly more often. Another interesting finding is provided by the comparison of gender-specific preferences of game genres, comparing how these affect EQ. The genres preferred by women are in the middle to lower range of this scale, while the genres preferred by men also appear in the upper middle range. This finding is also consistent with the results of the study, in which women on average have a lower EQ than men.

The average EQ of the gamers studied does not deviate fundamentally from the predefined and evaluated norm values of Satow's EI4. Consequently, digital gaming can neither be proven to have a positive effect on the development of emotional intelligence nor a negative one. No irregularities in the expression of the EQ and its sub-areas could be found in the study group.

FUTURE RESEARCH

It requires experts in the field of digital games who are aware of the opportunities and risks of their effects and can embed them in their educational work. This gives media education and media game education an important role. Further research is needed in the field of digital games and their influence on emotional development and emotional intelligence in order to be able to predict their effects with certainty. A combination of quantitative and qualitative long-term studies would be appropriate here, as they allow for more than just a cross-sectional comparison on emotional intelligence in the context of gaming. Emotional intelligence is becoming an increasingly important component of the current and future world of work. For the area of emotional self-hygiene, emotional intelligence continues to gain importance as self-protection against depression. Games, including digital games, create positive emotions and can help to develop a feeling for one's own emotions or those of other people.

The study showed that games can be equated with play. Furthermore, as this study showed there was no significant impact of playing digital games on empathy, which claims that they have a potential to endanger the development of empathy. Scientific proof was provided that gamers are not emotionally neglected or even emotionless. They are quite capable of acting with the emotional diversity of a game and creating a transfer to analogue reality. What conditions contribute to a successful transfer from in-game actions to actions outside the digital game? The answer to this question justifies further research in this field and can benefit the medium "digital game" in its public recognition as an opportunity rather than a risk. For a deeper understanding of the findings, long-term studies are needed to test statistical significances for correlations and clusters.

ACKNOWLEDGEMENTS

I would like to thank my research supervisors, Friederike Siller and Horst Pohlmann. Without their advice, this paper would have never been accomplished.

REFERENCES

Anderson, C. A., Carnagey, N. L., Flanagan, M., Benjamin Jr., A. J., Eubanks, J., Valentine, J. C. (2004). Violent video games: Specific effects of violent Violent video games: Specific effects of violent content in aggressive thoughts and behavior. In: Advances in experimental social psychology, S. 199–249. DOI: 10.1016/S0065-2601(04)36004-1.

Anderson, C. A., Dill, K. E. (2000). Video games and aggressive thoughts, feelings, and behavior in the laboratory and in life. In: Journal of Personality and Social Psychology 78 (4), S. 772–790. DOI: 10.1037/0022-3514.78.4.772.

Anderson, C. A., Gentile, D. A., & Buckley, C. E. (2007). Violent video game effects on children and adolescents. Oxford: University Press.

Andric, I. (2017). Runtastic into a SuperBetter Life! In: Denk, N., Pfeiffer, A., Rembold, É & Wernbacher, T. (Hg.): Welt der Spiele 360°. Sammelband des Zentrums für Angewandte Spieleforschung der Donau-Universität Krems. Krems an der Donau: Donau-Universität Krems, eBook.

Badawia, T. (2008). Mittendrin und/oder nur dazwischen? Identitätskonzepte von Migrantinnen und Migranten. In: Theunert. H. (Hg.): Interkulturell mit Medien. Die Rolle der Medien für Integration und interkulturelle Verständigung; [Beiträge aus Medienpädagogik, Kommunikationswissenschaft, Migrationsforschung, Jugendsoziologie, interkultureller Pädagogik und Jugendmedienschutz; basiert auf der Tagung "Interkulturell mit Medien", die am 30.11.2007 vom JFF - Institut für Medienpädagogik in Forschung und Praxis in München veranstaltet wurde. München: kopaed (Reihe Medienpädagogik, 14), S. 25–38.

Bonn, B., Johannes, K. (2019). eSport, Sport und Schule. Szenarien für die Auseinandersetzung von Schule mit eSports. In: merz / medien + erziehung 63. Jahrgang (Nr. 2 (April 2019)), S. 25–31.

Bosley, I., Kasten, E. (2018). Emotionale Intelligenz. Ein Ratgeber mit Übungsaufgaben für Kinder, Jugendliche und Erwachsene. Berlin, Heidelberg: Springer.

Bundesministerium des Innern, für Bau und Heimat (Hg.) (2020): Polizeiliche Kriminalstatistik 2019. Ausgewählte Zahlen im Überblick. Online verfügbar unter https://www.bmi.bund.de/SharedDocs/downloads/DE/publikationen/themen/sicherheit/pks-2019.pdf;jsessionid=DF3D46AC05D478F4F1DC14F9FEBB3C2D.1_cid287?__blob=publicationFile&v=10, zuletzt aktualisiert am Mai 2020, zuletzt geprüft am 16.10.2020.

Caillois, R.; Barash, M. (2001). Man, play, and games. 1. Illinois paperback. Urbana, Ill.: Univ. of Illinois Press.

Coeckelbergh, M. (2007). Violent computer games, empathy, and cosmopolitanism. In: Ethics Inf Technol 9 (3), S. 219–231. DOI: 10.1007/s10676-007-9145-3.

Deinet, U., Derecik, A. (2013): Das Konzept der sozialräumlichen Aneignung und die neuen Medien. In: Hartung-Griemberg, A., Lauber A. & Reißmann, W. (Hg.): Das handelnde Subjekt und die Medienpädagogik. Festschrift für Bernd Schorb. Unter Mitarbeit von Bernd Schorb. München: kopaed, S. 73–88.

Eichenberg, C., Wolters, C. (2012). Virtuelle Realität in der Behandlung psychischer Störungen. Eintauchen ins virtuelle System. Hg. v. Bundes-ärzte-kammer (Arbeitsgemeinschaft der deutschen Ärztekammern) und Kassenärztliche Bundesvereinigung (Deutsches Ärzteblatt). Online verfügbar unter https://www.aerzteblatt.de/pdf.asp?id=128395, zuletzt geprüft am 16.10.2020.

Ekman, P. (2010).: Gefühle lesen. Wie Sie Emotionen erkennen und richtig interpretieren. 2. Aufl. (Spektrum Taschenbuch).

Feierabend, S., Rathgeb, T., Reutter, T. (2019). JIM-Studie 2019. Jugend, Information, Medien. Basisstudie zum Medienumgang 12- bis 19-Jähriger in Deutschland. Hg. v. Medienpädagogischer Forschungsverbund Südwest. c/o Landesanstalt für Kommunikation Baden-Württemberg (LFK). Stuttgart. Online verfügbar unter https://www.mpfs.de/fileadmin/files/Studien/JIM/2019/JIM_2019.pdf.

Ferguson, C. J. (2018). Video Game Influences on Aggression, Cognition, and Attention. DOI: 10.1007/978-3-319-95495-0.

Fritz, J. (2011). Wie Computerspieler ins Spiel kommen. Theorien und Modelle zur Nutzung und Wirkung virtueller Spielwelten. Berlin: Vistas (Schriftenreihe Medienforschung der Landesanstalt für Medien Nordrhein-Westfalen, Bd. 67).

Fritz, J., Rohde, W. (2011). Mit Computerspielern ins Spiel kommen. Dokumentation von Fallanalysen. Berlin: Vistas (Schriftenreihe Medienforschung der Landesanstalt für Medien Nordrhein-Westfalen, 68). Online verfügbar unter http://lfmpublikationen.lfm-nrw.de/modules/pdf_download.php?products_id=255.

Geisler, M. (2019). Digitale Spiele in der Medienpädagogik. Einstellungen, Erfahrungen und Haltungen von Spielleitenden. München: kopaed.

Greitemeyer, T., Osswald, S. (2009). Prosocial video games reduce aggressive cognitions. Mannheim.

Greitemeyer, T., Osswald, S., Brauer, M. (2010). Playing prosocial video games increases empathy and decreases schadenfreude. In: Emotion (Washington, D.C.) 10 (6), S. 796–802. DOI: 10.1037/a0020194.

Groen, M., Witting, T. (2016). There Are No Girls on the Internet. Gender und Kommunikation in Online-Gaming-Szenen. In: Brüggemann, M., Knaus, T. & Meister, D. M. (Hg.). Kommunikationskulturen in digitalen Welten. Konzepte und Strategien der Medienpädagogik und Medienbildung. München: kopaed (Schriften zur Medienpädagogik, 52), S. 179–192.

Häder, M. (2015). Empirische Sozialforschung. Eine Einführung. 3. Aufl. Wiesbaden: Springer VS.

Hüther, G., Quarch, C. (2016). Rettet das Spiel! Weil Leben mehr als Funktionieren ist. München: Carl Hanser Verlag. Online verfügbar unter http://www.hanser-fachbuch.de/9783446447011.

Jarren, O., Klinger, U. (2017). Öffentlichkeit und Medien im digitalen Zeitalter: zwischen Differenzierung und Neu-Institutionalisierung. In: Gapski, H., Oberle, M.; Staufer, W. (Hg.) Medienkompetenz: Herausforderung für Politik, politische Bildung und Medienbildung. Bonn: Bun-deszentrale für politische Bildung, 33-42.

Klimmt, C., & Trepte, S. (2003). Theoretisch-methodische Desiderata der medienpsychologischen Forschung über die aggressionsfördernde Wirkung gewalthaltiger Computer- und Videospiele. Zeitschrift für Medienpsychologie, 15 (4), S. 114-121.

Lewinsohn, P. M., Biglan, A., Zeiss, A. (1976). Behavioral treatment of depression. In: Park O. Davidson (Hg.): The behavioral management of anxiety, depression and pain. New York: Brunner / Mazel, S. 91–146.

Liska, G. (2017). Das Spiel als Freiraum beim Denken und Fühlen. In: Denk, N., Pfeiffer, A., Rembold, É & Wernbacher, T. (Hg.): Welt der Spiele 360°. Sammelband des Zentrums für Angewandte Spieleforschung der Donau-Universität Krems. Krems an der Donau: Donau-Universität Krems, eBook.

McGonigal, J. (2011). Reality is broken. Why games make us better and how they can change the world. New York: Penguin Press, eBook.

Mößle, T., Wölfling, K., Rumpf, H.-J., Rehbein, F.; Müller, K. W., Arnaud, N. et al. (2014): Internet- und Computerspielsucht. In: Mann, K. (Hg.): Verhaltenssüchte. Berlin, Heidelberg: Springer Berlin Heidelberg, S. 33–58.

Picton, I., Clark, C., Judge, T. (2020). Video game playing and literacy: a survey of young people aged 11 to 16. Hg. v. National Literacy Trust. Online verfügbar unter https://cdn.literacytrust.org.uk/media/documents/Video_game_playing_and_literacy_repo rt_final_updated.pdf, zuletzt geprüft am 16.10.2020.

Salen, K., Zimmerman, E. (2010). Rules of play. Game design fundamentals. [Nachdr.]. Cambridge, Mass.: The MIT Press.

Salovey, P., Mayer, J. D. (1990). Emotional Intelligence. In: Imagination, Cognition and Personality 9 (3), S. 185–211. DOI: 10.2190/DUGG-P24E-52WK-6CDG.

Satow, L. (2012a). Emotional Intelligence Inventar (EI4). Test- und Skalendokumentation. Online verfügbar unter URL: http://www.drsatow.de.

Satow, L. (2012b). Emotional Intelligence Inventar (EI4). Testmanual und Normen. Online verfügbar unter http://www.drsatow.de, zuletzt geprüft am 30.09.2020.

Staudacher, N. (2019). Digitale Spiele und ihr Potenzial als Bildungs- und Lernräume - In: Magazin

Erwachsenenbildung.at 13 (2019) 35-36, 7 S. - Online verfügbar unter

http://nbn-resolving.de/urn:nbn:de:0111-pedocs-166730, zuletzt geprüft am 15.03.2021.

Süss, D. (2016). Digitale Medien als Lebens-, Genuss- und Suchtmittel für Jugendliche. In: Brüggemann, M., Knaus, T. & Meister, D. M. (Hg.): Kommunikationskulturen in digitalen Welten. Konzepte und Strategien der Medienpädagogik und Medienbildung. München: kopaed (Schriften zur Medienpädagogik, 52), S. 43–53.

ANALYZING USAGE PATTERNS IN ONLINE GAMES

Wilfried Elmenreich, Mathias Lux

A typical life cycle of an online game is reflected in its usage patterns. A game first builds a user base, then reaches an absolute peak, to then being played by a minimum number of dedicated fans at the end of its life. Apart from this development, extraordinary internal and external events can be observed as changes in usage in games, especially multiplayer and massive multiplayer ones. For the usage of video games, the COVID-19 pandemic has impacted usage as it had on the game business itself. However, research lacks data to investigate these relations further. Usage statistics of games are rarely accessible for researchers. In this paper, we relate usage statistics to viewership and popularity of a game using available data sources like online statistics or activity on Twitch.tv. In a first study, data from the online role-playing game (MMORPG) Eternal Lands is analyzed. Eternal Lands is a free, multiplayer, online game that was created already in 2002. The usage patterns show day/night cycles of players in the prime time of the time zones where most players are located and increased playing activity on weekends. A general trend over time shows a slowly diminishing user base over the years since its introduction. In April 2020, a significant rise in user activities can be observed, attributed to lockdowns in many countries due to the COVID-19 pandemic. This can be attributed to regular players investing more time playing the game during the lockdown and to new or recurring players, who have not played the game intensively before, were looking for a distraction during the lockdown. In a second study, we focus on complementary viewer statistics on the popular game streaming platform Twitch.tv. We can observe that the COVID-19 pandemic impacted the playing time, as mentioned earlier. We relate usage data to viewership and streaming statistics of popular games. With the example of Eternal Lands, being a game that never went viral, we discuss the possibility of approximating a game's popularity through game streaming and viewership.

Keywords: Video games, Online Games, User Analysis, COVID-19

INTRODUCTION

When it comes to games, just like with movies, people recall immediately the blockbusters and AAA games, as well as those that went viral. However, just like with music or movies, we can assume that popularity of games follow a long-tail distribution. Such an assumption means that besides the few blockbusters, each with a massive number of players, we have a vast number of less popular games, which still have an active community. Image 1 shows this for a specific time point with the current players using the Steam platform. With the long-tail distribution we can assume that while only three games exceed 200,000 active players, the cumulative number of players of the other games exceeds the player base for the popular games by far. Of course, a rank-frequency plot like the one in Image 1 can only give us a snapshot of a game's life cycle.

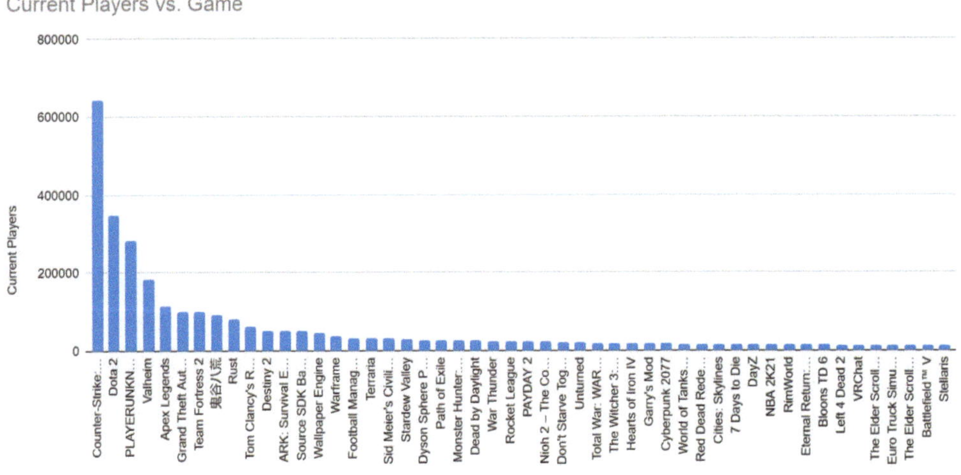

Image 1. Steam player statistics from Feb 19, 09:10 CET from the 50 most played games at that specific time. Taken from https://store.steampowered.com/stats/.

The full story of how a game is played, how the community forms, and its popularity diminishes over time is complex and lengthy. A simple example would be that people buy the game in the first week after release, play it and then archive it or sell it. However, early access, open betas, and viral games, just to name a few, do not follow this simple pattern. Moreover, data about game-playing activities is hard to get. Besides privacy concerns, playing statistics represent a vast amount of knowledge and a valuable resource not likely to be shared freely. Therefore, the possibilities for research are limited.

Game streaming, on the other hand, is by nature easily accessible. The entry barrier to open your own channel is small, and game streaming channels live from their popularity and discoverability. We hypothesize that the popularity of games in streaming is related to the popularity of games played.

While games have become more and more popular over the years, the COVID-19 pandemic in 2020 and 2021 presents a novel use case for gaming. In the COVID-19 pandemic context, many players resorted to online games to keep social interactivity up and talk and game with other people. So we chose to take a close look at the period around May 2020, often called the *first lockdown* in Europe, and focused on the questions:

1. Can we find indications that playing behavior changed in the COVID-19 crisis in Q1-Q2 2020?
2. Can we find indications that game streaming statistics show similar signs?

RELATED WORK

In general, the approach of analyzing players' behavior in games has been called *game analytics* (El-Nasr et al., 2016). Yee (2016) found that a model of different gamer types emerges from clustering data. In e-sports, researchers analyze past games to predict upcoming ones' outcomes (Schubert et al., 2016). For massive multiplayer online games (MMOs), researchers are interested in players' behavior to optimize network traffic and reduce latency (Suznjevic and Matijasevic, 2013). Moreover, the interrelations between game mechanics and player behavior have been investigated (Moll et al., 2020).

Johnson and Woodcock (2019) investigate the interrelation between game streaming on the popular streaming platform Twitch.tv and the game industry in three different scenarios. With statistical data and interviews, they investigate (i) how streaming impacts consumer choice, (ii) how it alters the lifespan of niche and older games, and (iii) how making the development of games public via streaming changes the opinions of players. However, Harpstead et al. (2019) identified the need for shared data and unified access in the use case of analyzing streaming statistics as many researchers draw conclusions on data they have gathered on their own.

DATA

Any approach for data-driven research is, of course, heavily relying on identifying and gathering relevant data. In the domain of digital commercial games, sales numbers are an obvious first choice. However, recent and detailed statistics of sales are typically hard to get by. Moreover, sales do not reflect the number of actual players. Some players might play the game only a short time, while others might re-visit the game regularly. For genres like MMOs, or battle royale games, numbers like subscriptions and online players are more relevant than sales. For online games or games with an online DRM component, the number of currently online players is often available through a web page or an API. Detailed, in-depth player statistics, including what the players are doing in the virtual world, are typically only available to developers and publishers. If needed, one has to scrape them from recordings or record them in controlled experiments, as Moll et al. (2020) did.

On the other hand, metadata is often not directly related to the actual game usage but can reflect the players' activity beyond the virtual world. We consider metadata of a game to be the data generated from the active community around the game. Metadata includes reviews from commercial reviewers as well as consumers and social media mentions like tweets or Reddit threads. Hubs for the communities around games are also a valuable resource. These hubs can be on Discord servers, in the Steam Community Hubs, Telegram, or any other online communication platform.

Game Streaming provides another wealth of data. Game streaming generally refers to one or more people playing a video game and broadcasting the game screen and a live camera stream of the player to the general public on the internet. The game stream typically has a lag of up to ten seconds. The degree of interactivity is often high, as streamers interact with their audience by reacting to chat messages. As Johnson and Woodcock (2019) researched, a general correlation between popularity in streaming and popularity of a game in terms of active players is generally assumed. Based on a survey among Twitch spectators, Gandolfi (2016) classifies streamers into three classes of personalities, each with different game streams. *Challenge* streams are commonly found in e-sports, where players interact less with the audience but engage in competition. *Exhibition* streams balance interaction with professional playing on a high level of player skill. *Companion* streams rely mostly on the interaction between streamers and the audience. These three types of streams also indicate that the causal relation between playing a game and viewing it in a game stream is influenced by the streamer's personality, i.e. in a way that viewers tend to watch gameplay of a game they are not interested just to follow one of their favorite streamers.

Still, game streaming is a huge phenomenon, which keeps getting bigger. According to May (2021), the number of hours viewed online on game streaming platforms rose 78.5% from 2019 to 2020 to 27.9 billion hours watched. The biggest streaming platforms are Amazon-owned Twitch.tv, Youtube from Google, and Facebook Gaming. Table 1 gives an overview of how many hours were streamed and watched on these three platforms.

Table 1. Comparison of the three biggest game streaming platforms based on their statistics in Q4 2020. Data from May (2021) with numbers in millions (rounded).

	live total h watched	total h streamed
Twitch.tv	5436	230M
YouTube Gaming	1924	10M
Facebook Gaming	901	15M

Besides the selection of which data to use, the source of the data is also of importance. While manual aggregation from web pages and APIs is a swiss army knife of data gathering, services and reports provide a look back into time. Besides quarterly reports like the one from May (2021), multiple web pages provide data aggregation services. Examples are *steamdb.info* or *steamspy.com*, which focus on aggregation from Valve's Steam platform. Aggregated Twitch.tv usage data is available on *sullygnome.com*, *twitchtracker.com*, and *twitchstats.net*. An ambitious example is *playtracker.net*, where cross-platform statistics are estimated.

SCENARIOS

With the vast landscape of games and the long tail of popularity, there are plenty of games to choose from for a close investigation of changes in playing behavior during the COVID-19 pandemic. We picked three games with a focus on maximizing the conceptual distance between the games.

Table 2. Games selected for the study in direct comparison. Active players according to playtracker.net, hours viewed for 365 days according to sullygnome.com as of Nov. 17 2020

	Eternal Lands	Drawful 2	CS:GO
Genre	MMO	Party Game	FPS for e-sports
Free to play	yes	no	yes
Platforms	PC, Mac, Android	Full range	Steam
Active players~	200	6.4M	59M
Hours viewed ~	2k	400k	681M

Eternal Lands

Eternal Lands is a free and online multiplayer game developed by Radu Privantu. The game went first online in 2003. It is available on PC, Linux, Mac, and Android and features an open-source client with a proprietary server and content. The game features 3D-graphics, which had been state-of-art at the time of creation. The game is set in a medieval fantasy world, which is typical for many MMORPGs. A significant point of attraction and long-time motivation for players is communicating via the chat system, which, in Eternal lands can be even put to focus while hiding the game graphics. The game is under ongoing development by the community, which communicates over an online bulletin board. We collected the statistics of online players from the web page starting from Mar 13, 2012, to Feb 24, 2021, with an interval of 1 hour.

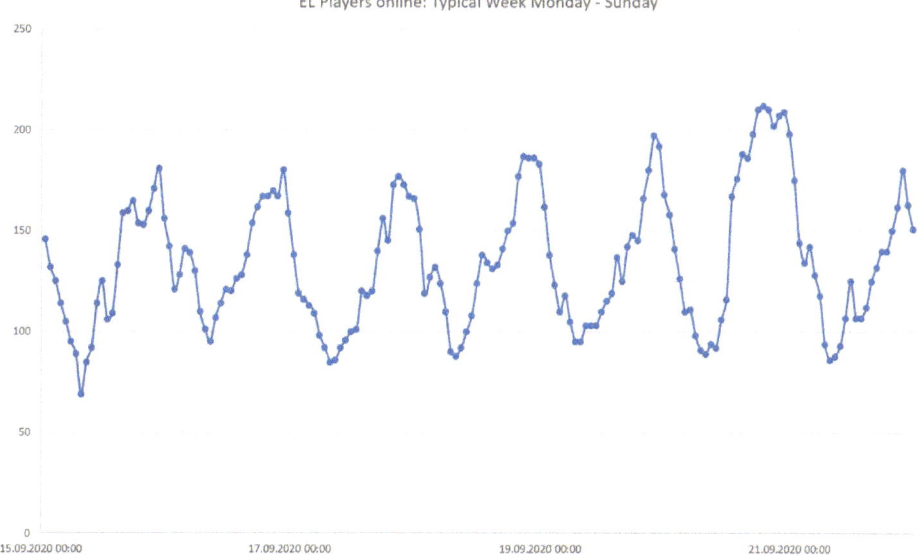

Image 2. Number of active players in Eternal Lands in a typical week from Monday to Sunday.

Eternal Lands as an MMO is small in terms of online players. Within the time of collecting statistics, we found a maximum of 497 players online and a weekly average ranging from 124 to 273 online players. The players are from similar time zones and typically meet at scheduled times of the day. Image 2 shows a sample of the data from a week in September 2020. The number of players shows a pattern with more players being online in CET prime time and a low number of players in the morning in the CET time zone.

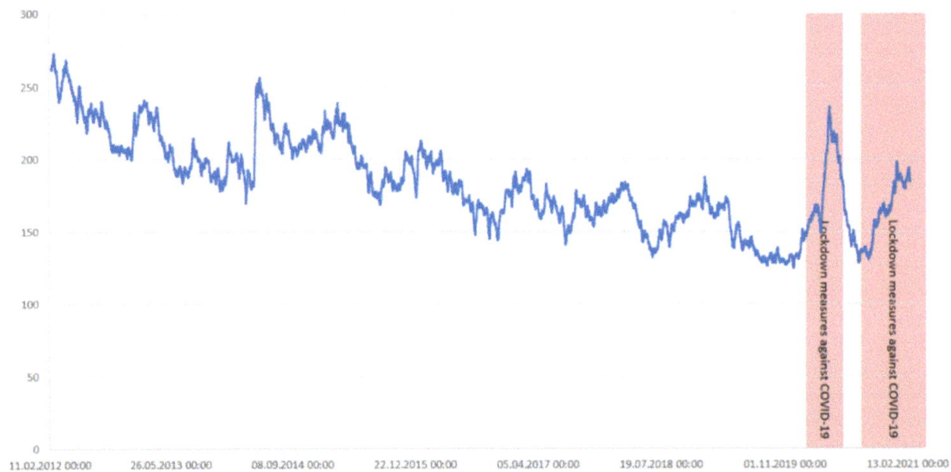

Image 3. Active players in Eternal Lands for the full sampled period.

Image 3 shows the number of active users declining until a steep rise in May 2020 (left marked area) with another increase in October 2020 (right marked area). This indicates that the pandemic correlates to an increase of online players, either due to more intensive play or new or re-visiting players. While the number of online players clearly show a trend correlated with the pandemic lockdowns, the number of hours viewed and streamed on Twitch.tv for Eternal Lands is so small that no rise in streaming hours can be argued for the time of the COVID-19 pandemic (see Image 4).

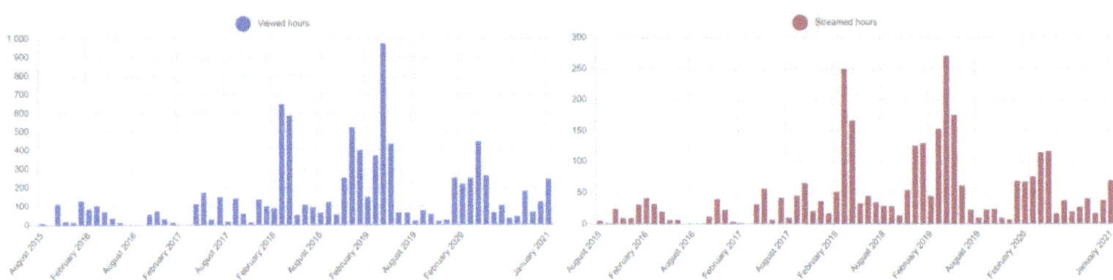

Image 4. Total hours watched (left) and streamed (right) on Twitch.tv. Taken from https://sullygnome.com/game/Eternal_Lands/longtermstats

Drawful 2

Developed by Jackbox Games, Drawful originally came out from a Jackbox Party Pack, game collections focusing on local multiplayer fun games with a lot of interaction between players. Drawful 2, being the second installment, was released in 2016 as a standalone game and adds additional streaming features, a second color, and the option to include spectators into the games' interaction. Only one player needs to own the game. All others can join in using their mobile phone or any web browser. Playing relies on communication: (i) players get phrases to draw, then (ii) guess what the others have drawn, and (iii) the guesses of players are presented as multiple choices. Points are awarded for good drawing and misdirection by entering smart answers in step (ii).

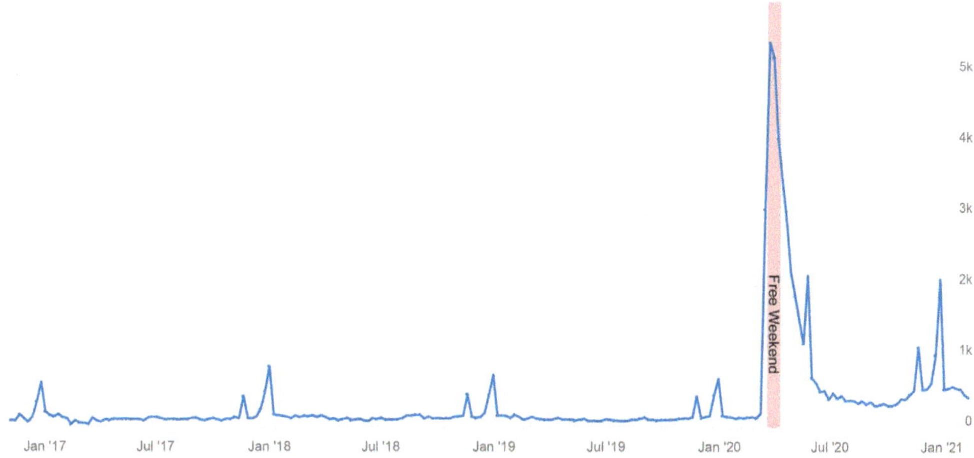

Image 5. Concurrent Drawful 2 players on Steam. Taken from
https://steamdb.info/app/442070/graphs/

Image 5 shows the concurrent Drawful 2 players registered on the Steam platform from Jan. 2017 to Jan 2021. A surprising pattern can be easily seen: Drawful 2 is more often played in the holiday season, as indicated by the spikes in the graph. In March 2020, in the midst of the first lockdown in Europe, there was a free weekend on Steam, leading to a spike in players on

Steam, although the game was featured as a free giveaway on the Epic Store too. In general, the time of the COVID-19 pandemic shows a higher number of concurrent users, also leading to a visually more prominent holiday season spike around December 2020 to January 2021.

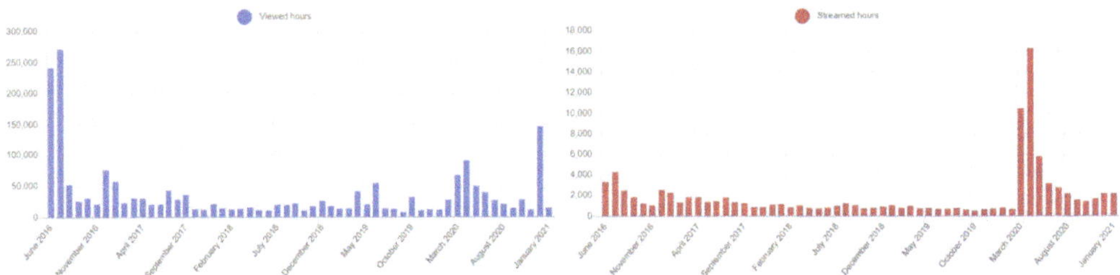

Image 6. Total hours watched (left) and streamed (right) on Twitch.tv. Taken from https://sullygnome.com/game/Drawful_2/longtermstats

Image 6 shows the total hours watched and streamed on Twitch.tv. Especially in the hours streamed the free weekend shows. Still, throughout the pandemics the bars are higher in both graphs, indicating higher interest in Drawful 2.

Counter-Strike: Global Offensive

Counter-Strike: Global Offensive (CS:GO) is a first-person shooter, where teams of four compete in a tactical scenario featuring terrorists and counter-terrorists. In general, it is played in rounds in a best of 30 fashion, and roles are switched after round 15. Besides the actual shooting, team-based tactical decisions and in-game economy are essential aspects of the game. CS:GO is a popular e-sports game with a stable viewer and player base. It's only available on Steam and is one of - if t not the - most popular games on Steam.

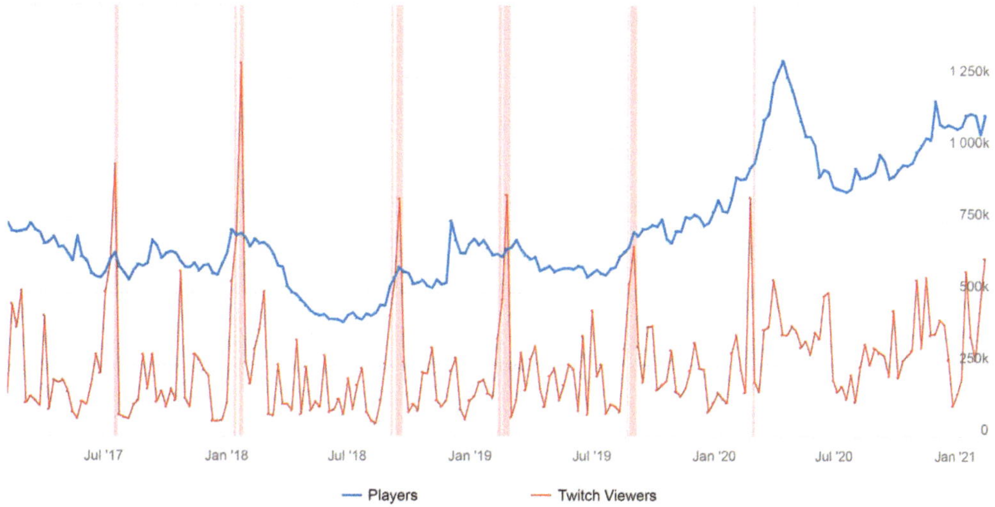

Image 7. Concurrent CS:GO players (blue) on Steam and Twitch viewers (red). The bars indicate time spans, where a prominent e-sports tournament took place, src. https://steamdb.info/app/730/graphs/

Image 7 shows the development of concurrent players and game stream viewers over time. The bars correlate with peaks in the number of viewers and relate to popular e-sports tournaments. With March 2020 a visually significant rise in the number of players can be seen in the image leading to an overall higher number of players throughout the rest of 2020, and, therefore, the COVID-19 pandemic.

CONCLUSION

For all three games we investigated, the impact of the COVID-19 pandemic can be found easily by visual inspection of the data in the graphs. However, the effect is different. While for Drawful 2, the rise in players is visible, we hypothesize that the free weekend on Steam and the giveaway on the Epic store at the same time acted as triggers for the rise in players and streamers. For CS:GO, it is interesting to see a visual correlation between the e-sports tournaments and the number of Twitch viewers. It is important to point out that the tournaments are exclusively streamed on Twitch.tv by the respective organizers. However, throughout the pandemic, the number of players rose to the highest number (1.3 million players, src. https://steamdb.info/app/730/graphs/) in the whole life cycle of CS:GO. These numbers are straightforward to interpret as CS:GO is only available on Steam, and the data on concurrent players are assumed to be accurate. Drawful 2, in contrast to that, is available on many platforms, including Playstation, XBox, Epic, and Apple TV, where we could not find estimations on the number of active players.

Regarding our first research question, we can answer that we can see indications of a change in player behavior throughout the pandemic for each game. With the accurate number of active players for Eternal Lands and CS:GO, active players' increase correlates well with the time of the lockdowns in Europe. For Drawful 2, the influence of the free weekend and the free giveaway are unclear. Future research will be to identify strong correlations between data and metadata, e.g., between active players and game streaming. That'd allow us to estimate the active player base of games like Drawful 2, available on multiple platforms with no data easily accessible. Moreover, the seasonal character of Drawful 2 and the events impacting the viewership of CS:GO are interesting patterns, which might be common in games of the same or similar genres. Last but not least, MMOs and online games often have problems with a declining user base. Investigating the correlations between active players, streamers, viewers, and events might bring some practical advice for game marketing.

REFERENCES

El-Nasr, M. S., Drachen, A., & Canossa, A. (2016). *Game analytics*. Springer London Limited.

Yee, N. (2016, October). The gamer motivation profile: What we learned from 250,000 gamers. In *Proceedings of the 2016 Annual Symposium on Computer-Human Interaction in Play* (pp. 2-2).

Schubert, M., Drachen, A., & Mahlmann, T. (2016). Esports analytics through encounter detection. In *MIT Sloan Sports Analytics Conference*. MIT Sloan.

Suznjevic, M., & Matijasevic, M. (2013). Player behavior and traffic characterization for MMORPGs: a survey. *Multimedia systems, 19*(3), 199-220.

Moll, P., Frick, V., Rauscher, N., & Lux, M. (2020, June). How players play games: observing the influences of game mechanics. In *Proceedings of the 12th ACM International Workshop on Immersive Mixed and Virtual Environment Systems* (pp. 7-12).

Johnson, M. R., & Woodcock, J. (2019). The impacts of live streaming and Twitch. tv on the video game industry. Media, Culture & Society, 41(5), 670-688.

Harpstead, E., Rios, J. S., Seering, J., & Hammer, J. (2019, October). Toward a Twitch Research Toolkit: A Systematic Review of Approaches to Research on Game Streaming. In *Proceedings of the Annual Symposium on Computer-Human Interaction in Play* (pp. 111-119).

Gandolfi, E. (2016). To watch or to play, it is in the game: The game culture on Twitch. tv among performers, plays and audiences. *Journal of Gaming & Virtual Worlds, 8*(1), 63-82.

May, E. (2021, January 31). *Streamlabs and Stream Hatchet Q4 Live Streaming Industry Report.* Streamlabs. https://streamlabs.com/content-hub/post/streamlabs-and-stream-hatchet-q4-live-streaming-industry-report

"WHERE DID YOU LEARN TO FIGHT?" GAMIFICATION OF AN ONLINE FIGHTING CLASS FOR STUDENTS AT GERMAN SPORT UNIVERSITY COLOGNE

Swen Koerner, Mario S. Staller

The spread of SARS-CoV-2 in Germany and the general restrictions on social contacts decided upon pose major challenges for institutional teaching and learning settings (Koerner & Staller, 2020a). For the German Sport University Cologne (GSU), one of the world's most renowned sport universities, due to the officially ordered shift from presence to online courses innovative and adapted solutions were in need, especially where the teaching and learning of motor and tactical skills had to be arranged. In this situation, especially for students of the GSU "self-defence" course of the summer semester 2020 the question aroused, "where do we learn to fight?"

The paper presents the conceptual framework of a regular university Krav Maga based self-defence course – "train2fight the virus" – developed specifically for this challenge, in which elements of gamification (Schell, 2017) play a driving role. Particularly, the story envisaged that besides weekly team challenges in the areas of "fitness", "growth", "fighting", "solidarity" and "knowledge" students had to prepare for a strenuous fight against an unknown and highly skilled end-boss via live online trainings.

Keywords: SARS-CoV-2, e-Learning, institutional teaching, self-defense, gamification

---

INTRODUCTION

The spread of SARS-CoV-2 in Germany and the general restrictions on social contacts decided upon pose major challenges for institutional teaching and learning settings (Koerner & Staller, 2020a). In March 2020, a collective helplessness quickly spread among many lecturers at the German Sport University Cologne (GSU), one of the world's most renowned sport universities, as a result of the ban on face-to-face teaching. While online based solutions were already in place for theoretical courses primary dealing with cognitive content, for the online teaching of practice-oriented sports and movement skills using electronic devices neither experience nor orientation was available.

In order for the practical online teaching to take place under conditions of COVID-19 innovative and adapted solutions were in need, which at the same time at least principally meet the demands of the respective university curriculum. In the case of GSUs` regular self-defence module PE 2.35, this meant that students had to be enabled to "understand and practically apply basic principles of self-defence" (GSU, 2020) by means of online teaching. The paper presents the conceptual framework of GSUs` "self-defence" course developed specifically for this challenge, in which elements of gamification (Schell, 2017) play a driving role.

First, we start by addressing the special challenge of teaching practice content in self-defence (2), followed by the idea of taking the pandemic as a development opportunity for both, the teacher and the students (3). Framed and informed by key elements of modern gamification (4), the concept of "train2fight the virus" is presented, representing an online self-defence class, which has been awarded with GSU´s e-learning prize in 2020 (5). With regards to the pending systematic analysis of student's views, we conclude by referring to related issues and questions of outcome according to recent literature on gamification. In sum, the article provides a conceptual proposal for a gamified design and delivery of online practice-based university courses adjusted to the contemporary requirements of physical distancing.

TRAIN2FIGHT THE VIRUS

Level one – challenge

The spread of SARS-CoV-2 in Germany and the resulting measures of physical distancing (BR, 16.3.2020) still impact many areas of society, not at least the teaching at schools and universities. For the German Sport University

Cologne (GSU) with its wide range of practice-oriented courses the ongoing corona pandemic addresses major challenges.

On the part of the university management for the field of practical fighting any form of face-to-face teaching for the summer semester 2020 was prohibited, including the self-defence course taught by the first author SK, which is an integral part of his teaching load and is anchored as PE 2.35 in many of GSU´s students degree programmes. At the same time, with the availability of WebEx, the universities` digital infrastructure met the technical requirements for online teaching in principle. But how should practice-oriented online teaching be delivered? Actually, nobody could answer this question reliably, simply because nobody had experience.

As if the notion of online teaching at universities using electronic devices and/or the use of asynchronous media formats (YouTube, Vimeo etc.) is still not unusual enough (Martin et al., 2019), many practical areas lacked ideas about what online-based teaching specifically could look like. Even if the lecturer can acoustically and optically interact with the students through microphone and camera, the teaching and learning of practical self-defence skills does not seem to fit into the realm of virtual learning due to its bodily foundation. Fighting essentially relies on processes of interpersonal synergy (Krabben et al., 2019) established through the dense interaction of human bodies.

In addition to the university's digital infrastructure, the collective teaching culture as well as demands established by the practice content itself the corona pandemic finally poses a major challenge to the lecturer. Similar to sports coaching (Muir et al., 2011) and police training (Staller, 2020), university teaching can be identified as a context of professional work, in which the practice is not only about the question of what to teach how to whom. As the corona pandemic impressively demonstrates, professional university teaching heavily depends on contextual factors as it is true for the sports (Staller & Koerner, 2020a) and police domain (Koerner & Staller, 2020a).

The pandemic altered key constraints of social live in general and has led to the mandatory limiting of physical proximity between citizens and therefore between learners and teachers. It's more than likely, that these contextual changes have impacted accustomed routines of design and delivery of many university teachers. As experienced by SK, being the head of the university department responsible for the teaching of martial arts and combat sports at GSU, this was definitely the case among many colleagues.

Corona and the associated measures of physical distancing may have irritated quite a few existing beliefs, values and attitudes for teachers and students on how to teach and learn martial arts and related practice-based movement and sports skills (Andreucci, 2020). The unfamiliar situation caused by the global pandemic has indicated, that teaching involves not solely questions of what to teach to whom in which way under which conditions, but foremost depends on the individual teacher´s mindset. Corona scratches on the underlying assumptions that guide one's actions in relation to teaching self-defence, ranging from that "nothing can be done" to the straight opposite.

Level two – idea

In this situation the idea arose to take Corona as a challenge for the exploration of new ways of teaching self-defence to students at GSU through the means of e-learning. As time between the announcement of the national lockdown and the start of the summer semester was scarce, there was a strong need for ideas especially in sports practice. At this time many students and teachers experienced in private amateur sports, competitive sports or at school what it feels like when training suddenly comes to a stop: no more instructed training, no more sports lessons.

The reference point for the planning and design of PE 2.35 on self-defence was not how the course would "normally" be run and how self-defence is or should "normally" been taught. The point of reference was the crisis itself, which excludes many of the usual pedagogical routines: "Imagine another lockdown in the future: As a learner, athlete or student - would you be happy about your club and trainers providing alternative ways of practice? As a trainer – who would you like to be then? The one who does something for your learners or the one who buries his or her head in the sand?" The students haven been asked these questions addressing their personal mindset at the beginning of the semester. The same questions were answered by the authors in advance (Staller & Körner, 2020; Koerner & Staller, 2020), resulting into a conceptual framework that was adapted to the Krav Maga course.

Questions of "normal" teaching make sense under "normal" conditions, whereas under corona conditions innovative solutions are needed (Demantowsky, 2020; Hanstein & Lanig, 2020), which at the same time at least principally meet the demands of professional teaching. These comprise of the balance of student´s expectations and needs (who-dimension, e.g. that the course takes place in a motivational manner) and the demands of the respective university curriculum (what-dimension). Concerning PE 2.35 the module description states, that during the semester students have to be enabled to "understand and practically apply basic principles of self-defence" (GSU, 2020). Against this background, the framework of gamification has been identified as a promising pedagogical approach for the conceptual design and delivery of GSU´s fighting class.

Level three – concept

While interesting academic debates on the question of what gamification is or is not do exist (Boendermaker, 2017; Fatta et al., 2019; Huotari & Hamari, 2017; Seaborn & Fels, 2015; Yohannis et al., 2014), the conceptualization of PE 2.35 followed a straight hands-on approach. According to widely used definitions, the gamification of GSU´s mandatory self-defence course required a) "the use of game design elements in non-game contexts" (Deterding et al., 2011) guided by b) a clear pedagogical intention, i.e. "to engage people, motivate action, promote learning and solve problems" (Kapp, 2012). In short, gamification was used to support GSU´s students online learning of genuine movement-based self-defence skills in times of Corona.

More specifically, for the process of detailed planning and reflection of design and delivery of a gamified Krav Maga e-learning course, Schell's (Schell, 2017) element tetrad of game design has been applied. The tetrad contains the following elements (see figure 1):

a) Story, defined as the narrative structure of the game / gamified context, including plots and characters involved, among others reinforced by

b) Aesthetics, including elements perceptible to the human senses, affording participants perception and action, resembling characters and narratives of the story.

c) Mechanics as the very core component of a game / gamified context, comprising demanding tasks and the set of rules, guiding and affording the players / learners task completion, linking them to the respective goal and providing immediate feedback on players / learner's decisions and actions.

d) Technology as the hardware and infrastructure of the game, in a broadest sense comprising all technical prerequisites and means that enable the game / gamified context to be set in place and played smoothly.

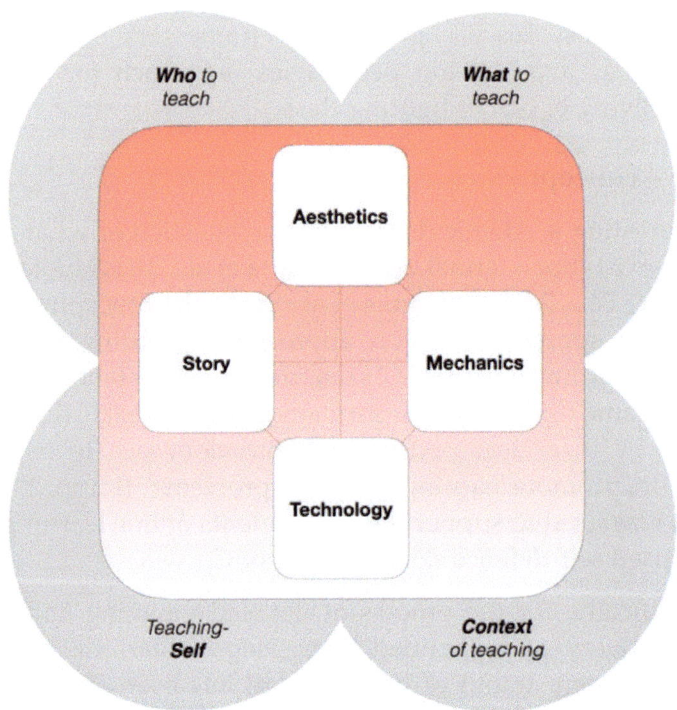

Figure 1. Elements and dimensions of gamification (Staller & Koerner, 2021).

Within the planning and reflection of a gamified e-learning environment each element of the tetrad requires single attention by the designer / teacher. It is important to note that all of them are interconnected, effect each other and have to be designed in a strong interactional relationship. For instance, elements of aesthetics have to take up and thematically stage the story, immediate feedback (mechanics) for results has to be made perceivable (aesthetics), and technological infrastructure has to be set according to the goals of the game / gamified learning environment, and so on. As the tetrad provides a conceptual framework for the design and delivery and therefore provides valuable solutions for the how-dimension of teaching, it is in turn deeply connected to the what- (what to teach), who- (who to teach), self- (the teaching self) and the contextual dimension mentioned before (see 2; figure 1). Next, we present how it has been adapted to the specific context of GSU`s PE 2.35 self-defence course and the challenge of online teaching.

Level four – application

Due to the daily topicality of Corona, the overall social situation in Germany during lockdown as well as the serious impact of the pandemic on various teaching and learning contexts in sport, the crisis itself provided the explicit point of departure for PE 2.35. At the beginning of the first online session using WebEx (WebEx training, Cisco) a fake headline from a fake daily newspaper ("The Gotham Times") has been presented to the students. The headline stated: "Krav Maga Classes cancelled - Clubs and Universities without concept". Triggered by the headline, which, although purely fictional, reflected the current status at GSU as well as individual experiences in many cases, an immediate discussion arose among the students. The discussion centred on what could be expected, what could be done from a learning and teaching perspective in times of Corona. At this point SK left it up to them to decide how to proceed.

As one and officially recommended option, a complete conversion of the course to the asynchronous provision of purely theoretical content has been offered. In this case the material would have been uploaded to the GSU e-learning platform. As part of this option, practical units would have been postponed until a possible end of the lockout. The alternative was to meet in WebEx on a weekly basis at course time. In this case, theoretical aspects as well as the joint training of practical self-defence skills would be the subject matter. The majority of the students voted for the latter. Thus, the way for the prepared gamified concept was paved, which was introduced next starting with the overarching narrative of the story: "Welcome, we are train2fight the virus! Corona is a challenging task for all of us. We are neither virologists nor politicians, but we will fight the pandemic with our strengths and means. Week after week you will receive challenges that you will work on as a team. Besides the team-work, a strenuous challenge is waiting for each of you: You will have to face a highly skilled end-boss, who already started his combat training. Your only option is to win this fight. You´d better prepare…"

The exposure of the story around the "fight against the virus" marks the beginning of something totally different to a normal self-defence course at GSU. PE 2.35 had just turned into a gamified learning environment and students became players. And, being typical for a gamified context, the story had been condensed in the title train2fight the virus! and had been visualized in a logo (aesthetics) of the same name (see figure 2). Title and logo clearly articulated the prospect of an ongoing series of missions the students had to face on a course-, team- and individual level (mechanics). Through the announcement of an end-boss, whose level of ability has been demonstrated by showing a short training clip (aesthetics; technology), the student's motivation for practice had been awakened at an early stage and a (first) indirect impulse to confront one's own identity was given by facing a tangible opponent character.

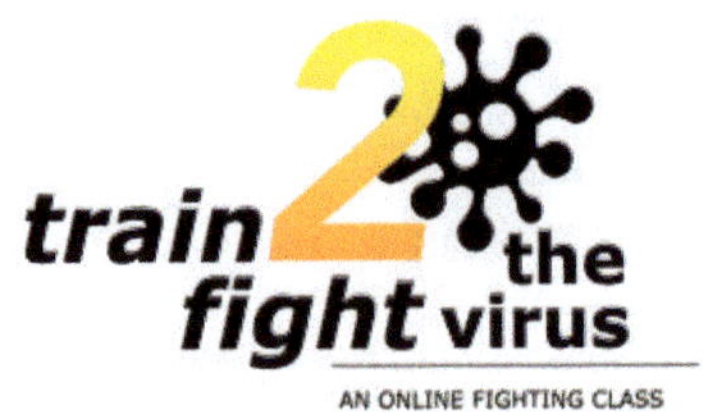

Figure 2. Logo-Design for PE 2.35.

In order to further stimulate the dynamic of the online-learning environment, its mechanic comprised weekly challenges within the areas of "fitness", "growth", "fight", "solidarity" and "knowledge", which had to be mastered on a team-level, with each team consisting of four randomly assigned students. The selection of areas and individual content was guided by GSU´s curriculum for PE 2.35, e.g., containing self-defence skills and related knowledge ("fight", "knowledge"), generally meaningful areas of learning for sport students ("growth", "knowledge", "fitness") as well as context-related aspects to the Corona pandemic feeding the cluster of "solidarity". For example, in week three the teams (and always each of its member) had to jump rope ("fitness"), to fight against invisible opponents ("fight"), to learn something they couldn´t do yet, e.g., juggling ("growth"), to support elder people with the completion of tasks in everyday life ("solidarity") and to carefully listen to a media interview, where SK explains the specialities of Batman's contemporary fighting style ("knowledge").

To allow for student's autonomy in decision-making, with regards to weekly task accomplishment teams could choose either to be "Virus Fighter", "Virus Hero" or "Virus Dominator", with each level differing in effort, difficulty and the expected reward. For example, a team that chose for each member to jump-rope on two days a week for 60 seconds twice reached the level of "Virus Fighter" and was awarded 15 points. Doing the exercise on two more days meant to be "Virus Hero" with 20 points on the score. Finally, to jump-rope on six days a week accomplished the Dominator-Level and resulted in 25 points. At the end of the semester the team that scores the most points won.

Teams were encouraged to consider their decisions carefully, since selecting a level had obliged them to complete all challenge areas in that level. For example, a team that decided to jump rope on six days, had also to help elderly people on six days, to practice juggling on six days and fought invisible intruders on six days. Finally, to become "Virus Dominator" they had to produce an audio Bat-Cast ("knowledge") on Batman's fighting style, while Virus Fighter, on the other hand, would only have to answer an online quiz (technology) on the related media interview. The mechanic of weekly challenges in relevant areas of student´s live together with the choice of different degrees of difficulty and effort resulting in different rewards, has been a key feature of gamifying PE 2.35.

During the first few weeks, the challenges were conceived by SK and presented at the end of each session in the form of a specially produced YouTube video. The use of video technology and internet-based ways of distribution goes along with several advantages. First, they allowed for multiple channels to explain the tasks to be done, e.g., by visualizing (aesthetics) how a fight against invisible intruders (a metaphor for corona and thereby part of the story) could look like. Second, using YouTube made the information accessible to all teams and members anytime and anywhere. Third, the challenge videos allowed for a further identification with the overarching story, since the videos addressed the "train2fight the virus" challenges and contained the game logo as well as the logo of the Justice League (aesthetics), as whose member SK had introduced himself in the opening session.

The respective weekly achievement was checked off by the teams in a specially designed (containing logo, aesthetics) challenge sheet (technology) and sent to SK. Points and ranking were presented in a table (aesthetics; technology) so that the teams could observe their current status and reflect their own development (possibilities) in the mirror of the others (mechanics). Consistent to SK´s character as a member of the Justice League, for the second session all teams were requested to create a signature name and logo suitable to the strengths and characteristics of their group, for which each of them was rewarded additional 25 points. "KravMagicians", "Shadow Warriors" and "Buffers" were now part of the game, together with "Quarantinos Unstoppable" and "SpoHomefighters" which made reference to the virus theme. From the fifth week of the semester onwards the teams took over the challenge part (see figure 3). Week after week one of the teams developed the challenges for the other teams and made them available via challenge sheet and video. By doing so, the students switched their role within the "game" and turned into learning designers responsible for the creation of demanding and motivating tasks and feedback.

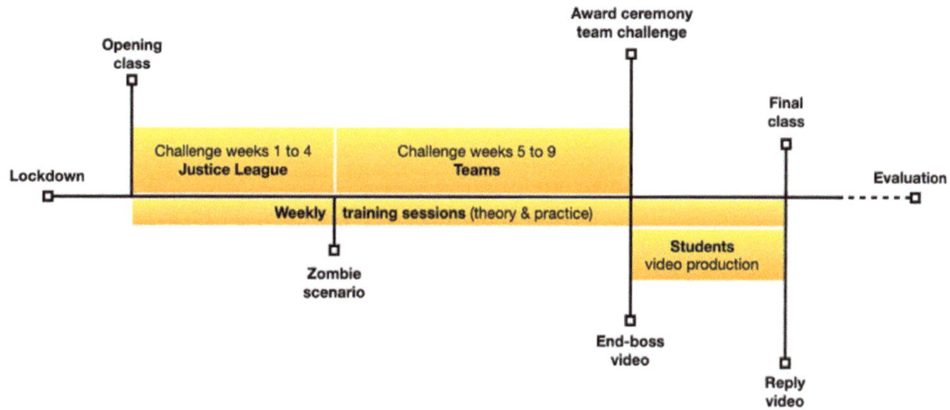

Figure 3. Timeline of PE 2.35.

In addition to the weekly challenges, designed for asynchronous processing, a synchronic training took place every week (see figure 3). Within these units at regular course time the participants were trained in Krav Maga based self-defence by SK, using the WebEx video function. In preparation of the fight against the fearsome end-boss (story), participants were taught "important" techniques. Instead of the usual material used for self-defence training, for example towels and pillows were used as punching pads and chairs or door frames became "opponents", where the students have learned to fight safely, i.e., without or with light contact on the surface.

Due to WebEx` video function (technology), a high amount of interactive teaching and fighting was allowed. For instance, the students were regularly asked to punch or kick against a target (a pillow or towel) held by SK. The task was to immediate stop attacking as soon as they see SK falling on the ground (mechanics), representing the end of a violent encounter. Due to the gallery-view SK was able to observe who stopped and who didn't and to provide immediate feedback ("Hey Mike, that would not have been self-defence anymore…"). Similarly, student's reaction to a straight punch attack could be observed and, according to the perceived individual action be physically answered or commented ("Good defence, Lucy. But what about that…?"). All principles and techniques treated week after week were listed and illustrated in an Excel spreadsheet (technology).

In order to prepare the students for their decisive fight challenge, regular training sessions altered with live scenarios. For instance, the fourth session consist of a zombie scenario, introduced by the thematic keynote slide and soundtrack based on the series "The Walking Dead" (story; aesthetic). The story (as a story within the overarching story) envisaged that SK had disappeared, while the students woke up in front of their cameras on the 8th floor of a hospital and had to fight their way out. On each level of the building, they were attacked by changing opponents (all played by with masks and further requisites by SK).

In the subsequent theory unit of the 4th week, the students were given the key to understanding that and why fighting works online via camera interaction even without contact (and thus online practical teaching of self-defence can work): In the sense of "ecological dynamics" (Koerner & Staller, 2020b; Seifert & Davids, 2016), motor performance is dependent on the continuous co-regulation of perception and movement. Students were asked to situations they had just experienced within the scenario, e.g.: "When the zombie at the camera just came towards you and wanted to choke you, what did you do? Then came answers such as "I dodged," "I ducked," "I punched," etc. These are exactly the same competences that also build the focus of a regular PE 2.35 course at GSU. The only difference: no direct physical contact.

After each training, different aspects of self-defence and human movement organisation (e.g., history of Krav Maga, prevalence of violence, legal aspects of self-defence, motor creativity, etc.) were discussed and reflected on the events just experienced. Week after week the Justice League had prepared a related keynote of about 20 minutes, which always contained a work assignment, which was first discussed in breakout sessions in the team and then together on course level. The keynote design included occasional references to DC`s superhero Batman (e.g., the weekly ranking slide with the Batcave in the background), reinforcing the story-based game character of the learning environment (aesthetics).

The meaning of the Batman references as part of a story within the story, was finally revealed to the students through the end-boss video presented in the last third of the semester and uploaded on YouTube (technology). The video was produced from the first-person perspective of the students and picked up different tasks, situations and characters from past online teachings (e.g., zombies) and culminates in a one-minute final fight against Batman (played by SK in a Batman suit; aesthetics). In a significant phase of the fight, when Batman becomes aware of what the students are capable of (depicted by Batman, who repeatedly suffers invisible attacks, stumbling and falling on the ground) he poses the striking question: Where did you learn to fight? The recognition resonating in the question refers to the initial issue of the course - how to teach and learn self-defence at GSU in times of Corona – and thus providing an answer.

Every single student had the task to choreograph an answer fight from what he or she had learned during the online-course and to shoot it within three weeks with not more than 60 seconds length (mechanics). The actions and reactions had to match the actions and reactions shown in the reference video (e.g., situationally appropriate kicking at the right distance with functional dynamics). The evaluation criteria (balance, appropriateness, distance, dynamic, variability) were dealt with in each session as part of the practical unit and were except of few concessions the same as in presence course of PE 2.35. In the end, the "Shadow Warriors" won the team challenge, with all teams scoring high. To some extend the students videos showed excellent fighting performances. In quite a few cases the story has been taken up and students appeared in the character of known opponents (e.g., by using aesthetics of the "Joker" figure).

CONCLUSION

The Corona pandemic generally addresses a major challenge for university teaching and learning at the moment. This especially has been true for the past mandatory self-defence course PE 2.35 at GSU, centrally dealing with the delivery of practical skills. Acknowledging the context-specificity of gamification (Hamari et al., 2014) the application of design elements of games appeared as a worthwhile and suitable way for GSU´s Krav Maga based self-defence course. The design of "train2fight the virus" as an online class on self-defence emerged out of the very special constraints of current physical distancing and the sanction of teaching in presence.

From our subjective perspective the gamified online-teaching (how-dimension) consisting of weekly team challenges and the preparation for a strenuous fight against an end-boss on the individual level succeeded in connecting student's needs (who-dimension) and curricular demands of PE 2.35 (what-dimension) within the conditions, the context of the SARS-CoV-2 pandemic prescribed (context-dimension).

Teaching self-defence online, shaped and guided by the element tetrad of game design (Schell, 2017) also offered a valuable opportunity for the re-development of teaching expertise (Staller & Koerner, 2020b). Importantly, the development of expertise in a so far unfamiliar teaching setting has not only been experienced by SK (self-dimension) but also by the students. During the course they took over the responsibility for the creation of motivating and demanding challenges for the others. Last not least, by addressing the challenge of "Solidarity" in times of Corona, which is no explicit issue within the university curriculum, a further stimulus for the reflection on action on the individual level (self-dimension) has been set.

The students view on PE 2.35 have been collected using a SoSci-Survey but have not been systematically evaluated yet. However, "train2fight the virus" has been nominated from among the students for the universities' teaching award. Finally, an independent jury rewarded "train2fight the virus" with GSU´s teaching prize in the autumn of 2020. In addition to this recognition, we hope that the analysis of the survey data will provide further insights on the effects of gamified online training from a participant's perspective. Has gamification been motivating, and if so, to which elements of game design motivation is attached? Or were there exceptions? Did participants mentally drop out of the game and still give the course a good rating? Did the students learn what they should and wanted to learn despite of the online handicap, or did the format counteract (some of) the goals and wishes? Within the next level of our "train2fight the virus" experience, answers to these and other questions are to be discussed in relation to current research (Walther, 2003; Rigby & Ryan, 2011; Rogers, 2017; Whitton & Langan, 2018).

REFERENCES

Andreucci, C. A. (2020). Gyms and Martial Arts School after COVID-19: When to Come Back to Train? Advances in Physical Education, 10(2), 114–120. https://doi.org/10.4236/ape.2020.102011

Boendermaker, W. J. (2017). Seroius Gamification. Motivating adolescents to do cognitive training. Ridderprint.

BR (Bundesregierung) (26.03.202020). Vereinbarung zwischen der Bundesregierung und den Regierungschefinnen und Regierungschefs der Bundesländer angesichts der Corona-Epidemie in Deutschland [Agreement between the Federal Government and the Heads of Government of the Federal States in view of the corona epidemic in Germany]

Demantowsky, M. (2020). Teaching between Pre- and Post-Corona. An Essay (2). Public History Weekly, 2020(4). https://doi.org/10.1515/phw-2020-16838

Fatta, H., Maksom, Z., & Zakaria, M. H. (2019). Game-based Learning and Gamification: Searching for Definitions. International Journal of Simulation: Systems, Science & Technology. https://doi.org/10.5013/ijssst.a.19.06.41

GSU (German Sport University) (2020). Module Handbook PE 2.35 „Martial Arts, Combat Sports and Self-Defence". DSHS Köln.

Hamari, J., Koivisto, J., & Sarsa, H. (2014). Does Gamification Work? - A Literature Review of Empirical Studies on Gamification. 3025–3034. https://doi.org/10.1109/hicss.2014.377

Hanstein, T., & Lanig, A. K. (2020). Digital lehren. 25–34. https://doi.org/10.5771/9783828875647-25

Huotari, K., & Hamari, J. (2017). A definition for gamification: anchoring gamification in the service marketing literature. Electronic Markets, 27(1), 21–31. https://doi.org/10.1007/s12525-015-0212-z

Koerner, S. & Staller, M.S. (2020a). Coaching self-defense under COVID-19: Challenges and solutions in the police and civilian domain. Security Journal. DOI: https://doi.org/10.1057/s41284-020-00269-9

Koerner, S. & Staller, M.S. (2020b). Police training revisited — Meeting the demands of conflict training in police with an alternative pedagogical approach, Policing: A Journal of Policy and Practice DOI: https://doi.org/10.1093/police/paaa080

Krabben, K., Orth, D., & Kamp, J. van der. (2019). Combat as an Interpersonal Synergy: An Ecological Dynamics Approach to Combat Sports. Sports Medicine, 49(12), 1825–1836. https://doi.org/10.1007/s40279-019-01173-y

Martin, F., Wang, C., Jokiaho, A., May, B., & Grübmeyer, S. (2019). Examining Faculty Readiness to Teach Online: A Comparison of US and German Educators. European Journal of Open, Distance and E-Learning, 22(1), 53–69. https://doi.org/10.2478/eurodl-2019-0004

Muir, M., Morgan, G., Abraham, A., & Morley, D. (2011). Developmentally Appropriate Approaches to Coaching Children. In I. Stafford (Ed.), Coaching Children in Sport (pp. 17–37). Routledge.

Rigby, S., & Ryan, R. M. (2011). Glued to Games: How Video Games Draw Us In and Hold Us Spellbound. Praeger.

Rogers, R. (2017). The motivational pull of video game feedback, rules, and social interaction: Another self-determination theory approach. Computers in Human Behavior, 73. https://doi.org/10.1016/j.chb.2017.03.048

Schell, J. (2017). The art of game design. Morgan Kaufmann.

Seaborn, K., & Fels, D. I. (2015). Gamification in theory and action: A survey. International Journal of Human-Computer Studies, 74, 14–31. https://doi.org/10.1016/j.ijhcs.2014.09.006

Seifert, L., & Davids, K. (2016). Ecological Dynamics: a theoretical framework for understanding sport performance, physical education and physical activity. 1–13. https://hal.archives-ouvertes.fr/hal-01291044/document

Staller, M. S. (2020). Optimizing Coaching in Police Training. Doctoral Thesis. Leeds Becket University.

Staller, M. S., & Körner, S. (2020a). Kampfsport Coaching in Corona Krisenzeiten - Herausforderungen, Möglichkeiten und Grenzen [Martial Arts Coaching during the corona crisis. Challenges, possibilities and limits]. Leistungssport 50(4), 13-16.

Staller, M. S., & Körner, S. (2020b). Regression, Progression and Renewal: The Continuous Redevelopment of Expertise in Police Use of Force Coaching. European Journal of Security Research. https://doi.org/10.1007/s41125-020-00069-7

Staller, M. S., & Körner, S. (accepted, 2021). Game on! - Spielen in der polizeilichen Lehre [Game on! – Gaming in police teaching]. Lehre.Lernen. Digital, accepted for puplication.

Walther, B. K. (2003). Playing and Gaming: Reflections and Classifications. Game Studies, 3(1).

Whitton, N., & Langan, M. (2018). Fun and games in higher education: an analysis of UK student perspectives. Teaching in Higher Education, 24(8), 1–14. https://doi.org/10.1080/13562517.2018.1541885

Yohannis, A. R., Prabowo, Y. D., & Waworuntu, A. (2014). Defining Gamification. 2014 International Conference on Information Technology Systems and Innovation (ICITSI), 284–289. https://doi.org/10.1109/icitsi.2014.7048279

ONE YEAR OF SNEAK GAMING. INSIGHTS INTO A PROJECT THAT AIMS TO BRING TOGETHER DEVELOPERS, GAMERS AND RESEARCHERS IN TIMES OF COVID-19

Simon Wimmer, Natalie Denk, Constantin Kraus

SNEAK GAMING is supposed to offer a low-threshold opportunity for developers to present, test and promote game projects of any kind. In the same way, players will have a low-threshold opportunity to try out unpublished and experimental projects, give feedback and talk to the developers. For game researchers, the event format provides an opportunity to stay up to date with the latest trends in emerging game production, the needs of players and the challenges and opportunities of contemporary game production. With the COVID-19 pandemic, the event concept faced a challenge. The concept had to be adapted for an online presence. This article shows the changes from the original concept of SNEAK GAMING and provides insights into how the challenges were tackled, how the project has evolved and what challenges it will have to face in the future.

Keywords: COVID-19, community-building, hands-on, digital showcase

———

INTRODUCTION

SNEAK GAMING was intended to be a low-threshold opportunity to bring together Austrian game developers, people interested in games and academics from the field of games research. As the project was planned for the end of 2019 and the launch was scheduled for spring 2020, it is obvious that this project (like pretty much everything) faced unexpected challenges. On the following pages we would like to present the project, but also talk about the adaptations, challenges, experiences and future plans.

The name SNEAK GAMING comes from the idea that the format is based on the concept of the "sneak preview". The term "sneak preview" is commonly used for a format of movie screenings in the cinema whose titles are not announced until the screening. This means that visitors do not know which movie they will see when they buy their ticket. Typically, movies are shown at a "sneak preview" before their official cinema release. This cinema format already exists in Austria and enjoys great popularity. "Sneak previews" are usually offered as a series of events, such as every Tuesday at the Apollo Kino (Cineplexx, 2013).

Somewhat anticipatory, it can be mentioned at this point that SNEAK GAMING was intended for submissions of games from Austria. Due to the planned venue, which was the MuseumsQuartier in Vienna, this criterion was not given too much attention in the early planning phase. It was assumed that primarily local game developers would present their games at the on-site sessions.

SIMILAR CONCEPTS AND DISTINCTION

Organising events to promote the local game developer scene is nothing new, nor are events to promote indie games. In fact, our cooperation partner Jogi Neufeld from SUBOTRON (2021) with its regular series of lectures is a central contact point for exactly such efforts in Austria. The monthly Game Dev Meetup Vienna should also not be overlooked, whose agenda they describe as follows: "Usually we have a keynote from 19 to 20, then a break with a couple of announcements followed by 2 games that people show off and discuss." (Hartinger, 2021) On a broader level, one can of course also mention the multitude of games fairs, which often have different orientations such as Play Austria (2021) which is also organised by SUBOTRON or events such as the ludovico's Button Festival (2021) and many more.

SNEAK GAMING should be distinguished by its "sneaky" character, where visitors do not know in advance which games will be presented. Furthermore, SNEAK GAMING should be characterised by the fact that staff from the Centre for Applied Games Studies at Danube University Krems attend the event and give feedback on the games from a game studies point of view. However, the latter criteria also apply to the events mentioned above.

REALIZATION

However, like so many things planned for 2019, the SNEAK GAMING project has been affected by the COVID-19 pandemic. But one thing after the other:

At the end of 2019, a concept was developed in which SNEAK GAMING was planned as an opportunity for developers to show their games - from prototypes to market-ready games - to an interested audience before the official release and, above all, to let them play and try them out.

Planned location

SNEAK GAMING was planned to take place before the regular events of the SUBOTRON "pro games" series in order to reach a broader audience than would be the case at a playtesting session announced by the respective developer studio. In addition, the events could have benefited from each other.

However, with the onset of restrictions due to the COVID-19 pandemic, it became clear that implementation in this form would not be possible for the first scheduled session on 16 April 2020.

While we were discussing how this first session could take place, the website www.sneakgaming.at was also created at the same time. The aim of the website is to present the project, to inform about the upcoming dates and to handle the submissions of the developers. Furthermore, all games presented at a SNEAK GAMING session will be featured in a showcase.

After a good year of COVID-19, one tends to forget how unforeseeable and new this situation was. In internal discussions, the question arose as to whether we should wait, since physical events would be possible again in the course of the year, or whether we should generally change the format to an online format. In August 2020, the educational YouTuber CGP Grey (2020) posted a video online in which he mentions two things that aptly describe the decisions around SNEAK GAMING. First, "there are no solutions, only trade-offs" (CGP Grey, 2020) and second, that decisions are made in the context of a fog of the future (CGP Grey, 2020).

The original SNEAK GAMING concept envisaged a low-threshold, shared gaming experience in a station setting where interested people could try out unpublished games for themselves and have an informal chat with the developers. In an online Zoom session, of course, such a format is not possible. This is particularly true with regard to a hands-on experience, but also an informal chat has it's limits in virtual space. This could be due to the audio mixing of several participants with the microphone switched on, the problem of destructive zoom bombers (especially at the beginning of the pandemic) or simply unstable internet connections. We also realised that a purely online format would present an extra challenge for analogue or hybrid games. In the end, we decided to give developers a certain time window for game presentations and a subsequent feedback session. This is despite the danger that it resembles a pitch. But, there are no solutions, only trade-offs.

Website

The website is addressing two target groups: developers and players. This duality is also reflected in the overall design. In the desktop version one half of the screen is used for players, where they are informed about SNEAK GAMING and always see the next appointment. On the other half of the screen, developers are invited to submit any form of unreleased game. Developers can submit their game description via a form and submit screenshots, trailers or even the whole game via cloud storage.

Another central element of the website is the showcase, where all games presented at SNEAK GAMING are archived in short form. Each game presented receives a virtual "poster" that is implemented as an animated Gif with central keywords and a small text field with a short description.

Figure 1. Example of virtual "poster" and overview of showcase

In addition, blog posts reflecting on past SNEAK GAMING sessions can be found on the website.

An outline of all SNEAK GAMING sessions so far

SNEAK GAMING Vol 1

The first SNEAK GAMING session has been scheduled for 16 April 2020 from 7pm to 10pm. Six games have been selected to present their games on this date. Starting at 7:15pm, half an hour has been allocated for each game.

SNEAK GAMING Vol 2

For the second SNEAK GAMING session, the number of presenters was reduced to five based on feedback, but also on our own experience. However, the time frame of 30 minutes per presentation including feedback was kept.

SNEAK GAMING Vol 3

The third online SNEAK GAMING session took place at the beginning of June. Due to further feedback, we decided to reduce the time per game to 20 minutes and to stay with five games per evening.

SNEAK GAMING Real Life Event

After a time window opened up in September in which on-site events were possible under certain conditions, the first on-site SNEAK GAMING session took place on 8 September. As in the original concept, we used the hour before the SUBOTRON "pro games" event. However, this did not take place in Room D of the MuseumsQuartier as usual. Also, the usual registration-free system could not be implemented, as Covid restrictions required a ticket system.

SNEAK GAMING XL

SNEAK GAMING XL was a special programme: SNEAK GAMING XL. Not only did this format start at 04:00 pm and included 15 games, but this time already published games could be submitted. Moreover, there was a new virtual format that allows further exchange around the games in addition to the usual Zoom session. During the SNEAK GAMING XL event (and also beyond) both text- and voice-channels of the participating games were available on the SUBOTRON Discord server. Furthermore, the Zoom session was also hosted on the SUBOTRON Twitch Channel. Conny Lee and Christian Stipkovits from the radio station "FM4" handpicked five of the submitted games, played them, and chatted with the devs live on the radio_fm4 Twitch Channel at their program "Spielekammerlshow" for two hours (FM4 Spielekammerlshow, 2021).

SNEAK GAMING Vol. 4

The latest SNEAK GAMING session was SNEAK GAMING Vol. 4 on May 7, 2021, a date that marked the one-year anniversary of the event series, with nine developers presenting new games. For the first time, one of them asked not to be named on the website or social media afterwards, as he did not want to reveal anything more until the final launch of his game.

Also, for the first time in SNEAK GAMING's history, there was an presentation by a SNEAK GAMING "alumni" who briefly mentioned a game already presented at SNEAK GAMING and talked about further developments.

SNEAK GAMING in numbers and graphics

So far, more than 30 games have been presented at SNEAK GAMING. If classifiable, the respective development status was indicated and documented at the time of submission. It is becoming clear that the request for unfinished and experimental games (such as games created in the context of game jams) has been met.

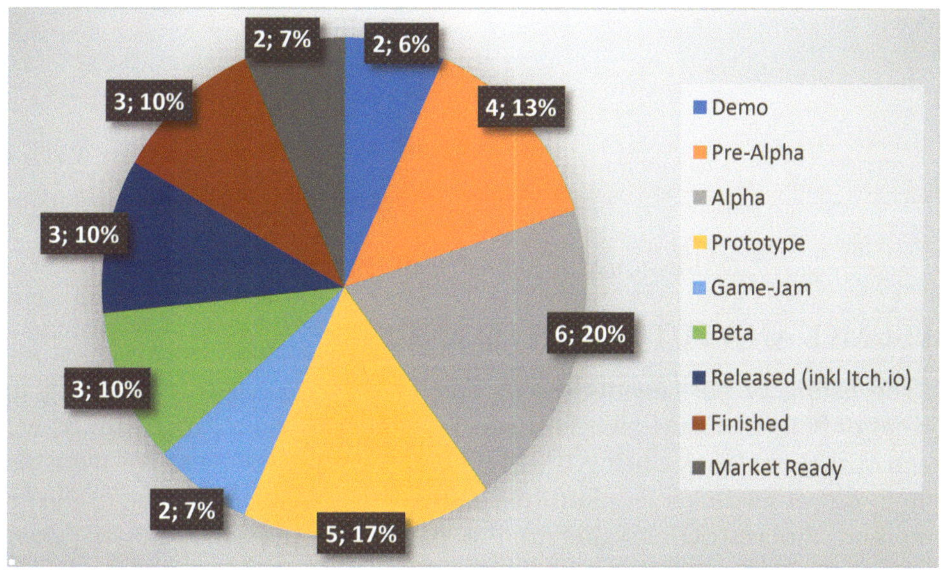

Figure 2. Distribution of the different stages of the presented games

The majority of the games presented so far were developed for Windows, Mac and/or Linux, followed by various virtual reality platforms and mobile games. However, it is worth noting that, as encouraged on the website, a board game and games for (nowadays) unconventional platforms like Atari ST were also presented. In the case of the board game, the developers helped themselves to a digital version of the game using the "tabletopia" (Tabletopia, 2021) tool.

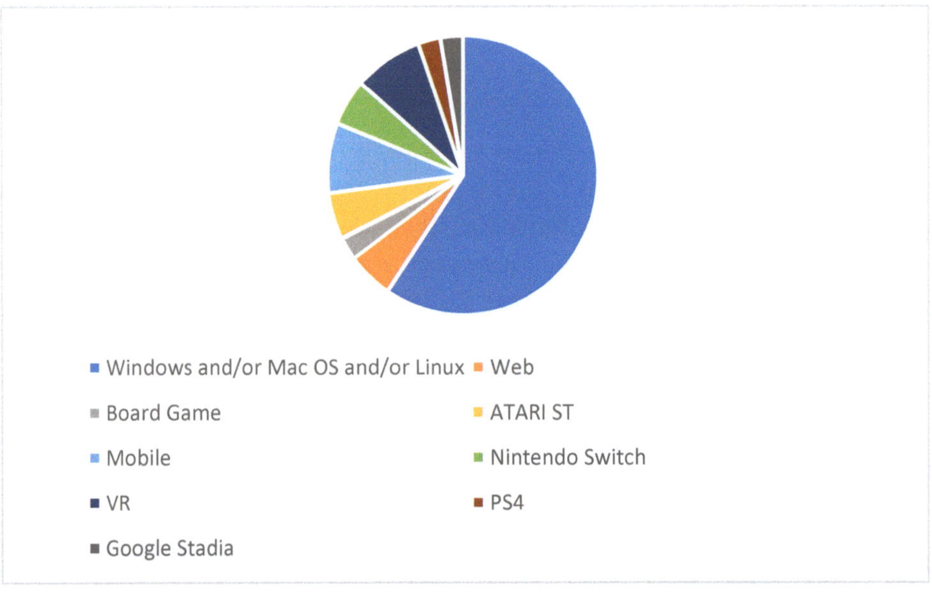

Figure 2. Distribution of the different platforms of the presented games

RESUMEE AND FUTURE CHALLENGES

The listing of past events shows that we were constantly adapting the format of SNEAK GAMING. This was based on our own experience as well as on verbal feedback during the SNEAK GAMING sessions. We were also able to access feedback on platforms such as Reddit. In the course of the last year and the restrictions due to the pandemic, the well-known "Zoom Fatigue" (Bailenson, 2021) could also be observed to a certain extent.

What the future SNEAK GAMING session will look like is open due to the "fog of the future". Vaccinations give hope for a return to the usual normality. And the often expressed wish of developers and audience at the end of a virtual SNEAK GAMING session to have a beer together show how much real gatherings are missed.

ACKNOWLEDGMENTS

Our thanks go to our sneaky cooperation partner Jogi Neufeld (SUBOTRON), to all the staff at the Centre for Applied Games Studies and of course to all the talented game developers who presented their games at SNEAK GAMING or plan to do so in the future – equally to all interested players who took part in SNEAK GAMING as visitors.

REFERENCES

Bailenson, Jeremy N. (2021) Nonverbal Overload: A Theoretical Argument for the Causes of Zoom Fatigue. Retrieved from https://tmb.apaopen.org/pub/nonverbal-overload/release/1 last checked 15.04.2021

Button Festival, (2021). https://www.buttonfestival.at/ last checked 15.04.2021

CGP Grey. (2020). Q&A With Grey: Thinking About Lockdowns Edition [Video]. Retrieved from https://www.youtube.com/watch?v=SVmEXdGqO-s&list=PLqs5ohhass_TBfFNkl6wUaPSBTZK4xTv8 last checked 15.04.2021

Cineplexx (2013). Sneak Preview OV Retrieved from https://www.cineplexx.at/film/sneak-preview/ last checked 15.04.2021

FM4 Spielekammerlshow, (2021). Retrieved from https://www.twitch.tv/radio_fm4 last checked 15.04.2021

Hartinger, M. (2021) Vienna Gamdev Meetup - Meet gamedevs, watch and discuss local creations, Retrieved from https://www.meetup.com/de-DE/Vienna-Game-Dev-Meetup/events/cssbprycchbgb/ last checked 15.04.2021

https://playaustria.com/ last checked 15.04.2021

SNEAK GAMING Reddit (2020). Retrieved from https://www.reddit.com/r/Austria/comments/k77cav/sneak_gaming_xl_zoom_event_spiele_made_in_austria/ last checked 15.04.2021

SUBOTRON, (2021). Retrieved from https://subotron.com/ last checked 15.04.2021

Tabletopia. (2021). https://tabletopia.com/ last checked 15.04.2021

AUTHORS

JOSEPHINE BAIRD is a lecturer at the University of Uppsala Game Design Department, and is a PhD candidate at the University of Vienna. She is a game designer and game design consultant, as well as a published scholar, author and visual artist - often producing work on issues related to gender and sexualities. She is also an actor and public speaker, and co-hosts a weekly podcast.

https://josephinebaird.com

NATALIE DENK has a degree in Educational Science and Game Studies. Her research focuses on Game-based Learning, Educational Game Design and the examination of age and gender aspects in relation to digital games and gaming cultures. Since 2014 she has been involved in several national and international research projects at the Center for Applied Game Studies at the Danube University Krems, Austria. Furthermore, she is responsible for the university courses "Game Studies", "Media Game Pedagogy", "Action Oriented Media Pedagogy" and "Game-based Media & Education". Since July 2019 she acts as Head of the Center for Applied Game Studies.

Twitter: @Natalie_Denk

JARED DERRY is an alumni from Texas A&M University with a B.S. in Visualization. He is currently a 3D Technical Artist and Generalist in pursuit of a career that is rich with opportunities to breathe life into inspiring concepts using computer graphics.

ALEXIEI DINGLI is a professor of AI at the University of Malta. He has been conducting research and working in the field of AI for more than two decades, assisting different companies to implement AI solutions. His work has been rated World Class by international experts and he won several local and international awards (such as those by the European Space Agency, the World Intellectual Property Organization and the United Nations to name a few). He has published several peer-reviewed publications and formed part of the Malta.AI task-force which was set up by the Maltese government, aimed at making Malta one of the top AI countries in the world.

WILFRIED ELMENREICH is professor for Smart Grids at the Institute of Networked and Embedded Systems at the Alpen-Adria-Universität

Klagenfurt, Austria. His research interests include intelligent energy systems, self-organizing systems and technical applications of swarm intelligence. Wilfried Elmenreich is a member of the Senate at the Alpen-Adria-Universität Klagenfurt, Counselor of the IEEE Student Branch, and is involved in the master program on Game Studies and Engineering. He is the author of several books and has published over 200 articles in the field of networked and embedded systems. Elmenreich researches intelligent energy systems, self-organizing systems, and technical applications of swarm intelligence.

https://elmenreich.alturl.com

KATHRYN FRIESEN is a recent graduate from Texas A&M University with a degree in Visualization. She currently works as a game designer at the University's Learning Interactive Visualization Experience Lab, where she was a systems designer for Virtual Disaster Day 2021, a live educational role-playing scenario that hosted 600+ people over three weekend sessions in March 2021. She is also the author of The Power of Legends series, of which two books are published and a third is to be released in late summer of 2021.

MARTIN GABRIEL studied history at the University of Klagenfurt and received his PhD for a thesis on the character of warfare during the 1878 Austro-Hungarian occupation of Bosnia and Hercegovina. Since 2012 he has been a lecturer in modern history at the University of Klagenfurt and has also taught at the University of Graz. His research interests and publications focus on 16th to 19th century global and imperial history (with a special interest in Spain, the United States, and Austria-Hungary), social and economic histories of colonialism, colonial violence as well as Star Trek as a cultural history phenomenon.

SONJA GABRIEL has a PhD in Educational Science, master degrees in Educational Media and Applied Game Science. Her research focuses on Game-Based Learning, values in digital games as well as media ethics and media literacy in general. Since 2018 she is professor for media literacy at University College of Teacher Education Vienna/Krems. She has already been involved in several national and international projects dealing with topics of digital game based learning and media literacy as well as evaluation of digital tools for learning.

RICARDA GOETZ-PREISNER is a political scientist with a strong focus on popular culture and gender. She works for the City of Vienna in the basic research section of the Women´s Department. As an independent researcher and writer she publishes in various media in the realm of cultural studies. She is interested in questions at the interface between (international) politics, culture and equality. She has relevant experience as an expert public speaker in topics ranging from popular culture to national and international politics. She is a member of different gremia e.g. the Bundesstelle für die Positivprädikatisierung von digitalen Spielen (BuPP).

SABINE HARRER is a senior lecturer at the department of game design Gotland, Uppsala University Sweden. Their research focuses on media studies and critical game design. They are the author of Games and Bereavement (transcript 2018).

DANIELA HAU has a degree in Media Game Pedagogy and Economics. She works as a secondary school teacher and as a coordinator for the "Service de Coordination de la Recherche et de l'Innovation pédagogiques et technologiques" (SCRIPT). Daniela Hau has been involved in the development of the national guide to digital game-based learning ('Digital spielend lernen') and the implementation of the national media competence framework ('Medienkompass'). She devotes a lot of time liaising with schools and teachers to review the effectiveness of digital game-based learning approaches and media competences in every-day learning settings.

STEVE HILBERT is a social pedagogue with a Game Studies Master's degree. For more than 20 years, he worked in a variety of settings including homelessness, unemployment, early childhood and youth. As a long-time experienced instructor of a secondary school class for potential dropouts, he has elaborated a school program to promote the search for identity through an educational use of digital games. Since 2019, he works for the Ministry of National Education, Childhood and Youth of Luxembourg's ""Resource Center for Non-Formal Education for secondary schools" as a policy maker, focusing on the promotion of digitalization on a national level.

MONA KHATTAB is a doctoral candidate at the University of Vaasa, Finland. Her research focuses on the impact of new digital media on communication, focusing on individual access and outreach. Khattab's research interests include social media, new technology, digital cultures, games and media studies.

BENJAMIN KIRCHENGAST is currently working on his master's degrees in Contemporary History and Media as well as Game Studies. His research focuses on the discourse in video game culture about gender, historical representation, and far-right politics. He is also working in the field of remembrance culture for Verein GEDENKDIENST, where has been involved in the conceptualisation of workshops about Austrian remembrance politics and historical representation of National Socialism and the Holocaust in popular media.

Twitter: @benkiga

SWEN KOERNER is since 2009 a Full Professor, currently heading the Department of Training Pedagogy and Martial Research at German Sport University Cologne. His current research focuses on training pedagogy, the professionalization of police training, martial arts studies and gamified learning designs.

CONSTANTIN KRAUS studied to become a teacher for elementary schools at the Kirchlich Pädagogische Hochschule in Graz. He is currently studying for a master's degree in MediaGamePedagogy at Danube University Krems. Besides working in an after-school program at an elementary school, he has been working as a sound designer, graphic artist and motion designer for several years.

BASTIAN KRUPP recently completed his M.A. in action-oriented media pedagogy at the Danube-University Krems. He wrote his master thesis on the connection between the development of emotional intelligence and gaming. As a trained educator and studied social worker, he is interested in the effect of digital games as a methodical instrument for initialising educational processes. Since 2017 he has been working as a social pedagogical director in open youth work where he has been able to realise numerous projects with and through games.

MATHIAS LUX is an associate professor at the University of Klagenfurt. His research areas are user intentions and retrieval in multimedia as well as interactive multimedia focusing on video games. In his scientific career he has (co-) authored more than 150 scientific publications and he is well known for the development of the award-winning and popular open-source tools Caliph & Emir and LIRE for multimedia information retrieval. In Klagenfurt, he is a co-founder of the game studies and engineering master program and has established a lively community of game developers and enthusiasts who meet at regular events and game jams.

JOSEY MEYER has a degree in Visualization: Animation and Game Design. Her research has pertained to game-based learning at the Texas A&M University Learning Interactive Visualization Experience Lab from 2017-2020. She worked on ARTé: Lumiere and ARTé: Hemut, an educational game art-history series, as a game designer. As of 2021 she works as a full-time associate dungeon designer on Diablo IV at Blizzard Entertainment

MONTY PATINO is an undergraduate Visualization student at Texas A&M University. She grew up playing video games like Zelda, Mario and Animal Crossing, and never really grew out of it. With a background in traditional art and illustration, Monty hopes to work in the video game industry as a concept artist, and use her art to give a voice to stories that have been generally under-represented in media. Her work has been shown around the United States in exhibitions by the College Board, VASE, and Texas A&M University.

ANDREW "ANDY" PHELPS is a professor at the Human Interface Technology Laboratory (HITLabNZ) in the College of Engineering at the University of Canterbury in Christchurch, New Zealand. He is also a professor of Film & Media Arts in the School of Communication, and a professor of Computer Science in the College of Arts & Sciences, both at American University in Washington, D.C. He is currently the director of the American University Game Center, and also serves as president of the Higher

Education Video Game Alliance (HEVGA). His latest game is Fragile Equilibrium available on Steam, Itch.io, and XBOX One.

https://andyworld.io

ALEXANDER PFEIFFER is currently working as a senior researcher at the center for applied game studies at Donau-Universität Krems, Austria. He is a recipient of a Max Kade Fellowship awarded by the Austrian Academy of Science to work at the MIT Education Arcade, USA in 2019/2020.

https://www.alexpfeiffer.at

EUGEN PFISTER leads the SNF-Ambizione research project "Horror - Game - Politics" (hgp.hypotheses.org) at the Hochschule der Künste Bern - HKB. He studied History and Political Sciences at the University of Vienna and the Université Paris IV – Sorbonne. PhD in co-tutelle at the Universita degli studi di Trento and at the Johann-Wolfgang-Goethe-Universität Frankfurt am Main.Founding Member of the research group "Geschichtswissenschaft und Digitale Spiele" (gespielt.hypotheses.org). Since 2015 he operates a research blog on cultural studies and video games called "Spiel-Kultur-Wissenschaft. Mythen im Digitalen Spiel" (spielkult.hypotheses.org)

MARIO PLATZER is a transport planner and has been working in the field of mobility research for more than 10 years. He holds a degree in civil engineering and sociology and is an expert in social behaviour theories, especially in measures for changeing mobility behaviour. He is currently researching how innovative communication technologies, such as social media, augmented reality and serious games force sustainable mobility behaviour. He is also highly experienced in qualitative and quantitative methods of data collection and analysis and has been responsible for the impact evaluation of several games.

FRANK POURVOYEUR is working as an Indie Game Developer and Media Artist in Austria. He graduated in Game Studies at the Danube University Krems where he is currently studying Game Based Media and Education at the Center for Applied Game Studies. His research interests are consciousness in video games and cooperatively shared digital experiences.

DORIS C. RUSCH is an associate professor / docent for game design at the Department of Game Design, Uppsala University. Her work focuses on transformative play. She has created numerous award winning projects, mainly on games on mental health and has been an international keynote speaker and presenter including Clash of Realities, DiGRA, Game Developers Conference, Meaningful Play, Nordic Game Conference, FDG and TEDx. She authored "Making Deep Games - Designing Games with Meaning and Purpose" (Taylor & Francis 2017).

ALEXANDER K. SEEWALD has a degree in Computer Science. His research focusses on basic and applied research in Machine Learning, Artificial Intelligence, Image/Video/Audio Analysis, Pattern Recognition, Robotics and Deep Learning. Since 2008 he has been involved in several national and international research projects at the Medical University Vienna, the University of Vienna, and the Center for Applied Game Studies at the Danube University Krems. Furthermore, he currently holds the university course "Future Media".

https://seewald.at

ALESHA SERADA is a researcher and a PhD student at the University of Vaasa, Finland. Inspired by Alesha's Belarusian origin, their research interests revolve around exploitation, violence, horror, deception, Machiavellian ethics and other banal and non-banal evils. Their dissertation, in work with support from the Nissi Foundation, discusses construction of value in games and art on blockchain. In their spare time, Alesha writes on late Soviet and post-Soviet visual culture to make it more accessible to the English-speaking academic audience.

TANJA SIHVONEN is professor of Communication Studies at the University of Vaasa, Finland. She is specialized in digital media, games, and participatory cultures on the internet. Her most recent work considers astroturfing, queer role-playing, and blockchain technologies.

https://www.univaasa.fi/en/profile/?view=1319353

MARIO S. STALLER is a professor at the University of Applied Sciences of Police and Public Administration North-Rhine Westphalia in Aachen,

Germany. His research focuses on the professionalization of police education concerning competencies to cope with the demands in the field. He works with recruits, regular officers and highly specialized personal. Furthermore, his projects focus on coaching, coach education and gamification in specific contexts.

Twitter: @mariostaller

TOBIAS UNTERHUBER studied modern German literature, comparative literature and study of religions at LMU Munich and at the University of California, Berkeley. In 2018, he earned his PhD with his thesis on the works of Swiss author Christian Kracht. He is a post-doc for literature and media studies at the Leopold-Franzens University Innsbruck. In addition to pop literature, literary theory, discourse analysis, literature & economics and gender studies, his research focuses on video games from a cultural studies and media history perspective. He is an editor of the game studies journal PAIDIA.

Twitter: @letobili

PASCAL MARC WAGNER is an M.A. cognitive and cultural linguist with a B.A. in English Studies and German Media and Human Rights Law from the Ludwig-Maximilians-Universität Munich. At the time of writing he is developing a video game-based language diagnosis test for primary school children at the Goethe-Institut Munich and teaching English linguistics at the Johannes Gutenberg-Universität Mainz. He founded the blog and ludolinguistics resource page languageatplay.de to further advance the field of linguistics into the study of digital games. In 2020, he co-founded the antifascist network "Keinen Pixel den Faschisten!".

https://languageatplay.de

THOMAS WERNBACHER works as a senior scientist at the Centre for Applied Game Studies at Danube University Krems. He studied Psychology at Karl-Franzens University of Graz and Game Studies as well as Game-based Media & Education at Danube University Krems. In his work, he researches the use and effects of playful approaches in the context of mobility, education and health. His expertise includes behavioural concepts in the form of gamification and nudging with a special focus on incentive models. The current focus of his research is on communicating sustainable development goals in the context of urban development and climate change.

SIMON WIMMER holds a Bachelor's degree in Cultural and Social Anthropology and a Master's degree in Game Studies. Since 2017 he works at the Center for Applied Game Studies at the Donau-Universität Krems. His focus is on storytelling, conceptual- and narrative design.